COASTAL SEDIMENTATION

Benchmark Papers in Geology

Series Editor: Rhodes W. Fairbridge
Columbia University

Volume

1 ENVIRONMENTAL GEOMORPHOLOGY AND LANDSCAPE CONSERVATION, Volume 1: Prior to 1900/*Donald R. Coates*
2 RIVER MORPHOLOGY/*Stanley A. Schumm*
3 SPITS AND BARS/*Maurice L. Schwartz*
4 TEKTITES/*Virgil E. Barnes and Mildred A. Barnes*
5 GEOCHRONOLOGY: Radiometric Dating of Rocks and Minerals/*C. T. Harper*
6 SLOPE MORPHOLOGY/*Stanley A. Schumm and M. Paul Mosley*
7 MARINE EVAPORITES: Origin, Diagenesis, and Geochemistry/*Douglas W. Kirkland and Robert Evans*
8 ENVIRONMENTAL GEOMORPHOLOGY AND LANDSCAPE CONSERVATION, Volume III: Non-Urban/ *Donald R. Coates*
9 BARRIER ISLANDS/ *Maurice L. Schwartz*
10 GLACIAL ISOSTASY/*John T. Andrews*
11 GEOCHEMISTRY OF GERMANIUM/*Jon N. Weber*
12 ENVIRONMENTAL GEOMORPHOLOGY AND LANDSCAPE CONSERVATION, Volume II: Urban Areas/*Donald R. Coates*
13 PHILOSOPHY OF GEOHISTORY: 1785-1970/ *Claude C. Albritton, Jr.*
14 GEOCHEMISTRY AND THE ORIGIN OF LIFE/ *Keith A. Kvenvolden*
15 SEDIMENTARY ROCKS: Concepts and History/*Albert V. Carozzi*
16 GEOCHEMISTRY OF WATER/*Yasushi Kitano*
17 METAMORPHISM AND PLATE TECTONIC REGIMES/*W. G. Ernst*
18 GEOCHEMISTRY OF IRON/*Henry Lepp*
19 SUBDUCTION ZONE METAMORPHISM/*W. G. Ernst*
20 PLAYAS AND DRIED LAKES: Occurrence and Development/*James T. Neal*
21 GLACIAL DEPOSITS/*Richard P. Goldthwait*
22 PLANATION SURFACES: Peneplains, Pediplains, and Etchplains/*George F. Adams*
23 GEOCHEMISTRY OF BORON/*C. T. Walker*
24 SUBMARINE CANYONS AND DEEP-SEA FANS: Modern and Ancient/*J. H. McD. Whitaker*
25 ENVIRONMENTAL GEOLOGY/*Frederick Betz, Jr.*
26 LOESS:Lithology and Genesis/*Ian J. Smalley*

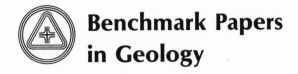

27 PERIGLACIAL PROCESSES/*Cuchlaine A. M. King*
28 LANDFORMS AND GEOMORPHOLOGY: Concepts and History/ *Cuchlaine A. M. King*
29 METALLOGENY AND GLOBAL TECTONICS/*Wilfred Walker*
30 HOLOCENE TIDAL SEDIMENTATION/*George deVries Klein*
31 PALEOBIOGEOGRAPHY/*Charles A. Ross*
32 MECHANICS OF THRUST FAULTS AND DÉCOLLEMENT/*Barry Voight*
33 WEST INDIES ISLAND ARCS/*Peter H. Mattson*
34 CRYSTAL FORM AND STRUCTURE/*Cecil J. Schneer*
35 OCEANOGRAPHY: Concepts and History/*Margaret B. Deacon*
36 METEORITE CRATERS/*G. J. H. McCall*
37 STATISTICAL ANALYSIS IN GEOLOGY/*John M. Cubitt and Stephen Henley*
38 AIR PHOTOGRAPHY AND COASTAL PROBLEMS/*Mohamed T. El-Ashry*
39 BEACH PROCESSES AND COASTAL HYDRODYNAMICS/*John S. Fisher and Robert Dolan*
40 DIAGENESIS OF DEEP-SEA BIOGENIC SEDIMENTS/*Gerrit J. van der Lingen*
41 DRAINAGE BASIN MORPHOLOGY/*Stanley A. Schumm*
42 COASTAL SEDIMENTATION/*Donald J. P. Swift and Harold D. Palmer*
43 ANCIENT CONTINENTAL DEPOSTIS/*Franklyn B. Van Houten*
44 MINERAL DEPOSITS, CONTINENTAL DRIFT AND PLATE TECTONICS/ *J. B. Wright*
45 SEA WATER: Cycles of the Major Elements/*James I. Drever*
46 PALYNOLOGY, PART I: Spores and Pollen/*Marjorie D. Muir and William A. S. Sarjeant*
47 PALYNOLOGY, PART II: Dinoflagellates, Acritarchs, and Other Microfossils/*Marjorie D. Muir and William A. S. Sarjeant*

Benchmark Papers in Geology / 42

A BENCHMARK® Books Series

COASTAL SEDIMENTATION

Edited by

DONALD J. P. SWIFT

Atlantic Oceanographic and Meteorological Laboratories-National Oceanographic and Atmospheric Administration

HAROLD D. PALMER

Dames & Moore
Washington, D. C.

Dowden, Hutchinson & Ross, Inc.
STROUDSBURG, PENNSYLVANIA

Benchmark Papers in Geology, Volume 42
Library of Congress Catalog Card Number: 78-18696
ISBN: 0-87933-330-8

80 79 78 1 2 3 4 5
Manufactured in the United States of America.

LIBRARY OF CONGRESS CATALOGING IN PUBLICATION DATA
Main entry under title:
Coastal sedimentation.
(Benchmark papers in geology ; v. 42)
Bibliography: p.
Includes indexes.
1. Coasts—Addresses, essays, lectures.
2. Sedimentation and deposition—Addresses, essays, lectures. I. Swift, Donald J. P. II. Palmer, Harold Dean, 1934-
GB451.2.C63 551.3'6 78-18696
ISBN 0-87933-330-8

Distrubuted world wide by Academic Press, a subsidiary of Harcourt Brace Jovanovich, Publishers.

SERIES EDITOR'S FOREWORD

The philosophy behind the "Benchmark Papers in Geology" is one of collection, sifting, and rediffusion. Scientific literature today is so vast, so dispersed, and, in the case of old papers, so inaccesible for readers not in the immediate neighborhood of major libraries that much valuable information has been ignored by default. It has become just so difficult, or so time consuming, to search out the key papers in any basic area of research that one can hardly blame a busy man for skimping on some of his "homework."

This series of volumes has been devised, therefore, to make a practical contribution to this critical problem. The geologist, perhaps even more than any other scientist, often suffers from twin difficulties—isolation from central library resources and immensely diffused sources of material. New colleges and industrial libraries simply cannot afford to purchase complete runs of all the world's earth science literature. Specialists simply cannot locate reprints or copies of all their principal reference materials. So it is that we are now making a concerted effort to gather into single volumes the critical material needed to reconstruct the background of any and every major topic of our discipline.

We are interpreting "geology" in its broadest sense: the fundamental science of the planet Earth, its materials, its history, and its dynamics. Because of training and experience in "earthy" materials, we also take in astrogeology, the corresponding aspect of the planetary sciences. Besides the classical core disciplines such as mineralogy, petrology, structure, geomorphology, paleontology, and statigraphy, we embrace the newer fields of geophysics and geochemistry, applied also to oceanography, geochronology, and paleoecology. We recognize the work of the mining geologists, the petroleum geologists, the hydrologists, the engineering and environmental geologists. Each specialist needs his working library. We are endeavoring to make his task a little easier.

Each volume in the series contains an Introduction prepared by a specialist (the volume editor)—a "state of the art" opening or a summary of the object and content of the volume. The articles, usually some thirty to fifty reproduced either in their entirety or in significant extracts, are selected in an attempt to cover the field, from the key papers of the last century to fairly recent work. Where the original works are in foreign lan-

guages, we have endeavored to locate or commision translations. Geologists, because of their global subject, are often acutely aware of the oneness of our world.The selections cannot, therefore, be restricted to any one country, and whenever possible an attempt is made to scan the world literature.

To each article, or group of kindred articles, some sort of "highlight commentary" is usually supplied by the volume editor. This should serve to bring that article into historical perspective and to emphasize its particular role in the growth of the field. References, or citations, wherever possible, will be reproduced in their entirety—for by this means the observant reader can assess the background material available to that particular author, or, if he wishes, he too can double check the earlier sources.

A "benchmark," in surveyor's terminology, is an established point on the ground, recorded on our maps. It is usually anything that is a vantage point, from a modest hill to a mountain peak. From the historical viewpoint, these benchmarks are the bricks of our scientific edifice.

RHODES W. FAIRBRIDGE

PREFACE

This book contains a selection of twenty papers that illustrate directions of research in coastal sedimentation over the past sixty years. Four main themes are sampled: the coastal equilibrium profile, coastal deposits, studies of fluid motion, and studies of substrate response. A fifth theme, surf zone studies, has been dealt with in companion volumes. We have supplemented the references cited by our authors with citations of our own that serve to bridge the inevitable gaps in a small sample of a long-lived research field. Our collection is "front-end loaded" in that we stress the research of the sixties and the seventies, when synergism became apparent between the fields of classical geology and physical oceanography.

We would like to take this oppurtunity to thank our respective institutions for extending library and other facilities to us during the preparation of this volume. We would especially like to thank Claire Ulanoff for secretarial assistance, and all those on the editorial staff of Dowden, Hutchinson & Ross who in their various capacities have been responsible for the production of the book.

DONALD J. P. SWIFT
HAROLD D. PALMER

CONTENTS

Series Editor's Foreword vii
Preface ix
Contents by Author xv

Introduction 1

PART I: THE COASTAL EQUILIBRIUM PROFILE

Editors' Comments on Papers 1 Through 7 8

1 **FENNEMAN, N. M.:** Development of the Profile of Equilibrium of the Subaqueous Shore Terrace 14
J. Geol. **10**:1–2, 22–32 (1902)

2 **FISCHER, A. G.:** Stratigraphic Record of Transgressing Seas in Light of Sedimentation on Atlantic Coast of New Jersey 26
Bull. Am. Assoc. Pet. Geol. **45**(10):1656–1666 (1961)

3 **BRUUN, P.:** Sea-Level Rise as a Cause of Shore Erosion 37
J. Waterways and Harbors Division, Proc. Am. Soc. Civ. Eng. **88**: 117–130 (1962)

4 **DIETZ, R. S.:** Wave-Base, Marine Profile of Equilibrium, and Wave-Built Terraces: A Critical Appraisal 51
Geol. Soc. Am. Bull. **74**:971–990 (1963)

5 **MOORE, D. G., and J. R. CURRAY:** Wave-Base, Marine Profile of Equilibrium, and Wave-Built Terraces: Discussion 73
Geol. Soc. Am. Bull. **75**:1267–1273 (1964)

6 **DIETZ, R. S.:** Wave-Base, Marine Profile of Equilibrium, and Wave-Built Terraces: Reply 80
Geol. Soc. Am. Bull. **75**:1275–1281 (1964)

7 **WRIGHT, L. D., and J. M. COLEMAN:** River Delta Morphology: Wave Climate and the Role of the Subaqueous Profile 87
Science **176**:282–284 (1972)

PART II: COASTAL DEPOSITS

Editors' Comments on Papers 8 Through 12 92

8 **CURRAY, J. R.:** Transgressions and Regressions 97
Papers in Marine Geology, R. L. Miller, ed., Macmillan, 1964, pp. 175–203

9 **PILKEY, O. H., and D. FRANKENBERG:** The Relict-Recent Sediment Boundary on the Georgia Continental Shelf **126**
Bull. Georgia Acad. Sci. **22**(1):37–40 (1964)

10 **CURRAY, J. R., F. J. EMMEL, and P. J. S. CRAMPTON:** Holocene History of a Strand Plain, Lagoonal Coast, Nayarit, Mexico **130**
Lagunas Costeras, Un Simposio, A. A. Costonares and F. B. Phelger, eds., Mexico: Universidad National Autónoma de Mexico, 1967, pp. 63–100

11 **SWIFT, D. J. P., R. B. SANFORD, C. E. DILL, Jr., and N. F. AVIGNONE:** Textural Differentiation on the Shore Face During Erosional Retreat of an Unconsolidated Coast, Cape Henry to Cape Hatteras, Western North Atlantic Shelf **169**
Sedimentology **16**:221–250 (1971)

12 **SHERIDAN, R. E., C. E. DILL, Jr., and J. C. KRAFT:** Holocene Sedimentary Environment of the Atlantic Inner Shelf off Delaware **199**
Geol. Soc. Am. Bull. **85**:1319–1328 (1974)

PART III: STUDIES OF FLUID MOTION

Editors' Comments on Papers 13 Through 16 **210**

13 **MURRAY, S. P.:** Bottom Currents near the Coast during Hurricane Camille **213**
J. Geophys. Res. **75**(24):4579–4582 (1970)

14 **PALMER, H. D., and D. G. WILSON:** Nearshore Current Regimes in a Linear Shoal Field, Middle Atlantic Bight, USA **217**
IXme Congres International de Sedimentologie, Nice, France, 1975, pp. 137–141

15 **CASTON, V. N. D.:** A Wind-driven Near-bottom Current in the Southern North Sea **222**
Estuarine and Coastal Mar. Sci. **4**:23–32 (1976)

16 **CSANADY, G. T.:** Wind-Driven and Thermohaline Circulation Over the Continental Shelves **233**
Effects of Energy Related Activities on the Atlantic Continental Shelf, B. Manowitz, ed., Brookhaven National Laboratories, 1976, pp. 31–47

PART IV: STUDIES OF SUBSTRATE RESPONSE

Editors' Comments on Papers 17 Through 20 **252**

17 **INMAN, D. L., and G. A. RUSNAK:** Changes in Sand Level on the Beach and Shelf at La Jolla, California **255**
Beach Erosion Board TM-82, U. S. Army, Corps of Engineers, Washington, D. C., July 1956, pp. 1–30

18 **COOK, D. O., and D. S. GORSLINE:** Field Observations of Sand Transport by Shoaling Waves **285**
Mar. Geol. **13**:31–55 (1972)

19 **LUDWICK, J. C.:** Tidal Currents, Sediment Transport, and Sand Banks in Chesapeake Bay Entrance, Virginia **310**
Estuarine Research: Vol.II. Geology and Engineering, L. E. Cronin, ed., Academic Press, 1975, pp. 365–380

20 **LAVELLE, J. W., P. E. GADD, G. C. HAN, D. A. MAYER, W. L. STUBBLEFIELD, and D. J. P. SWIFT:** Preliminary Results of Coincident Current Meter and Sediment Transport Observations for Wintertime Conditions on the Long Island Inner Shelf **326**
Geophys. Res. Lett. **3**(2):97–100 (1976)

Author Citation Index **331**
Subject Index **335**

About the Editors **341**

CONTENTS BY AUTHOR

Avignone, N. F., 169
Bruun, P., 37
Caston, V. N. D., 222
Coleman, J. M., 87
Cook, D. O., 285
Crampton, P. J. S., 130
Csanady, G. T., 233
Curray, J. R., 73, 97, 130
Dietz, R. S., 51, 80
Dill, C. E., Jr., 169, 199
Emmel, F. J., 130
Fenneman, N. M., 14
Fischer, A. G., 26
Frankenburg, D., 126
Gadd, P. E., 326
Gorsline, D. S., 285
Han, G. C., 326
Inman, D. L., 255
Kraft, J. C., 199
Lavelle, J. W., 326
Ludwick, J. C., 310
Mayer, D. A., 326
Moore, D. G., 73
Murray, S. P., 213
Palmer, H. D., 217
Pilkey, O. H., 126
Rusnak, G. A., 255
Sanford, R. B., 169
Sheridan, R. E., 199
Stubblefield, W. L., 326
Swift, D. J. P., 169, 326
Wilson, D. G., 217
Wright, L. D., 87

COASTAL SEDIMENTATION

INTRODUCTION

Geologists have always found a certain fascination in shorelines, for it is in this region that three of the four ancient elements meet in a common interface. This juxtaposition of earth, air, and water form a planetary "triple point," and nowhere is the demonstration of interactions between geological agents and the earth's response more dramatically revealed than at the junction of the marine, atmospheric, and terrestrial realms.

Early geologists were drawn to shorelines because the work of marine processes provided a variety of features or situations relevant to their pursuits. Many came with singular purpose—to collect fossils or to examine structure exposed in cliffs—but others found a spectrum of geological features that lead to a broader understanding of terrestrial rhythms. One of the most perceptive and eloquent was Hugh Miller, and in his classic *Old Red Sandstone* of 1841, he describes the northern coast of the Moray Firth in Scotland as follows:

> I know not a more instructive walk for the young geologist than that furnished by the two miles of shore along which the section extends. Years of examination and inquiry would fail to exhaust it. It presents us, I have said, with the numerous organisms of the Lower Old Red Sandstone; it presents us also, towards its western extremity, with the still more numerous organisms of the Lower and Upper Lias; nor are the inflections and faults which its strata exhibit less instructive than its fossils or its vast denuded hollow. I have climbed along its wall of cliffs during the height of a tempestuous winter tide, when waves of huge volume, that had begun to gather strength under the night of the Northern Ocean, were bursting and foaming below; and as the harder pebbles, uplifted by the surge, rolled by thousands and tens of thousands along the rocky bottom, and the work of denudation went on, I have thought of the remote past, when the same agents had first begun to grind down the upper strata, whose broken edges now projected high over my head on the one hand, and lay buried far under the waves at,my feet on the other [Miller, 1859, p. 200].

Geologists' concern with degradation and aggradation along shorelines was a natural extension of the geomorphologists' studies of river deposits, channel morphology, and lacustrine features. In fact, much of the early theory regarding erosion and deposition in near-shore marine processes originated from G. K. Gilbert's (1890) scrutiny of the ancient shorelines of Lake Bonneville hundreds of miles from any sea. This concern pervades most writings in the early twentieth century, as shown by Barrell's statements regarding the continental shelf:

> Valuable studies on the principles controlling the character of the shore *line* have been made by Davis and Gulliver, but what is here emphasized is the study of the water *bottom,* both near shore and offshore, its sedimentary character and its form . . . . In such studies distinction must be drawn between the subaqueous profiles of aggradation and degradation [Barrell, 1915, p. 5].

We have collected a series of papers that addresses the subject of *Coastal Sedimentation.* However, this title requires some amplification, for it is one of several similar themes in this series. How does this volume differ from its companion volume, *Coastal Hydrodynamics and Beach Processes* (Fisher and Dolan, 1977)? What do we mean by sedimentation? In order to answer the first question, we must anticipate some of our subject material from Part III of this volume entitled, "Studies of Fluid Motion." From the point of view of coastal sedimentation, it is convenient to think of the coastal water mass as consisting of three hydrodynamic zones (Figure I.1). In an innermost zone of wave-dominated flow, water movements are driven by the regime of shoaling and breaking waves. Oscillatory wave surge is the dominant fluid motion and the source of energy for all the other motions, such as the wave-driven littoral current, rip currents, surf beat, and edge waves. The long-period motions characteristic of deeper water are damped out by friction in this shallow zone. The seaward extent of this zone depends of course on the prevailing wave regime, but on most coasts it extends to at least ten meters depth.

Further seaward lies a zone of wind-driven, friction-dominated flow. Oscillatory wave surge is still strongly felt on the sea floor when any sort of sea is running, but these rapidly fluctuating motions are almost always superimposed on a slowly varying component of flow that tends to occur during the same meteorologic event. This slowly varying component is wind induced and may be of the same intensity as wave surge, so that the resultant flow is a unidirectional, pulsating flow rather than a reversing oscillation. In the equation of motion describing the slowly varying flow, a wind

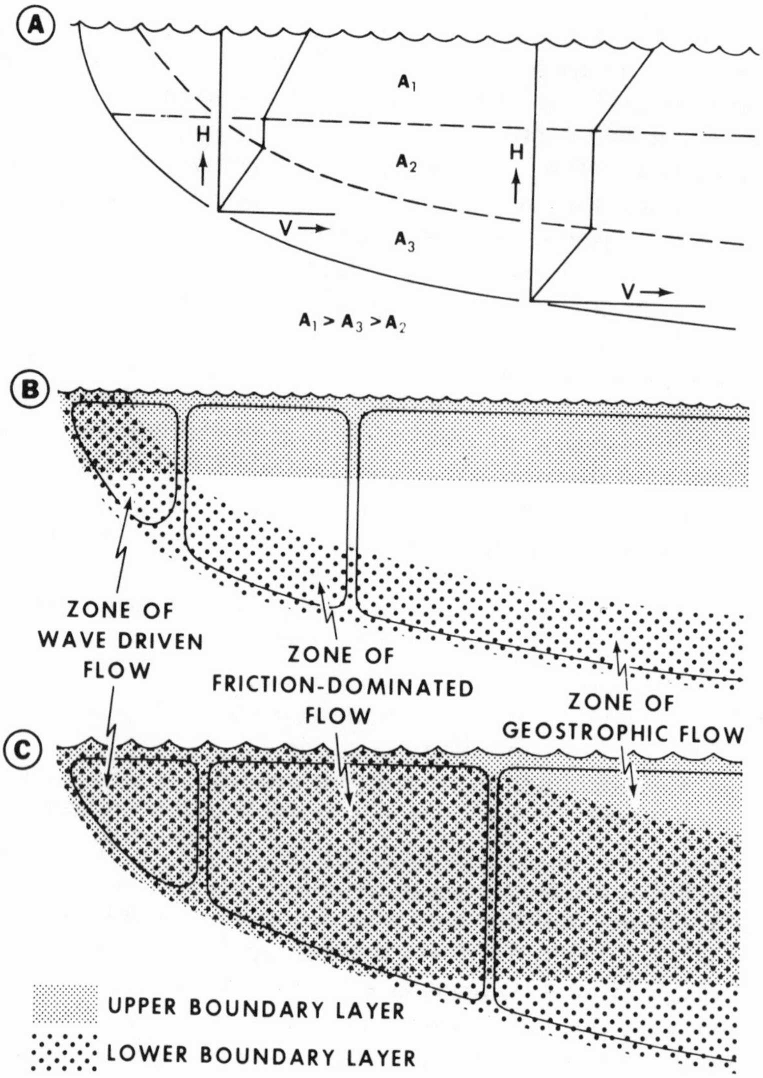

Figure I.1 Generalized velocity structure on an open marine coast. Section A: Velocity profiles throught the upper boundary layer, interior flow, and lower boundary layer, and relative values of eddy viscosity (A). H is depth; V is velocity. Section B: Velocity structure during a period of relatively mild flow. Section C: Velocity structure during peak flow. Source: Stanley, D. J., and Swift, D. J. P. (eds.), *Marine Sediment Transport and Environmental Management* (New York: John Wiley and Sons, 1974), p. 263.

stress term is set equal to a term representing bottom friction; hence, the term *friction-dominated flow.* The velocity profile is Couette-like in that it tends to consist of a uniform top-to-bottom gradient that shows little separation into distinct top and bottom boundary layers. The zones of wave-dominated and friction-dominated flow are together approximately equivalent to the coastal boundary layer (see Csanady's Paper 16).

An outer hydrodynamic zone of geostrophic flow extends from the zone of friction-dominated flow out to the oceanic front at the shelf edge. Here the water is sufficiently deep that Ekman's three-layer structure tends to arise in response to sustained wind stress. Upper and lower boundary layers develop vertical velocity gradients in response to wind friction at the surface or frictional retardation by the bottom, respectively. But flow in the central layer responds primary to a pressure gradient resulting from setup of the sea surface against the coast; velocity tends to be coast-parallel and uniform through this central range of depth. This flow is geostrophic in the sense that in the equation of motion, the frictional term is subordinate, and the pressure term is balanced primarily by the Coriolis term. These flow zones may be modified in areas of strong tidal currents, although tidal flows themselves divide into similar onshore and offshore zones.

On most continental shelves the behavior of flow zones and their dimensions change on a time scale measured in hours. They can be identified and mapped only by expensive arrays of current meters. Still, these are the patterns of fluid motion that govern sediment transport, and they comprise natural regions for the study of sedimentation. Despite the apparent similarity in titles, Fisher and Dolan's *Coastal Hydrodynamics and Beach Processes* deals exclusively with the innermost zone of wave-dominated flow. Our volume also considers this zone, but deals primarily with the second zone of friction-dominated flow. For the sake of categorization we might deal exclusively with this zone and title our volume "Near-shore sedimentation," which would not be very practical, however. The zone of friction-dominated flow is a transitional zone that actively exchanges sediment with the zone of wave-dominated flow and cannot be studied in isolation; hence, our adoption of the term *coastal.* We examine both of the inner zones and are therefore looking at "coastal hydrodynamics" on a larger spatial scale than does our companion volume.

By "sedimentation" we refer to sediment transport. We assume that the goal of the student of sediment transport is to determine the sources, pathways, and sinks of sediment in the system

being studied, to determine volume or weight rates of erosion, flux, and deposition, and to do this by means of direct measurement. The term *sedimentation* should refer to such studies, but it has instead come to mean inferences about such sediment movement based on classical geological examination of the sedimentary pile. The tools required for direct measurement are only now beginning to become available, and most of our important papers have reached their conclusions through such geological inference. Only a few papers herein, such as those of Ludwick (Paper 19) and Lavelle and others (Paper 20), are based on the kind of direct measurements that, as they become more widespread, must inevitably require a revision of this volume.

The papers in this volume are those that we and our colleagues have used most frequently in our own studies. They seem to fall naturally into a series of themes. The oldest and one of the most clearly defined is the theme of the *coastal equilibrium profile,* which is presented in Part I. The first scientists to consider the floor of the sea were geomorphologists who had only maps as data. As marine geology developed its own tools, *coastal deposits* (Part II), in terms of both surficial sediments and coastal stratigraphy, became subjects of study.

This classically geological line of development has been paralleled by two other hydrodynamically oriented lines of investigation. The evolution of beach and surf zone studies, as guided by the coastal engineering tradition, is the subject of other Benchmark volumes (Schwartz, 1972, 1973; Fisher and Dolan, 1977) and does not concern us here. The evolution of *studies of fluid motion* (Part III), in the tradition of fluid dynamics and physical oceanography is of particular interest as a field whose impact is only now beginning to be felt in the study of nearshore sedimentation. A final theme, presented as Part IV, *studies of substrate response,* consists of sedimentological papers informed by the fluid dynamical tradition.

REFERENCES

Barrell, J. 1915. Factors in the movements of the strandline and their results in the Pleistocene and Post-Pleistocene. *Am. J. Sci.* (4th ser.) **40**:1–22.

Fisher, J., and Dolan, R. 1977. *Coastal Hydrodynamics and Beach Processes.* Stroudsburg, Pa.: Dowden, Hutchinson & Ross, 451 pp.

Gilbert, G. K. 1890. Lake Bonneville. *U. S. Geological Survey Monograph 1,* 438 pp.

Miller, H. 1859. *The Old Red Sandstone,* or new walks in an old field. Reprint of the 7th ed. Boston: Gould and Lincoln, 427 pp.

Schwartz, M. L. 1972. *Spits and Bars.* Stroudsburg, Pa.: Dowden, Hutchinson & Ross, 452 pp.

——. 1973. *Barrier Islands.* Stroudsburg, Pa.: Dowden, Hutchinson & Ross, 451 pp.

Stanley, D. J., and Swift, D. J. P. (eds.). 1974. *Marine Sediment Transport and Environmental Management.* New York: John Wiley and Sons, 602 pp.

Part I

THE COASTAL EQUILIBRIUM PROFILE

Editors' Comments on Papers 1 Through 7

1 **FENNEMAN**
Development of the Profile of Equilibrium of the Subaqueous Shore Terrace

2 **FISCHER**
Stratigraphic Record of Transgressing Seas in Light of Sedimentation on Atlantic Coast of New Jersey

3 **BRUUN**
Sea-Level Rise as a Cause of Shore Erosion

4 **DIETZ**
Wave-Base, Marine Profile of Equilibrium, and Wave-Built Terraces: A Critical Appraisal

5 **MOORE and CURRAY**
Wave-Base, Marine Profile of Equilibrium, and Wave-Built Terraces: Discussion

6 **DIETZ**
Wave-Base, Marine Profile of Equilibrium, and Wave-Built Terraces: Reply

7 **WRIGHT and COLEMAN**
River Delta Morphology: Wave Climate and the Role of the Subaqueous Profile

One of the central threads of coastal sedimentation—the problem of the coastal equilibrium profile—begins in Paper 1. Most coastal profiles approach exponential curves, concave side up, with a shorter, steeper, nearshore limb. What is the origin of this profile, and what is its significance in coastal sedimentation?

Fenneman was a member of a school of classical geomorphologists that included Penck in Europe and Davis and D. L. Johnson in America. He was the first to deal specifically with the problem of the coastal profile, although the component ideas of equilibrium

morphological processes were in the air and were shared by his colleagues. Modern students exposed to the style of Victorian geomorphologists for the first time must find it rather startling; the rolling and sonorous phrases are forced to bear a heavy load. There are few illustrations and no equations. The effect of the latter rises to claustrophobic proportions in a long section, here omitted, that presents contemporary wave theory in entirely qualitative terms.

The literary style reflects a philosophic stance. The Victorian geomorphologists were inductive theoreticians. They observed geometric relationships on maps, conducted "thought experiments" in which scenarios were devised to explain the observed geometries, and inferred general relationships from these mental exercises. Perhaps the most that we can bring away from this first exploration of the problem is that the near-shore profile may be thought of as an equilibrium response to the wave climate and that coast erosion or coastal aggradation or both might play a role in its maintenance. These simple ideas were to have a profound and lasting impact in the history of coastal sedimentation.

The concept of the equilibrium profile reappears in Paper 2 after intermediate expression by three other writers, D. W. Johnson (1919), Ph. H. Kuenen (1950), and W. A. Price (1954). Paper 2 doesn't mention the equilibrium profile at all, but its observations nevertheless are crucial to the solution of the problem. Here we again see far-reaching conclusions drawn from relatively scanty data: Fischer has collected mollusc shells from the New Jersey beaches and from them infers the genesis of coastal stratigraphy. Fischer has an advantage denied previous students of the problem, for the school of Dutch coastal studies has begun to flourish and Fischer can draw upon the observations of Van Straaten and Kuenen as well as his own.

Fischer's central observation is that the shells of New Jersey barrier beaches are a mixed assemblage. Fresh-appearing shells of species indigenous to the shallow inner shelf are mixed with diagenetically altered lagoonal species. He infers (Paper 2, Figure 3) that the barrier islands have been retreating landwards in response to postglacial sea-level rise and that as they have done so, lagoonal deposits have been first buried by the advancing landward edge of the subaqueous barrier and then exhumed at the foot of the eroding barrier face.

The implications of this model for transgressive sedimentation become apparent in Paper 3. Bruun, a coastal engineer, postulates that a rise in sea level will cause a landward and upward translation of the equilibrium profile in which the shoreface undergoes ero-

sional retreat and the adjacent sea floor is aggraded to the new equilibrium depth by the debris eroded from the shoreface (Paper 3, Figure 4). The model is a kinematic or "black box" model; the concept of the equilibrium profile is assumed, not explained, and very little is said about the processes that move the sediment about. Bruun's Figure 4 lends an increased significance to Fischer's Figure 3, although Bruun does not appear to have been aware of Fischer's article.

While Fischer and Bruun were contemplating the nature of coastal erosion, David Moody, a graduate student at Johns Hopkins University, was intensively studying the morphology and behavior of the Delaware coast. His doctoral dissertation (Moody, 1964), not reproduced here due to space limitations, offers interesting additional confirmation of the Bruun hypothesis. On the Delaware coast the shelf floor seaward of the north-south–trending shoreface is not smooth but is corrugated, with sand ridges trending obliquely offshore toward the northeast. Careful study of past bathymetric surveys of the National Ocean Survey (then the U. S. Coast and Geodetic Survey) showed Moody that this was a very dynamic area. The shoreface was retreating, and the ridges were migrating to the southeast and extending their crest lines so as to maintain contact with the shoreface as they did so. Careful assessment of the sequence of bathymetric maps indicated a rough mass balance had prevailed on the Delaware coast between 1929 and 1961, in that shoreface erosion was compensated by deposition on the adjacent sea floor, which is Bruun's principle of erosional shoreface retreat. Furthermore, the time series, including the Great Ash Wednesday storm of 1962, was sufficiently detailed to give some insight into the dynamics of the retreat process (Moody, 1964). "The barrier face steepens over a period of years to a critical slope during which time the shoreline remains stable." However, a major storm such as the Ash Wednesday storm flattens the profile, drives it landward, and transfers the eroded sediment to the offshore sea floor.

Other field or laboratory studies that support the Bruun model for erosional shoreface retreat have been published by Ziegler and others (1964), Schwartz (1965, 1967, 1968), in Fisher and Dolan (1977), and Dillon (1970; reprinted in Schwartz, 1973). See Zenkovitch (1967) and references cited by him for the Russian point of view. Hayden and others (1975) present a statistical approach to the problem of inner shelf morphology.

The problem of the equilibrium profile appears once again in

Papers 4, 5, and 6 with a brisk exchange of argument and rebuttal between R. S. Dietz on one hand and D. G. Moore and J. R. Curray on the other. The issue at stake is whether or not the equilibrium profile concept of the old inductive school of geomorphologists can hold up in view of the vast amount of information available to marine geologists by 1963. Much of the debate is not directly relevant to us, because the proponents are considering the applicability of the equilibrium profile concept to the shelf as a whole. However, the coastal profile is an essential part of this discussion; hence, the papers do belong in our collection.

Dietz is here the iconoclast. He describes the concepts of wave base and the marine profile of equilibrium as largely erroneous. He agrees that a marine profile of equilibrium is developed in the near-shore zone, where it is associated with "a migrating lens of sand," but describes the adjacent shelf as drowned and relict. Moore and Curray argue that the concepts of wave base and the marine profile of equilibrium should be modified rather than rejected outright. Wave base is a climatic concept, and morphology is a time-averaged response to the wave climate. The marine profile of equilibrium has perhaps not yet been attained everywhere.

All members of the debate are clearly influenced by the ideas set forth by Shepard (1932) and Emery (1952) concerning the relict nature of shelf surfaces. No one seems to notice that Dietz' Figure 13 in Paper 4, cited as an example of a submerged relict fluvial surface extending almost to the beach, looks in fact very much like the dynamic, mobile inner shelf sand ridge topography that Moody was describing at that time on the Delaware coast.

We should mention at this point a thread that belongs to the more general theme of the equilibrium profile—That is, the null line concept, first proposed by Cornaglia in 1898 (reprinted in Fisher and Dolan, 1977). It envisages shoreface dynamics in terms of a Newtonian balance of forces experienced by a sand particle on the shoreface, in which the downslope (offshore) component of gravitational force is opposed by a net fluid force (averaged over a wave cycle) that is directed upslope (onshore) as a result of the asymmetry of surge experienced by a bottom beneath shoaling waves. As the shoreline is approached and the water becomes shoaler, wave surge becomes more intense, so that successively coarser particles may attain equilibrium at successive "null lines." Shoreface slopes as well as grain size would be controlled by this mechanism.

The hypothesis has been expressed in its most complete form

by Johnson and Eagleson (1966) and has been field tested by Miller and Ziegler (1958, reprinted in Fisher, 1977;1964). However, recent studies suggest that the null line mechanism is not of major importance, because slopes are not sufficiently steep over much of the shoreface, and because it tends to be overwhelmed by other mechanisms that are more important in determining sediment size gradients and bottom slope. These aspects are summarized in Swift (1976, p. 264–65) and in Komar (1976, p. 309–14). Several recent studies (Wells, 1967; Noda, 1972; Logvinenko and Barkov, 1973) have examined other ways in which oscillatory wave surge may interact with a cohesionless substrate so as to determine its slope, but these studies deal expressly with bottom slope near or in the surf zone and have not yet been applied to the shoreface as a whole.

In the final paper dealing with the equilibrium profile, we at last see classical geological and fluid dynamical concepts blending in the proportions required by the subject matter. We see the emergence of advanced computational techniques that are state-of-the-art at the time of this introduction. An algorithm for computing a coastal wave climate in terms of wave power in ergs per second per centimeter of wave crest is used (Paper 7, Table 2), and the results are compared with a series of subaqueous delta profiles (Paper 7, Figure 1). The implicit, qualitative insights of Fenneman (Paper 1) and Price (1954) are finally presented in a systematic quantitative form, although the authors do not appear to be aware of the history of their topic. The gain in insight is dramatic, but the statistical approach cannot take us all the way to the goal: The definitive *analytical* model of shoreface response to wave process is yet to be presented.

REFERENCES

Dillon, W. P. 1970. Submergence effects on a Rhode Island barrier and lagoon, and inferences on migration of barriers. *J. Geol.* **78**:94–106.

Emery, K. O. 1952. Continental shelf sediments of Southern California. *Geol. Soc. Am. Bull.* **63**:1105–08.

Fisher, J. S., and Dolan, R. 1977. *Coastal Hydrodynamics and Beach Processes.* Stroudsburg, Pa.: Dowden, Hutchinson & Ross.

Hayden, B., Felder, W., Fisher, J., Resio, D., Vincent, L., and Dolan, R. 1975. Systematic variations in inshore bathymetry. *Office of Naval Research Geography Programs,* Tech. Rept. No. 10, 51 pp.

Johnson, D. W. 1919. *Shore Processes and Shoreline Development.* New York: John Wiley & Sons, 584 pp.

Johnson, J. W., and Eagleson, D. S. 1966. Coastal processes, p. 404–92. In A. J. Ippen (ed.), *Estuary and Coastline Hydrodynamics.* New York: McGraw-Hill, 744 pp.

Komar, P. D. 1976. *Beach Processes and Sedimentation.* Englewood Cliffs, N. J.: Prentice-Hall, 429 pp.

Kuenen, Ph. H. 1950. *Marine Geology.* New York: John Wiley and Sons, 508 pp.

Logvinenko, N. V., and Barkov, L. K. 1973. On the Dynamics and Prediction of the Relief of an Underwater Shore Slope. *Oceanography* **14**: 238–244.

Miller, R. L., and Zeigler, J. M. 1958. A model relating dynamics and sediment pattern in equilibrium in the region of shoaling waves, breaker zone, and foreshore. *J. Geol.* **66**:417–41.

———. 1964. A study of sediment distribution in the zone of shaling waves, p. 133–53. In R. L. Miller (ed.), *Papers in Marine Geology—Shepard Commemorative Volume.* New York: Macmillan, 530 pp.

Moody, D. W. 1964. Coastal morphology and processes in relation to the development of submarine sand ridges off Bethany Beach, Delaware. Unpublished Ph.D. thesis, Johns Hopkins University, Baltimore, 167 pp.

Noda, E. K. 1972. Equilibrium beach profile scale-model relationship. *J. Waterways, Harbors & Coastal Engineering Div.,* No. 9367, WW4, Nov., pp. 511–528.

Price, W. A. 1954. Dynamic environments: Reconnaissance mapping, geologic and geomorphic, of continental shelf of Gulf of Mexico. *Trans. Gulf Coast Assoc. Geol. Soc.* **4**:75–107.

Schwartz, M. L. 1965. Laboratory study of sea level rise as a cause of shore erosion. *J. Geol.* **73**:528–534.

Schwartz, M. L. 1967., The Brunn theory of sea level rise as a cause of shore erosion. *J. Geol.* **75**:76–92.

Schwartz, M. L. 1968. The scale of shore erosion. *J. Geol.* **76**:508–517.

Schwartz, M. L. 1973. *Barrier Islands.* Stroudsburg, Pa.: Dowden, Hutchinson & Ross, 451 pp.

Shepard, F. P. 1932. Sediments on continental shelves. *Geol. Soc. Am. Bull.* **43**:1017–1039.

Swift, D. J. P. 1976. Continental shelf sedimentation. pp. 311–350, In Stanley, D. J., and D. J. P. Swift, eds., *Marine Sediment Transport and Environmental Management.* New York: John Wiley & Sons, 602 pp.

Wells, D. R. 1967. Beach equilibrium and second order wave theory. *J. Geophys. Res.* **72**:497–509.

Zeigler, J. M., Tuttle, S. D., Geise, G. S., and Tasha, H. J. 1964. Residence time of Sand composing the beaches and bars of outer Cape Cod. In *Proc. IX Conf. Coastal Engineering Am. Soc. Civil Engineers,* pp. 403–416.

Zenkovitch, V. P. 1967. *Processes of coastal development.* New York: John Wiley, 738 pp.

1

Reprinted from *J. Geol.* **10**:1–2, 22–32 (1902)

DEVELOPMENT OF THE PROFILE OF EQUILIBRIUM OF THE SUBAQUEOUS SHORE TERRACE

N. M. Fenneman

THE profile of a shore as seen at any one time is a compromise between two forms. One of these is the form which it possessed when the water assumed its present level; from this form it is continually departing. The other is the form which the water is striving to give to it; toward this form it is continually tending. There is a profile of equilibrium which the water would ultimately impart, if allowed to carry its work to completion. The continual change of shore line and the supply of new drift are everchanging conditions with which no *fixed* form can be in equilibrium. There are, however, certain adjustments of current, slope and load which, when once attained, are maintained with some constancy. The form involved in these adjustments is commonly known as the *profile of equilibrium*. When this profile has once been assumed the entire form may slowly shift its position toward or from the land, but its slope will change little or not at all. It may be compared to a stream channel which has reached grade but not base level.

The force which the water exerts is derived ultimately from the wind. The immediate agencies in the work are waves and currents. It will be convenient to consider these first as acting independently of the wind which caused them, and second, as acting under its continuous influence. It is also desirable to consider waves first in their free forms, while meeting no resistance and hence doing no external work. This condition is found in deep water. The various ways in which the bottom or shore may offer resistance and be subject to work may then be discussed.

[*Editors' Note:* Material has been omitted at this point.]

PROFILES RESULTING FROM FORCES DISCUSSED ABOVE.

In the actual operation of the forces discussed above, the resulting action on a sloping bottom may be outward at all places, or inward at all places, or outward over one part and inward over another. Forces in either direction may be gradually augmented or diminished. The different forces are capable of different combinations. Each set of conditions will lead to certain features of profile. If there be no change of condition, a permanent profile of equilibrium may be reached. The constant supply of load constitutes an ever shifting condition. Equilibrium as commonly realized depends on the uniformity of this supply.

Factors in profile-making.—The agencies which shape the marginal bottom may be treated in three groups, (1) oscillatory wave action and undertow, carrying material from shore; (2) on-shore currents and translatory wave action, carrying the material toward the shore; (3) currents alongshore. The tendency of the first group is to steepen the slope from the water's edge to the line at which its erosive power ceases, and deposition begins and to reduce the slope beyond that line. There is also for the second group a line of maximum power on the bottom, within which their effect is to steepen the profile by accumulation at the water's edge, and beyond which the slope is reduced by cutting down. Currents alongshore will be introduced later.

Conflict between on-shore and off-shore action.—The first two pairs of agencies are in conflict as to the direction in which bottom materials are to be moved. If all the water which moves shoreward must return over the same area and as a bottom current, this current would seem to have greater efficiency than the one above, moving in the opposite direction. This is certainly the case where translatory waves are not favored, as where the off-shore slope is steep. Where slope is gentle and translatory waves are well developed, they have one decided advantage. They are short as compared with the distance from wave to wave; hence all the shoreward movement of the water is concentrated into a small portion of the entire time. Divers are said to feel the passing of one of these waves as a sudden jerk between intervals of quiet. The undertow, on the other hand, has a steady flow except as interrupted by these sudden reverses.[1] The laws of energy give to these concentrated movements a much greater efficiency than to the same amount of motion more evenly distributed in time. On many shores of gentle slope, sand is worked landward, and in this process the agency just mentioned is doubtless important. The effect here referred to is that of waves of translation and is therefore inside the breaker line. It might accumulate sand on-shore but not in off-shore barriers. The dominance of shoreward action is essentially temporary (omitting currents alongshore from consideration). Its effect is to steepen by narrowing the slope. This steepening, in turn, is adverse to waves of translation.

Laws of equilibrium; eroding currents.—Ignoring the presence of a bank and the load derived from it, a current of uniform power tends to reduce the bottom to a level surface, that is, to require equal depth throughout. Equilibrium cannot exist on a level bottom where the power of the current is unequal at different places. In such cases, the depth must suffer a corresponding change until the power of water on the bottom is

[1] HENRY MITCHELL, "On the Reclamation of Tide-Lands and its Relation to Navigation," *Report of the U. S. Coast and Geodetic Survey, 1869*, Appendix 5, p. 85. In this paper Mr. Mitchell takes the extreme view that the sea restores to the continent "all the material washed from its bluffs and headlands." Certain exceptions are made for islands.

everywhere the same. A current of uniformly increasing power requires a uniformly increasing depth, that is, a plane slope. The opposite is true for a current of uniformly diminishing power. A current whose power is augmented at an *increasing rate*, as, for example, in geometrical ratio, requires a descent to deep water on a curve which is convex upward. Increase of power at a diminishing rate requires concavity. Loss of power at increasing rates, and loss at diminishing rates, require concavity and convexity respectively.

Uniform cutting or building.—If a uniform current on a level bottom has eroding power, the whole will be cut down at the same time, and the bottom will remain level while depth increases. In this case the load is furnished at all points equally, and is all carried forward at the same rate. If load be furnished in excess of carrying power, and at all points uniformly (as from top or sides), then the level surface of the bottom would be preserved while depth would decrease.

Load derived from the shore.—To make the case applicable to undertow, the excessive load must be supposed to be furnished at the end where the current enters upon the bottom in question. In this case deposition will first reduce the load at the end upon which it enters and at the same time reduce the depth and thus constrict the current, increasing its power. The latter influence will determine a higher level to which the bottom will be built; a level at which the power of the water is sufficient to carry the load which before was excessive. Filling will then advance forward over the bottom, the filled and unfilled portions both being level, the former growing while the latter diminishes, and the two being separated by a slope, mentioned below. It is evident that the depth at which this slope begins is determined jointly by the power of the water, the amount of the load, and the size of the fragments which make up the load.

The front.—The shape of the slope which intervenes between the area which has been filled and the bottom beyond, will be determined by the rate at which the power of the current decreases. If the loss of power were instantaneous, the slope

would be simply the subaqueous earth slope. If it be in any arithmetical progression, the slope will be a plane whose steepness will vary with the rate of decrease, the slope being steeper when the rate is higher. If the loss of power be in some other manner than by arithmetical difference, the slope will show a curve which will be convex or concave according as the rate of decrease is augumented or diminished. In actual deposition by a current advancing into deep water, the decrease of power is at an increasing rate, as may be seen from the following. If a plane slope be assumed, so that depth increases in arithmetical ratio, then the velocity of the current will decrease in similar ratio, but transporting power varies as the square of the velocity, hence its rate of decrease is progressively augmented. This will require convexity of slope, a feature generally observed at the edge of embankments and subaqueous terraces. The general law of equilibrium, as given above for an eroding current still applies; current power is uniform over all parts of the bottom, if by the term *current power* is understood *power with referrence to load* and the current considered is the *resultant of all conflicting currents*. In this case, while the current is acually losing power, the loss is balanced by the coincident loss of load, and the uniformity of power in comparison with load is maintained.

Presence of a bank; equilibrium on a slope.—The presence of a bank fixes not only a horizontal limit to the bottom in question, but determines that at this limit the depth shall be zero. This involves a slope. If equilibrium is to exist on this slope in harmony with the general law stated above, the advantage in power due to shallower water on one side must be balanced in one of four ways, (1) the equality of transporting power in deep and shallow water may be partially maintained by the participation of more water where the depth is great than where it is small. In the case of undertow this has been shown to be true; (2) currents in both directions may be stronger, so that the resultant motion in one direction may be more in shallow water than in deep water, it may even be zero or it may be in the opposite direction. The factors of translatory wave motion and on-shore

currents may occasion this condition; (3) the excessive power of the water on the shallow bottom may be employed in the transporting of a greater load or even in erosion. This is quite generally true; (4) the material may be heterogeneous, the larger stones coming to rest in the shallower water because of their ability to withstand the greater agitation at a higher level. Of all these reasons, it will be seen that only the first can provide for a permanent slope: the others depend upon a continual supply of fresh drift.

Necessity of a continuous supply of load.—Suppose now that a short section of coast line be enclosed between perfectly resistant walls or piers perpendicular to the shore line, and extending out to deep water. The transportation of material alongshore will thus be prevented. If the shore also be supposed to be perfectly resistant, so that no new drift can be furnished to the waves, then the profile of equilibrium, toward which the bottom will tend, is a steep descent from the water line to the depth at which undertow becomes ineffective, and then a low slope outward, following the base of effective undertow. This base is necessarily on a slope because of the increasing volume of undertow with distance from shore.

Effect of a supply of drift.—If now, drift be supplied at the shore line at a given rate, filling will occur at the foot of the steep descent leading down from the water line, until the bottom has risen to a level at which the power of the water is sufficient to transport the material at the rate at which it is furnished, and this filling will advance off-shore, ending in a convex front as shown above.

At the shoreward boundary of this filling area is an angle made by the plane of deposition, with the steeper descent leading down from the water's edge to the line at which deposition becomes possible. In an actual case, where the material of the shore yields to erosion, the water's edge is carried landward, and the first descent is not only far from vertical, but in weak material, is very gentle; probably always steeper, however, than the slope made by deposition farther out. This may be observed

on almost any of the coastal charts of the United States Coast and Geodetic Survey. The east coast of Florida furnishes typical illustrations.

Normal profile; cutting coast.—The normal profile then, of a shore where the resultant of transporting power is outward, is a compound curve, which is concave near the shore, passing through a line of little or no curvature, to a convex front. Where this front rests upon the bottom below the reach of currents, the descent merges into the more level bottom by another concave curve, due to deposition from suspension. If the supply of material from the shore be cut off, the entire shelf will be cut down and its slope reduced and it will necessarily be separated from the shore by a steeper slope than before. If, on the other hand, the supply of material be suddenly increased, a smaller shelf will grow from shore on the surface of the older, for the reason that the new load, being greater, is in equilibrium with the currents at a higher level than before. The greater the load, the nearer will the surface of deposition approach that of the water. On the Atlantic coast of the United States, the depth at which the concave curve merges into the plane of deposition varies from three fathoms near the mouths of some rivers, to ten or twelve fathoms where the lead is smaller. On some parts of the Pacific coast, where the lead is small, the concave curve descends to twenty or thirty fathoms.

Normal profile; building coast.—If the resultant of shore action be to carry material landward, the general character of the resulting curve cannot be very different, since this process also produces steepening near shore. In general the velocity of shoreward motion increases with nearness to land. If the effectiveness of this motion increases with its velocity, there is no accumulation until the shore is reached. The shore is then progressively steepened by accumulation, until the force which acts shoreward can no longer carry material up against the growing component of gravity. This landward urging of sediments is commonly thought to be one of the factors in the production of off-shore barriers. It is plain, however, that

unless the power of inward transportation is decreased before reaching the shore, no barrier can form. This decrease may, at times occur, for carrying power will depend not only on the velocity, but on the agitation of waves at the bottom. It has been seen that waves are rapidly reduced in size and vigor in the act of breaking. It is possible, therefore, that when the slope is so gentle that waves recover their form after breaking, thereby showing that oscillatory wave motion has been much reduced, deposition may take place along the line of wave reduction, which is essentially the breaker line. With these conditions alone, however, the growth of this feature would probably be confined to narrow limits by the undertow. It would, moreover, be a very transient feature, a mere incident in the process of shoreward transportation. The steepening of the shore, to which this process is incidental, would rapidly remove the conditions of the incident.

Variations of the compound curve.—The compound curve will be more marked in proportion as the surface of deposition is broad and its slope is gentle. Where it is narrow its significance may not appear from a profile drawn from widely spaced soundings.[1] If all the waste from the land be carried alongshore, the marginal terrace is of the cut type purely, in which the compound curve is not noticeable, the only prominent angle being that where the surface of cutting intersects the original steeper bottom.

Currents alongshore.—If the effect of currents alongshore were the same at all distances from land, they might be ignored as a factor in profile making. Their variation in strength at different distances from shore produces important results. It has been stated above that *for any one current* the power at the bottom with respect to the load must remain constant. It may also be shown that of *two currents*, each of which is furnished with load to its full capacity, the stronger, which may be supposed to dissipate gradually, will be in equilibrium with its load at the smaller depth. Hence if transportation alongshore be

[1] This is illustrated at many places on the Pacific coast of the United States.

distinctly greater in a zone adjacent to the land, a smaller terrace will rest upon the larger. If transportation parallel to the shore line be distinctly greater in a zone off-shore, and the supply of drift be at hand, a ridge will be built along the line of this more effective current.

Barriers.—It has been shown above that when the off-shore slope is too low for equilibrium, and there are no currents alongshore, steepening is effected, in the main, by accumulation at the water's edge, though there may be some small tendency to accumulation at or just within the breaker line. When currents are flowing, they have a zone of greater efficiency along this same line or just outside. This is because the material which they transport is more agitated by wave action, and is to some extent lifted into the current. Excessive transportation along this zone initiates the ridge which may continue to grow until it assumes the functions of the beach. It this then called a barrier.

The essential function of the barrier is to steepen the bottom slope by carrying the shore line farther out. If the slope is not abnormally low, the barrier is not needed; nor are the conditions present which make its formation possible, one of these conditions being that the agitation on the bottom at the breaker line should exceed that nearer shore. It was seen above that this condition is present, only on a deficient slope.

The slope may become deficient in several ways. The currents themselves might be the cause; or it may result from the sediments delivered by streams, as at many places on our Atlantic coast; or the gentle slope may have belonged to the original bottom over which the waters rose, as seems to have been the case with Lake Michigan in its former extension in the vicinity of Chicago. Doubtless far the most frequent occasion of deficient slope is the falling of the water level or the rising of the shore. That the immediate off-shore slope should in this case be too low, is the necessary consequence of the concavity of the normal slope near shore. The slope from the Atlantic shore line, where well removed from rivers, as on the

east coast of Florida, is perhaps ten fathoms in the first two miles, but if the sea level should fall ten fathoms, or the land should rise by that amount, the new ten-fathom line would lie many miles off-shore, and new barriers might be expected. On some of the small lakes of Wisconsin, especially those without outlet, as Silver Lake of the Oconomowoc group, the falling level has found a deficient slope and barriers are constructed.

The front of the marginal shelf.—If the marginal shelf be a pure wave-cut terrace with no addition by deposit, its limit will be marked by an angle where the plane of the shelf meets the original bottom. The depth of the shelf at this edge will constantly approach wave-base, for it may be safely assumed that wherever waves can agitate, there will be sufficient current to transport. If there are currents strong enough to erode below wave-base, the shelf may be cut still lower. The hardness of the rock can make no permanent difference. This is well illustrated even in so young and small a body as Lake Mendota at Madison, Wis., where the sandstone shelves southwest of Governor's Island and Maple Bluff are cut to the same depth as the clay shelves west of Picnic Point and Second Point.[1]

If the shelf is being broadened at the same time by materials carried across and deposited on its front, there will be, between its upper surface and its steep front, a curve convex to the sky as shown above. This steeper slope begins, not at the depth where the power of the water ends, but at the depth at which the power of the water becomes insufficient to carry the entire load. From this depth the slope becomes progressively steeper to the depth at which the movement of the water is ineffective. Off the Atlantic coast of the United States, the depth at which the slope begins to steepen is usually fifty or sixty fathoms, but the maximum of steepness is not attained until a much greater depth is reached. The depth familiarly assigned to wave-base along this coast is one hundred fathoms, and this figure expresses fairly well the horizon at which the maximum steep-

[1] See hydrographic map issued by the Wisconsin Geological and Natural History Survey.

ness is reached. This would mean that currents become unable to carry the *whole load* at fifty or sixty fathoms, and at one hundred fathoms or less, become unable to transport anything except in suspension. If the factor of transportation in suspension did not enter, the front of such a shelf should show the subaqueous earth-slope.

It is commonly assumed as above, that undertow and wave agitation lose their efficiency at the same point, the limit of the former being determined by that of the latter. Probably this is very generally true; moreover, since wave oscillation decreases with depth in geometrical ratio at a high rate, and the decrease of its agitating power is at a rate measured by the square of this same ratio, it may readily be seen that there is a somewhat definite horizon below which wave action is ineffective. Such a condition is signalized by a somewhat definite limit to the sedimentary shelf.

Transportation beyond wave-base.—The undertow may, however, be constricted laterally and preserved from dissipation, as when the water drifts into a re-entrant curve of the shore; or deep currents may result from a system of rebounds. By either of these means the power of the lower water may be increased, so that at depths greater than that of wave-base sand or even gravel may be transported.[1] In such cases no break in the profile may be seen at wave-base. Broad sheets or streaks of sand may cover the bottom to depths far beyond this line. Such troughs as those of the great lakes, in which all the surface water may be drifted simultaneously in one direction, should especially favor vertical circulation with vigorous movements below. Wave-base of Lake Michigan, where revealed by a sharp angle at the edge of a marginal terrace, is sixty or seventy feet below the surface; yet around much of its margin, a sand covered or gravel covered bottom, concave upward, extends outward to several times this depth with little or no evidence of change of slope at wave-base.[2] This is to be expected from the

[1] See H. C. KINAHAN, "The Beaufort's Dyke off the coast of the Mull of Galloway," *Proceedings of the Royal Irish Academy*, Third Series, Vol. VI, No. 1.

[2] Charts of Lake Michigan, War Department.

necessarily powerful undertow. In Lake Mendota, where wave-base is not lower than twenty feet, sands and even heavy gravels are irregularly distributed over the bottom at depths frequently approaching fifty feet. Some lie at the bases of steep slopes which gravity may have helped them to descend, but others are far from slopes and plainly illustrate the erosive power of currents resulting from a concentration of movement along certain lines.

2

Reprinted from *Bull. Am. Assoc. Pet. Geol.* **45**(10):1656–1666 (1961)

STRATIGRAPHIC RECORD OF TRANSGRESSING SEAS IN LIGHT OF SEDIMENTATION ON ATLANTIC COAST OF NEW JERSEY[1]

A. G. FISCHER[2]
Princeton, New Jersey

ABSTRACT

The presently transgressing sea has a marginal low and intermediate energy zone of tidal marshes and lagoons, protected from the wave energy of the open ocean by barrier beaches. Peats and muds are being deposited in the tidal marshes, sands and muds in the lagoons. The sands are largely derived from the ocean, and are deposited as tidal deltas, which start at gaps in the barrier beaches, and grow across the lagoon toward the mainland. The lagoons have biotas of variable character, depending on bottoms and salinity, but distinct from those of the open sea.

With the rise in sea-level, the lagoons transgress over the land, and the barrier beaches follow behind them, transgressing over older lagoonal deposits. The waves breaking at the face of the barrier are cutting into older lagoonal sediments, and are reworking "fossil" lagoonal shells into the sands of the open-shore and near-shore bottoms. The aragonitic members of the lagoonal fauna have been diagenetically eliminated, but distinctive calcitic lagoonal shells are thus being mixed into the open-sea assemblage. It is not known how much, if any, of the lagoonal sedimentary record escapes this erosion by the advancing surf zone.

Transgressive seas of the past, advancing over a lowland surface, are likely to have had a marginal belt of brackish or saline lagoons. The sediments which come first to be deposited on the former land surface are low-energy sediments of non-marine or semi-marine character. Insofar as these primary deposits of the transgression are not entirely removed by the surf zone advancing behind them, their remnants come to be disconformably overlain by the higher-energy, offshore deposits of the main seaway. Thus transgressive sequences are likely to contain a marked intra-sequential disconformity.

Transgressive sequences of this sort, and matching regressive sequences, which are longer and lack the disconformity, are found in the Cherokee (Pennsylvanian) cyclothems of southeastern Kansas and adjacent parts of Oklahoma and Missouri.

The classical Pennsylvanian cyclothems of the Illinois basin can also be interpreted in a parallel manner. Here the middle and upper shales are considered as the sediments of transgressive and regressive lagoons respectively—lagoons in an extended sense, namely, more or less brackish seas occupying the whole of the Illinois basin, and barred in part tectonically, in part by limestone deposits (calcarenites and algal banks).

Introduction

The geologist working with sedimentary rocks is constantly challenged by problems of paleoecology—under what conditions and in what manner did a sediment accumulate? The general approach to this problem is a search for intrinsic clues—mineral composition, fabric, sedimentary structures, and biotic content, matched to recent sediments deposited in known environments. There is, however, another line of evidence which may be used to good advantage, to check and supplement conclusions based on intrinsic evidence. The relations of a body of sediment to its lateral facies, and to its preceding and succeeding rock types, can offer powerful clues to its origin, for only certain ecologic settings can lie next to each other, and only certain environments can succeed one another. Even disconformities, occurring regularly in certain rock sequences, offer important evidence of the sequence of events, and can thereby provide clues for the interpretation of underlying and overlying strata.

We should therefore go to sites of modern deposition not only with a view toward finding out what sorts of sediments are accumulating in certain settings, but also to determine how these settings have shifted and are shifting, in post-glacial time, and what sort of stratigraphic relationships have been, are being, and will be brought about by the shifts taking place. Detailed investigations along these lines have been undertaken in the Gulf Coast region by various individuals and groups, particularly by Fisk and co-workers, by other petroleum company research groups, and by Shepard and co-workers in API project 51. Similar studies are needed elsewhere, on other coasts, in different tectonic settings.

This paper records some preliminary observations on the transgressive post-glacial record on the coast of New Jersey, and goes on to a contemplation of older transgressive and regressive stratigraphic sequences, particularly the cyclothemic Pennsylvanian deposit of the Cherokee type and of the Illinois type.

[1] Manuscript received, December 20, 1960.

[2] Princeton University.

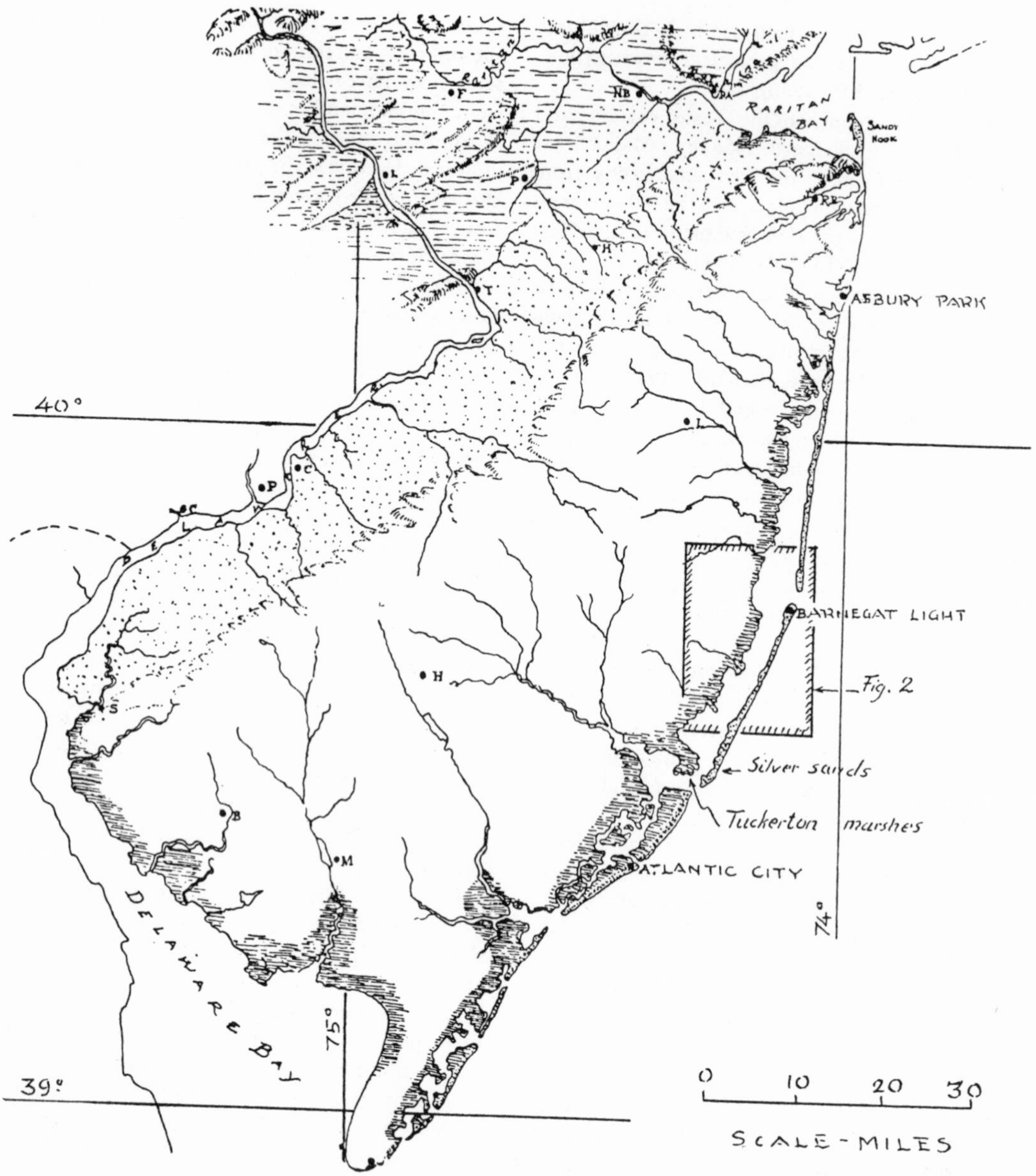

FIG. 1.—Index map showing lagoons and tidal marshes of New Jersey. Taken from Physiographic Map of New Jersey by W. Goodwin, and reproduced with author's permission.

Special thanks are due to Franklyn B. Van Houten, who read the manuscript critically, and guided the writer to much of the pertinent literature.

SEDIMENTATION ON NEW JERSEY COAST

The coast of New Jersey fronts on the open Atlantic Ocean, but also features large bodies of restricted water: the estuaries of Raritan Bay and Delaware Bay, and, along the intervening coast, lagoons such as Barnegat Bay (Figs. 1, 2) separated from the Atlantic Ocean by narrow barrier beaches. These lagoons are more or less brackish; they are fringed by tidal marshes, and contain marsh islands. They bear some resemblance to the Wadden Sea, along the coast of the Nether-

lands and Germany, which has been the subject of many ecological and geological investigations (Van Straaten, 1954, 1956).

Bays and lagoons on the Texas coast (Ladd and others, 1957, Shepard, 1956) and Maine coast (Bradley, 1957) have received some study from geologic-ecologic viewpoints, but the only geological study concerned with sediments of the New Jersey lagoons is Lucke's (1934a, 1934b, 1935) pioneer investigation of the Barnegat inlet tidal delta. And while the common species of shell-bearing invertebrates may readily be determined from such handbooks as Richards (1938) and Abbott (1954), the structure and distribution of the organic communities remain to be worked out.

Our knowledge of the Quaternary stratigraphy is even more deficient. Borings, study of underwater outcrops, and geochemical dating will be necessary to establish adequately the relations of the present depositional and ecologic setting with the stratigraphical record of past events.

During the past three years, the writer has visited the New Jersey bays and barrier beaches repeatedly, with students. Our scattered observations are no substitute for systematic research; yet they shed some light on the kind of geological record being developed in this area, and lead to thoughts about transgressive sequences in general. These observations, and the discussions which resulted from them, are the basis for this paper.

Barriers and inlets.—The barrier beaches (Johnson, 1919) are composed of sands derived in part from local sea-floor erosion and in part from coastal cliffs of the Atlantic Highlands at the north. These sands have been piled up by the breaking waves, and redistributed in dunes by wind action. The lagoons thus formed communicate with the ocean through gaps (inlets) in the barrier beaches. Some of these gaps (for example, Barnegat inlet) are narrow, and are swept by swift tidal currents.

Lagoons.—The lagoons are several miles wide and tens of miles long. Fresh water enters them from various rivers, sea water via the inlets. The general composition of the lagoonal waters is thus brackish, and shows salinity gradients decreasing from the inlets into the remoter parts. The faunas reflect these gradients: starfishes, for example, occur near Barnegat inlet, as does *Mytilus*, but neither of these forms penetrates far into the bay.

Extensive areas in the lagoons are floored by sand. Most of this contains some admixture of silt and clay, and some of it is pebbly. Some of this sand may be carried into the lagoon by rivers from the mainland, and some is probably blown in from the dunes of the barrier beaches at times of storm. However, the great bulk of this sand is carried into the lagoons from the seaward side, to form tidal deltas. A classic delta of this type is the one at Barnegat inlet (Fig. 2), described by Lucke (1934a, 1935). On the rising tide, swift currents flow through the inlet into the lagoon, where they spread out in distributary channels. They carry with them a load of sediment, which is dropped as their velocity decreases. On the ebb tide currents flow in the opposite direction, and carry sediment back out, but the amount returned to the sea is less than the amount brought in, and the net result is a delta-like accumulation of sediment, with its head at the inlet, and its growth directed laterally into the lagoon, and across the lagoon toward the mainland. Lucke showed that the sands in this delta are coarsest and cleanest near the head, and become finer and muddier toward the land. The bottom topography of the New Jersey lagoons suggests that they are largely floored by such tidal deltas, each of which was actively growing so long as it was supplied through an inlet; as the position of inlets changed with time, so did the location of active tidal deltas (Lucke, 1934a, 1934b).

Finer muds are less abundant in the investigated parts of the lagoons. Lucke found organic-rich muds in pockets of channels of the tidal delta, and the writer has observed superficial accumulations of mud on shallow bottoms near ship-channels—possibly the result of winnowing of bottom sediment during dredging or by ship traffic. Muds may be more important in the fresher, less tide-current-swept parts of the lagoons, in which the waters are also notably more turbid, but there has been no systematic sampling. One of Lucke's boreholes, located close to the mainland, opposite Barnegat inlet (Lucke, 1934a), found 7 feet of sand underlain by 18 feet of uniform gray clay, without reaching the bottom of this unit.

Tidal marshes.—Fringes of tidal marsh accompany the lagoons on the mainland and on the barrier beach side, and tidal marsh likewise covers the islands which have been developed in the lagoons by the growth of tidal deltas. These marshes are densely overgrown by a variety of halo-

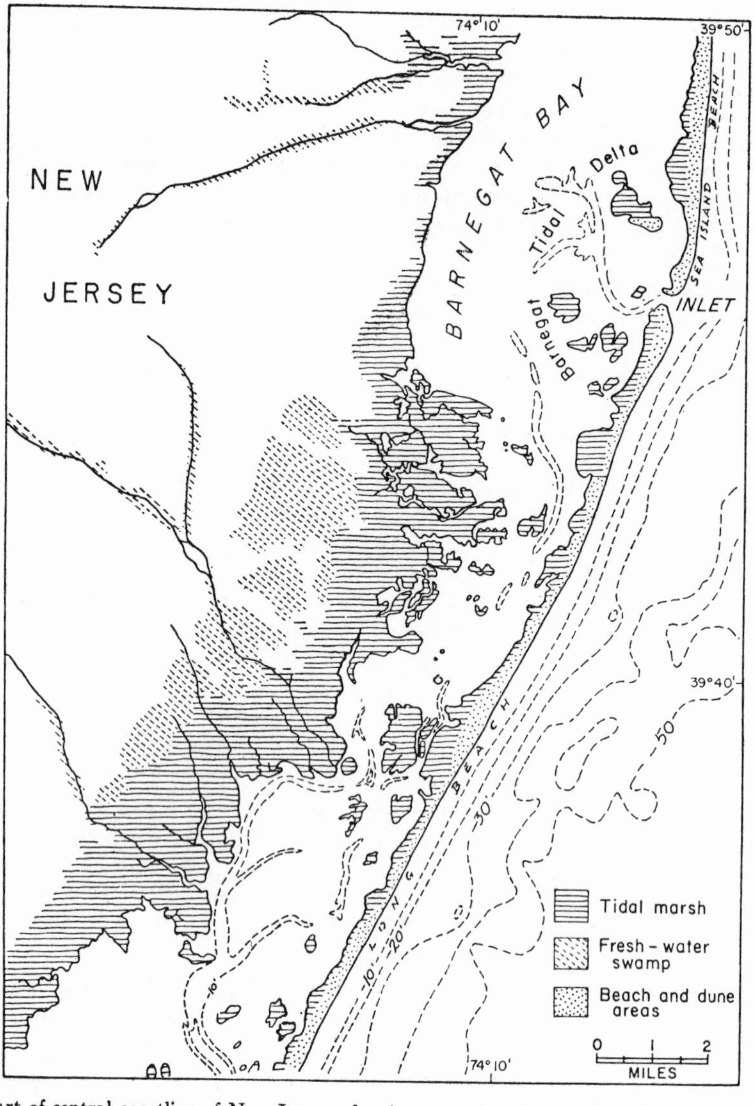

FIG. 2.—Part of central coastline of New Jersey, showing coastal swamps and tidal marshes, lagoon with islands of tidal marsh, barrier beach with inner fringe of tidal marsh and outer belt of sand dunes and beach, and ocean. Largely from New Jersey Dept. of Conservation and Development Atlas Sheet 33.

phytic plants, in particular members of the sedge genus *Spartina*. Commonly the ground between sedge plants is covered by mats of red or green filamentous algae. On the mainland side, the salt marsh tends to grade landward into fresh water marshes or wooded swamps. The width of the tidal marsh commonly exceeds a mile, despite the fact that the normal tidal range on the mainland side of Barnegat Bay is only about one foot. In places large expanses of the marsh are unbroken; elsewhere the marshes are laced by channels—the estuarine mouths of sluggish streams, or tidal channels separating marsh islands. Much of the tidal marsh has been modified by drainage ditches and dredging and filling for houses, boat channels, and anchorages. Sediments in the channels are sand. In some areas the tidal marsh sediments proper are also sandy. This occurs where the marsh is growing out over intertidal sand flats, such as at the end of the barrier beach south of

Silver Sands (Fig. 2) where tidal range is greater than normal for the lagoons; it is also the case where dunes encroach upon the marsh, and where storms cause oceanic waters to spill over the barrier beach and to discharge westward into the lagoon—also events which have been observed south of Silver Sands. Some storms stir the lagoon sufficiently to cause the development of temporary sandy beachlets, and to drive sand for some distance into the marshes. This occurred at Waretown, opposite Barnegat inlet, in the minor hurricane of 1960.

Yet, whereas sand is the dominant sediment in the lagoon, it is subsidiary in the marsh. Here peat and mud are dominant, and even pure clays may be deposited. The reasons for clay deposition on tidal flats and tidal marshes have been discussed by Van Straaten and Kuenen (1957, 1958). In Barnegat Bay, some clay is presumably brought in by rivers, some by tidal currents from the sea. At high tide, a thin sheet of water with suspended solids floods the marshes. Here the water is kept quiet by the baffle-work of plants, and the fine silt and clay particles settle out. This process may be further aided by entrapment of clay in the abundant mats of filamentous algae, and by the action of filter feeding organisms as suggested by Van Straaten and Kuenen—in our tidal marshes chiefly the horse mussel, *Volsella*. Most of the fine suspended material in the lagoonal waters settles to the lagoon floor rather than onto the tidal marshes; but due to reworking of the lagoonal sediments by waves, tidal currents, and human activity such as dredging operations and boat traffic, a large share of this fine material is recycled into the waters, and eventually comes to rest on the tidal marshes. Thus clay is abstracted from the lagoon, and concentrated in the marshes.

Most of the clay in the tidal marshes appears to be mixed with organic matter in a kind of "muck." Pure, putty-like clay, blue-gray when fresh and buff when oxydized, underlies fibrous peats at the southern end of the Tuckerton marshes, and at Sandy Hook, in a relationship suggestive of the underclay-coal sequence of the Pennsylvanian cyclothems.

Peat is the most characteristic sediment of the tidal marshes, although it may be quantitatively subordinate to clay-rich muck. Fibrous brown sedge peat, 2–3 feet thick, forms intertidal cliffs at various places along the inner margins of the barrier beaches, as well as at the southern end of the Tuckerton marshes.

Over extensive areas on the western inland side of the lagoons, clays and peat of tidal marsh origin are found on the lagoon bottom, below present tidal range, thus providing evidence of rising sea level.

FAUNAS, THEIR RELATIVE PRESERVATION, AND THEIR MIXING

The distribution of some of the common molluscan species is shown in Table I. Some occur only in the tidal marsh; others are known only from the lagoon; others are found in the lagoon and on the open ocean beach as well, and some are limited to the open beach. Ronai (1955) has listed Foraminifera dredged from Barnegat Bay.

Mytilus, the blue mussel, has a rather distinct distribution pattern. It grows in large numbers in the tidal channels near the inlets, and also lives associated with *Volsella* in the tidal marshes very near the inlets. It does not range far into the lagoons, possibly lacking tolerance for brackish water. It occurs rather abundantly on the oceanic beaches, but there is some question whether the shells here are derived from populations living beyond the normal surf line, or whether they are derived from the inlets, and, being light in weight and made even more buoyant by their tangled byssal masses, are carried along the open beaches by currents and wave action.

TABLE I. ECOLOGIC DISTRIBUTION OF MORE COMMON MOLLUSK SHELLS ON CENTRAL NEW JERSEY COAST

Species	*Tidal Marsh*	*Lagoon*	*Open Atlantic Beach and Adjacent Bottoms*
Volsella demissa	Living		
Littorina irrorata	Living		
Mya arenarea		Living	
Solemya velum		Living	
Ensis directus		Living	
Nassarius obsoletus		Living	
N. trivittatus		Living	
Pecten irradians		Living	Fossil only
Crassostrea virginica		Living	Fossil only
Anomia simplex		Living	Living and fossil
Mercenaria mercenaria		Living	Living
Anadara ovalis		Living	Living
Tagelus plebeius		Living	Living
Busycon canaliculatum		Living	Living
Polynices duplicatus		Living	Living
Lunatia heros		Living	Living
Spisula solidissima			Living
Barnea costata			Living
Mytilus edulis	Living*	Living*	Living?*

* In the lagoons and salt marshes *Mytilus* is largely restricted to the regions near inlets, where salinity is near normal. Shells on beaches may be dead shells carried from the inlet areas.

The distribution of *Anomia*, *Crassostrea*, and *Pecten* poses special problems. *Anomia*, the familiar silvery jingle shell, is abundant in Barnegat Bay, and such silvery shells are also common on the open beach. Here there are found, in addition, black shells of this species.

Crassostrea virginica, the common oyster, is fairly abundant in the bay (though not nearly so abundant as some decades ago). Its shells litter the open beaches in great numbers, but these shells on the ocean side are generally black and dull, and thus do not appear to be derived from living oyster banks. Some fresh-looking shells of *Crassostrea* have been found on the beach at Silver Sands, but of these specimens one was grown to a valve of *Volsella*, making it quite clear that these shells were somehow brought from the bay, by people or other extraneous agencies.

Pecten irradians, the Atlantic bay scallop, is likewise represented on the open beaches by abundant dull black valves. In addition to these there are rare orange to brown-colored valves, in which, as in the black ones, the stain permeates the entire shell. These brownish specimens are not generally as dull as the black ones, but do not appear fresh. No scallops with the white, shiny interiors that characterize live shells have been found by us on the open beach. It seems therefore, that these shells are not derived from populations living offshore at the present time.

It seems most reasonable to interpret these stained shells as old shells, which have lain buried under reducing conditions long enough to have undergone some carbonization of the proteinaceous material which makes up an appreciable proportion of the molluscan shell. Of the three species found in this state of preservation, the scallop and the oyster appear to be lagoon-dwellers, and only *Anomia* thrives both inside and outside the barrier beaches. Apparently then, a fossil lagoonal fauna is being reworked on the present beach.

But a glance at Table I will show that only a fraction of the lagoonal assemblage is being reworked on the beach. One might expect to find similarly blackened shells of all lagoonal species, especially the common and massive *Mercenaria mercenaria*, but they are not represented. The reason for this absence is a mineralogical one: of all the mollusks in the lagoon, only *Crassostrea*, *Anomia*, and *Pecten* have shells composed of calcite. All of the others are dominantly aragonitic. Presumably shells are removed by solution in this setting, in tens to hundreds of years after burial, so that only the calcitic shells are preserved when waves at the beach-front cut into older lagoonal deposits.

Such selective leaching of aragonite has been recognized throughout the geological column. Every paleontologist working with mollusks is familiar with rocks in which scallops, oysters, and certain mytiloids are preserved intact, as are brachiopods, while associated gastropods and pelecypods belonging to groups having aragonitic shells are represented by molds only.

Evidence for rapid dissolution of shells may be seen in the tidal marshes, where the shells of *Volsella* become deeply etched during the life of older individuals, and where dead shells are generally either soft and crumbly, or completely decalcified. In these peats and clays any evidence of shells is limited to the upper surface. Within the body of the peat or clay there are no shells, only cavities where *Volsella* may once have lived. Shells of *Nassarius* dug out of lagoonal sands are strongly etched, and Wagstaff (B.A. thesis, Princeton, 1959) found evidence of etching in suites of the calcareous foraminifer *Streblus*.

Lucke (1934a, 1935) also appealed to solution of shells in order to explain the lack of fossils noted in boring through the sands of the Barnegat inlet tidal delta. Van Straaten (1954) has called attention to shell solution in the Wadden Sea, in a very similar setting.

SUMMARY OF PRESENT SEDIMENTARY PATTERNS AND RECENT EVENTS

Transgression of the sea over a gently sloping coastal plain has resulted in a marginal belt of low-energy sedimentation—the lagoon and associated tidal marshes—separated from the open sea by barrier beaches of high-energy sedimentation. The lagoons are flanked on both sides by tidal marshes in which clays and peat accumulate; islands of similar salt marsh occur as well. The bottoms of the open lagoon are floored partly with sand, spread from the seaward side on tidal deltas.

These sediments must be present in a vertical, transgressive sequence below the barrier beach, where they are buried by sand dunes (Fig. 3). As sea-level rises, the belts of sedimentation shift landward, and wave attack on the seaward face of the barrier beach destroys part or all of the la-

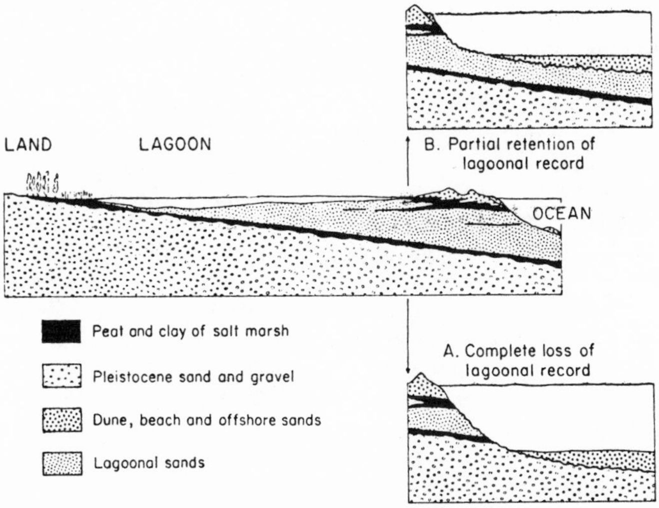

FIG. 3.—Diagram of marginal stratigraphic record developed by marine transgression. Tidal marsh peats and clays are followed by lagoonal sediments, these in turn by tidal marsh peats and clays, and these by dune sands and beach sands of the barrier beach. Further rise in sea-level and landward shift of barrier beach involve complete (A) or partial (B) destruction of marginal record, before overlap by offshore sands.

goonal sediments which become exposed there. These sediments are converted into new offshore bars, sand dunes, and tidal deltas, and their fossil fauna—diagenetically altered, with elimination of aragonitic shells—is being mixed with the Recent fauna of the open shore and adjacent bottoms.

How much, if any of the sediment of the lagoonal side escapes this destruction is not known.

IMPLICATIONS FOR SEDIMENTOLOGY AND STRATIGRAPHY

Several of the foregoing relationships are given little consideration in the course of stratigraphical work, or run counter to current generalizations, and may be misinterpreted when encountered in the geological column.

1. The setting of an advancing sea includes a complex array of marginal sediments formed in the low-energy near-shore environments.

2. These marginal sediments are likely to be partly or completely reworked by the succeeding higher energy zone. In the sedimentary record of a transgressing sea we may therefore expect to find either of two possible conditions (Fig. 3). (A) The marginal deposits have been completely reworked, and the marginal record of the transgression has been lost. The Cretaceous cyclic deposits of the Rocky Mountain region described by Young (1957) appear to be of this type. Or (B) the marginal low-energy record has been partly preserved, and is separated from the succeeding sediments of the more open, agitated waters by a wave-truncated surface (disconformity).

In transgressive-regressive sequences the transgressive part of the record is commonly thinner than is the regressive one. Wells (1960) has suggested that this reflects the time involved: that transgressions occurred more rapidly than did regressions. This is indeed reasonable, if one subscribes to the view that tectonic motion of the basins was slowly and steadily downward, and that transgressions and regressions were largely caused by eustatic changes of sea-level, for in that case tectonic sinking would have speeded the invasion and slowed the retreat of the seas. But regardless of these considerations, the transgressive record is also incomplete, thinned by its built-in disconformity, while the regressive record is more completely preserved unless it is truncated at the top.

3. The basal units of the high-energy record may be expected to contain fossils reworked from the marginal sediments which may otherwise have been completely obliterated. The fossils preserved in the reworked state are likely to be selected on chemical, even more than physical, characteristics.

4. The initial sediment of a major transgression may be very fine-grained, as pointed out by Van

Straaten and Kuenen (1957). In the New Jersey setting the finest sediments are those of the tidal marshes.

Little has been learned about the tidal ranges of ancient oceans and epicontinental seas. It does seem likely that many of those ancient seas which were broadly open to the oceans had appreciable tides. The lagoons of such seas may have been comparable with those of the New Jersey coast, subject to considerable tidal currents and to the construction of tidal deltas, growing from the sea toward the land. A fossil delta of the Barnegat type might easily lead the unwary stratigrapher to postulate a landmass in the Atlantic Ocean!

In largely landlocked seas, such as those which occupied the interior basins of North America during the late Paleozoic, tides must have been very weak, lagoons were probably very quiet, and the likelihood of tidal deltas is very small.

6. When an advancing sea develops barrier beaches or extensive spits, it advances not with one but with three shorelines. In the case of epicontinental seas with only moderate tides at best, and with limited fetch, and therefore with little wave action compared with oceanic shores, erosion at the outer beach might well spare most of the lagoonal sediments, possibly even those deposited at the inner margin of barrier beaches or spits. A stratigraphic section through such a transgressive sequence (Fig. 3) would show a transition from lagoonal shoreline deposits to deeper lagoonal deposits, then up into lagoonal shoreline deposits, disconformably overlain by the sediments of the beach and of the open sea. In this way, a single transgression could result in the formation of two separate coal seams, and other features which the stratigrapher might interpret as the result of oscillating sea-level.

APPLICATION TO CYCLOTHEMIC DEPOSITS

The Pennsylvanian-Lower Permian cyclothemic deposits of the North American interior record many transgressions and retreats of the sea. A number of different kinds of sequences were formed, in different places and at different times. Two of these sequences are here discussed: the Cherokee type of cycle, and the classical Illinois-type cyclothem.

Cherokee cycle.—The Cherokee group comprises the basal sediments formed during the mid-Pennsylvanian advance over the southwestern flank of the Ozark dome. The stratigraphic character and rock types have been described by Howe (1956). The rock types include the following five types. (1) *Underclay* (perhaps comparable with the intertidal marsh clays of Barnegat Bay). (2) *Coal*, presumably formed in fresh to brackish shoreline swamps (comparable with the shoreline peats of Barnegat Bay?). (3) *Shale*, ranging from gray to black; partly calcareous, partly phosphatic, and commonly containing clay-ironstone concretions. Some of these shales contain a marine fauna, whereas others appear to be non-fossiliferous. Howe interprets these shales as deposits "of a lagoonal or near-shore environment, differing from that of the coal only by the presence of sea water having moderate depth." It is here suggested that some of these shales may have been formed essentially at sea-level, in brackish to fresh lakes and bays or in the intertidal zone. (4) *Sandstone*, generally well sorted and strongly cross-bedded, and containing shale chips in the basal part. Some of the sands have yielded marine fossils; others grade into marine limestones. The long, narrow shape of these sandstone bodies in the subsurface of Kansas and Oklahoma has led to the name "shoestring sands." They are generally regarded as barrier beach and offshore bar deposits (Rich, 1923, Bass and others, 1937). (5) Impure *limestone*, ranging from apparently non-fossiliferous to highly fossiliferous, and probably deposited in various lagoonal settings.

Figure 4 illustrates the two lower cycles of the Cherokee Group in southeastern Kansas. In the light of what we have learned from the Atlantic coast of New Jersey, the sequence of events in Oklahoma may have been as follows: the mid-Pennsylvanian seas encroached over a Mississippian terrane covered with weathered chert gravel (chat), and showing little relief. A zone of surf and currents formed spits and barrier beaches, which separated a marginal low-energy zone of lagoons and shoreline swamps from the main sea. During comparatively rapid subsidence, these environments rapidly shifted shoreward in a transgressive phase. During slow subsidence or more abundant sediment supply, they gradually shifted seaward in a regressive phase. Tides were probably weak, hence tidal deltas are not to be expected.

The non-fossiliferous shales and coal of the Riverton Formation represent the initial shoreline swamps. The thin lagoonal sediments and

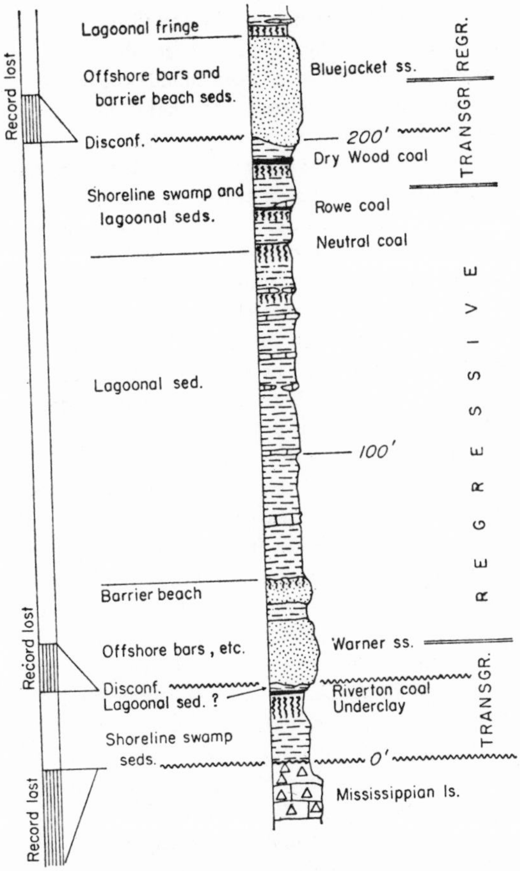

FIG. 4.—Lower part of Cherokee Group (Pennsylvanian), in southeastern Kansas. After Howe (1956), with the writer's interpretation added.

barrier beach deposits which migrated over them were largely or entirely destroyed by the advancing surf zone, and the lower part of the Warner Sandstone represents sand deposited at the seaward margin of this turbulent zone. This unit marks the peak of the transgression and extends into the regressive phase, for the sandstone at its top, containing plant roots, is best interpreted as subaerial barrier beach deposit. It is succeeded by thick lagoonal shales containing thin limestones. The cycle ends again in non-fossiliferous shales and coals—the shoreline marsh sediments which mark the peak of the regression. The succeeding cycle, in which the Bluejacket Sandstone represents the offshore deposits, is of a similar nature. In both cycles the transgression destroyed most of its own lagoonal record, whereas the sediments deposited in the regressive phase are more completely preserved. There may be a break in sedimentation at the peak of regression, but the most striking disconformity lies within the transgressive phase—at the base of the offshore sands.

Illinois-type cyclothem.—The classical cyclothems of Illinois are of a different nature (Fig. 5.). They show a basal alluvial portion, contain more limestone and no marine sandstone, and have a regressive phase of non-fossiliferous shale. The main unconformity separates the regressive from the succeeding transgressive phase. They have been dealt with in many papers, among them Wanless (1955), Weller (1956, 1957), Wheeler and Murray (1958), and Wells (1960). The general view (Elias, 1937) has been that the normal cyclothemic sequence leading from basal non-marine sediments through coal to brackish-water and marine shales and finally to clean marine limestones is the result of a simple increase in water depth, possibly coupled with variation in sediment supplied (Weller, 1956, 1957).

The observations in Barnegat Bay suggest a somewhat different interpretation. As sea-level rose, and the waters encroached on a land of very low topography, dotted with large subsiding areas such as the Illinois and Michigan basins, these negative areas first became huge swamps, and later vast bays, separated from the main sea-

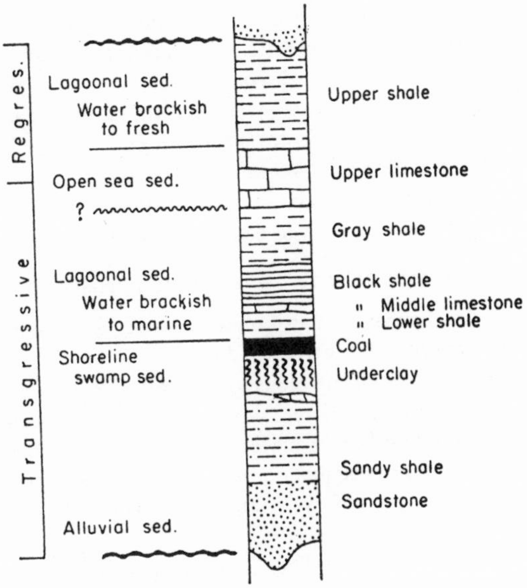

FIG. 5.—Idealized Pennsylvanian cyclothem of the Illinois type. After Weller (1956), with the writer's interpretation.

ways by tectonic barriers (sills, peninsulas) and by spits and barrier beaches. Here the terrigenous muds were trapped, and shales and occasional fine-grained limestones, with brackish to marine faunas, were deposited. When such areas were very effectively silled, temperature or salinity stratification developed, and led to the deposition of black mud.

The spits and barrier beaches of these Pennsylvania seas of the Mid-Continent region, as postulated here, remote from sources of terrigenous debris, were generally not composed of quartz sand, but of locally generated shell debris. The limestones represent the transgressive sediments of the open seaways deposited over lagoonal sediments when, as a result of rising sea-level, the barriers were drowned or shifted, and the former lagoonal areas became incorporated into the more open seas. Locally, and especially in their upper regressive portions, these limestones may incorporate remnants of calcareous barrier beach deposits.

The preservation of the transgressive part of the lagoonal record, as contrasted to its destruction in the Cherokee cycle, is due to the much greater scale—in width and depth—of the Illinois "lagoons."

The limestones of the Illinois cyclothem are abruptly followed by non-fossiliferous shales representing the main part of the regressive phase. They contrast sharply with the varied and fossiliferous lagoonal sediments found in the comparable part of the Cherokee cycle. This non-fossiliferous, monotonous nature of the regressive part of the Illinois cyclothem has not been satisfactorily explained. Weller (1957) suggests increased influx of mud, but does not offer plausible reasons for the close correlation of this event with the disappearance of fossils. A different hypothesis is here presented.

In the Cherokee setting, regression brought about a seaward retreat or relocation of sand barriers. The establishment of new barriers was dependent on the supply of new sand; hence, the lagoons of the regressing sea are likely to have been barred less effectively than the lagoons of the transgressing sea, in which the sand itself tended to migrate with the bars.

In the seas of the Illinois cyclothems, the sediments of the open seaway were calcareous, precipitated *in situ*. As subsidence slowed or ceased and the waters shallowed, extensive algal banks developed. These stromatolitic banks and associated pisolite beds form parts of the characteristic terminal members of the cyclothemic limestones in Kansas, Missouri, and Nebraska (Moore and others, 1951; Harbaugh 1959, 1960). It is here proposed that such algal banks, extending to sea-level, were developed during each regression, primarily on the arches which separated individual basins. Such banks barred the peripheral basins much more effectively than did the barrier beaches of the transgressive phase, so that the marginal basins were converted into vast lagoons of very low salinity. One can visualize the entire Illinois basin as a fresh water or nearly fresh lake, barred from the seas on the west by algal banks extending northward from the flanks of the Ozark dome. With further regression the Kansas area also became such a brackish lagoon.

Accordingly, both during transgression and regression extensive lagoons acted as traps for terrigenous clay. The barrier beaches formed during transgression permitted much water exchange between sea and lagoon, and marine and brackish-water faunas thrived in the latter. The great belts of algal banks which formed during regression barred the lagoons so effectively that these became nearly fresh, and did not support such lagoonal faunas of shell-bearing invertebrates.

LITERATURE CITED

Abbott, R. T., 1954, American seashells: Van Nostrand, 541 p.

Bass, N. W., Leatherock, C., Dillard, W. R., and Kennedy, L. E., 1937, Origin and distribution of Bartlesville and Burbank shoestring oil sands in parts of Oklahoma and Kansas: Am. Assoc. Petroleum Geologists Bull., v. 21, p. 30–311.

Bradley, W. H., 1957, Physical and ecologic features of the Sagadahoc Bay tidal flat, Georgetown, Maine, *in* Treatise on Marine ecology and paleoecology (H. S. Ladd, ed.): Geol. Soc. America Mem. 67, v. 2, p. 641–682.

Elias, M. K., 1937, Depth of deposition of the Big Blue (Late Paleozoic) sediments in Kansas: Geol. Soc. America Bull., v. 48, p. 403–32.

Fisk, H. N., and McFarlan, E. Jr., 1955, Late Quaternary deltaic deposits of the Mississippi River, *in* Crust of the earth (ed., A. Poldervaart), Geol. Soc. America Special Paper 62, p. 278–302.

Harbaugh, J. W., 1959, Marine bank development in Plattsburg limestone (Pennsylvanian), Neodesha-Fredonia area, Kansas: Kansas Geol. Survey Bull. 134, pt. 8, p. 289–331.

——— 1960, Petrology of marine bank limestones of Lansing Group (Pennsylvanian), southeast Kansas: Kansas Geol. Survey Bull. 142, pt. 5, p. 190–234.

Howe, W. B., 1956, Stratigraphy of pre-Marmaton Desmoinesian (Cherokee) rocks in southeastern Kansas: State Geol. Survey Kansas Bull. 123, 132 p.

Johnson, D. W., 1919, Shore processes and shoreline development: 584 p., New York, Wiley.
Ladd, H. S., Hedgpeth, J. W., and Post, R., 1957, Environments and facies of existing bays on the Central Texas Coast, *in* Treatise on marine ecology and paleoecology (H. S. Ladd, ed.): Geol. Soc. America Mem. 67, v. 2, p. 599–640.
Lucke, J. B., 1934a, A study of Barnegat Inlet, New Jersey, and related shoreline phenomena: Shore and Beach, v. 2, no. 2, p. 1–54.
——— 1934b, A theory of evolution of lagoon deposits on shorelines of emergence: Jour. Geology, v. 42, p. 561–548.
——— 1935, Bottom conditions in a tidal lagoon: Jour. Paleontology, v. 9, p. 101–107.
Moore, R. C., Frye, J. C., Jewett, J. M., Lee, W., and O'Conner, H., 1951, The Kansas rock column: Kansas Geol. Survey Bull. 89, 1932 p.
Rich, J. L., 1923, Shoestring sands of eastern Kansas: Am. Assoc. Petroleum Geologists Bull., v. 7, p. 103–113.
Richards, H., 1938, Animals of the seashore: 273 p., Boston, Humphries.
Ronai, P. H., 1955, Brackish water Foraminifera of the New York Bight: Cushman Lab. Foram. Research Contr., v. 6, pt. 4, p. 140–150.
Shepard, F. P., 1956, Late Pleistocene and Recent history of the central Texas coast: Jour. Geology, v. 64, p. 56–69.
Van Straaten, L. M. J. U., 1954, Composition and structure of Recent marine sediments in the Netherlands: Leidse Geol. Mededelingen, v. 19, p. 1–110.
——— 1956, Composition of shell beds formed in tidal flat environment in the Netherlands and in the bay of Arcachon (France): Geol. en Mijnbouw, n. s., v. 18, p. 209–226.
——— and Kuenen, Ph. H., 1957, Accumulation of fine-grained sediment in the Dutch Wadden Sea: Geol. en Mijnbouw, v. 19, p. 329–354.
——— 1958, Tidal action as a cause of clay accumulation: Jour. Sed. Petrology, v. 28, p. 406–413.
Wanless, H. R., 1955, Pennsylvanian rocks of the eastern interior basin: Am. Assoc. Petroleum Geologists Bull., v. 39, p. 1730–1820.
Weller, J. M., 1956, Diastrophic control of Late Paleozoic cyclothems: Am. Assoc. Petroleum Geologists Bull., v. 40, p. 17–50.
Weller, J. M., 1957, Paleoecology of the Pennsylvanian period in Illinois and adjacent States, *in* Treatise on marine ecology and paleoecology (H. S. Ladd, ed.): Geol. Soc. America Mem. 67, pt. 2, p. 325–364.
——— Wheeler, H. E., and Murray, H. H., 1958, Cyclothems (discussion): Am. Assoc. Petroleum Geologists Bull., v. 42, p. 442–447.
Wells, A. J., 1960, Cyclic sedimentation: A review: Geol. Mag., v. 97, p. 389–403.
Wheeler, H. E., and Murray, H. H., 1957, Base-level control patterns in cyclic sedimentation: Am. Assoc. Petroleum Geologists Bull., v. 41, p. 1985–2011.
Young, R. E., 1957, Late Cretaceous cyclic deposits, Book Cliffs, eastern Utah: Am. Assoc. Petroleum Geologists Bull., v. 41, p. 1760–1774.

3

Reprinted from *J. Waterways and Harbors Division, Proc. Am. Soc. Civ. Eng.*
88:117–130 (1962)

SEA-LEVEL RISE AS A CAUSE OF SHORE EROSION

By Per Bruun,[1] F. ASCE

SYNOPSIS

It is an established fact that sea level is rising slowly and irregularly. Also, it seems to be true that erosion on most seashores built up of alluvial materials greatly exceeds accretion. The paper attempts to relate the two phenomena; rise of sea level and erosion.

RISE OF SEA LEVEL AND ITS RELATION TO SOLAR RADIATION

According to R. W. Fairbridge,[2] preliminary conclusions from studies of solar variations and climatic change indicate that "ice ages" are produced by normal variations in observed solar-controlled metereological effects that are reinforced, probably aperiodically in geological time, by terrestrial, topographic "accident."

According to Fairbridge, long-term metereological and historical records disclose multiple cycles (for example, 11-yr, 22-yr, 40-yr, 80-yr, 189-yr and 567-yr periods) related, in part, to sunspots, geomagnetic phenomena, and planetary motions. Mean, annual temperatures in temperate belts vary 2°C-3°C. The longer the period, the greater the effect in terms of snowfield accumulations or glacier ice melting. This concerns the world hydrologic balance and, thus, is reflected on world tide gages. Whereas

snow and ice accumulation advances directly in short steps, melting shows marked retardation. Major astronomic cycles with period of 21,000 yr, 40,000 yr, and 92,000 yr control absolute and hemispheric variations in effective solar radiation. Mid-latitude temperatures will vary 3°C-5°C. and will mainly influence the northern hemisphere, which contains 95% of the world's sensitive mountain glaciers.

The earth will remain relatively cool while Antarctica lies symmetrically about the South Pole. However, temperate belts will continue to be alternately glaciated and deglaciated every 40,000 yr to 90,000 yr.

HYPOTHESES RELATING TO DISPLACEMENT OF LAND AND SEA

Sea-level changes in the Quaternary have evoked some ingenious theories. It is now established that during the half-million-year Quaternary, the sea level has oscillated in a manner as rapidly and extreme as was ever before observed in geological history. During the warm, interglacial phases, the shorelines advanced inland leaving erosion cliffs, terraces, and platforms behind. During the cool, glacial phases the shorelines advanced seaward and new land was built up leaving ridges and plateaus emerged for later submergence.

The ice-age oscillations involved withdrawals of huge water masses from the sea on the polar land masses. The present volume of sea water is approximately 1370 x 10^6 cu km, but there is still a considerable quantity of water locked in the present-day continental ice caps and glaciers. For the Northern Hemisphere the glacier area is estimated to somewhat over 2 x 10^6 sq km; for the Southern Hemisphere 13 x 10^6 sq km, thus a total of 15 x 10^6 sq km or approximately 37.5 x 10^6 cu km of ice averaging 2.5 km thick. Because the area of ocean surface is 360 x 10^6 sq km, a total melting of 37.5 x 10^6 cu km of ice would cause a sea-level rise of 95 m (315 ft). Due to oceanic crustal lowering, marginal to rising continental areas, and to the fact that the rising sea would spill over enormous lowlands greatly expanding the present ocean area, the final level of the ocean might be perhaps only approximately 50 m above the present datum. (Present as used herein is in a geological rather than literal sense.)

The area of the last glaciation maximum has been determined at approximately 40 x 10^6 sq km or to approximately 40 x 10^6 sq km ice. This corresponds to the maximum measured fall of sea level, 100 m (330 ft), or to "the Wisconsin" glacial period in the United States. However, owing to the progressive build-up of Antarctica over four or five glacial cycles of the Pleistocene, the total removal of water may approach 200 m (660 ft).

Fig. 1, from Fairbridge,[2] gives an impression of these grand-scale fluctuations in sea level. It may be seen that the "deglacial geoid" continues downward, which phenomena can only be explained by an overall sea level lowering controlled by the tectonic shape of the ocean basins, or polar shift. The drowned sea mounts of the Pacific Ocean may be taken as an example.

The development of sea-table elevation during approximately the past 20,000 yr is depicted in Fig. 2 from Fairbridge.[2] Approximately 10,3000 years ago began the Finiglacial retreat of Scandinavia and the late Valders retreat in North America. Sea level rose rapidly at 20 mm to 30 mm a year until it reached 15 m below present datum. The movement apparently came

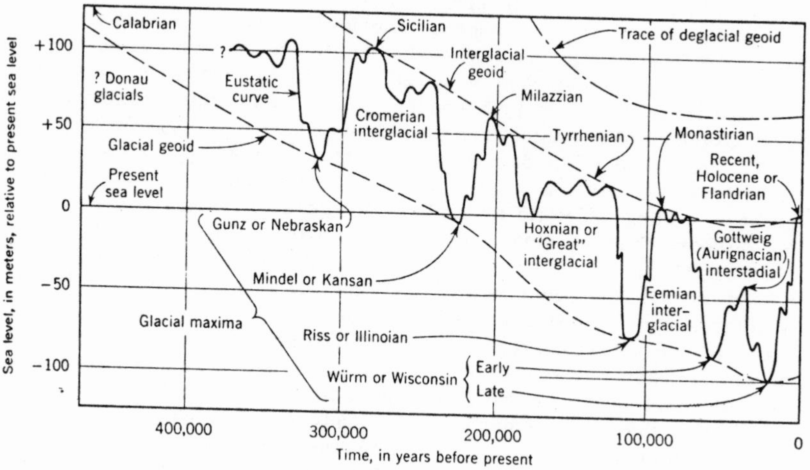

FIG. 1.—QUARTERNARY EUSTATIC OSCILLATIONS

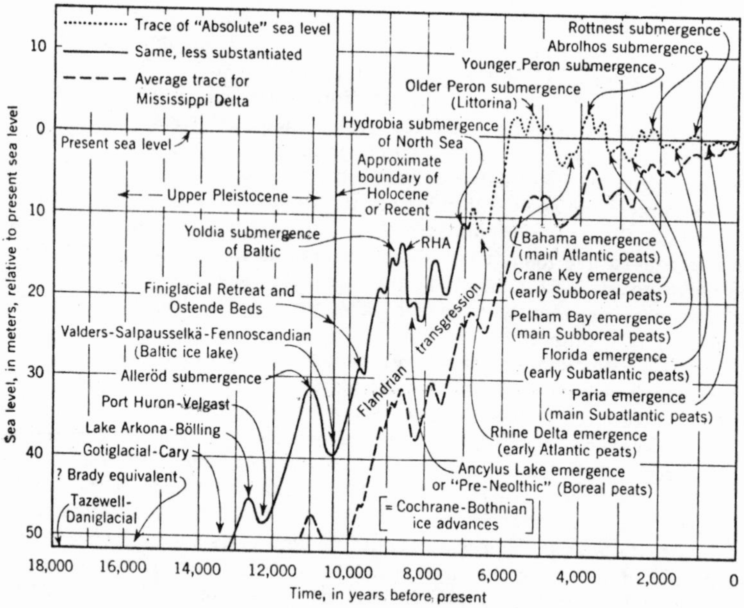

FIG. 2.—EUSTATIC OSCILLATIONS DURING THE POST-GLACIAL TRANSGRESSION

in sharp, jerky steps. The Mid-Recent or Middle-Littorina submergence is the culmination of the universal climatic and oceanic warming that began before the end of the Pleistocene. The level of the ocean rose at a remarkable rate. The highest position reached by this sea in most places was 3 m to 5 m above the present and occurred 5,000 yr to 6,000 yr B.P. Certain peat formations from 1.5 m to 2.0 m below surfaces in the Everglades south of Lake Okeechobee are dated fromtthis period.

The so-called "Florida Emergence" that occurred approximately 2,100 to 1,600 years ago had sea-level elevations of approximately 2 m below present, probably adding 1/8 mile to 1 mile to the general Florida shores, depending on offshore bottom slope. It coincides with a slight advance of northern glaciers. From a historic point of view, it is interesting to note that this period covers the Roman Era and that data from Britain, Italy, and the Mediterranean suggest a low sea level at this time. Apart from notoriously unstable and volcanic areas, there is widespread evidence of the "drowning" of Roman coastal structures. The deep foundations of some ancient habor works may not have been so difficult to construct as they seem today. Climatically, the Florida Emergence coincides with a universal cool phase.

In examining the sea-table fluctuations of the past 100 yr, note that average world rise for the 1900-1950 period was 1.2 mm, annually, corresponding to the average rise since the Roman Era of approximately 2,000 mm. Meanwhile deglaciation can result in sea level rising at 25 mm or approximately 1 in. per yr. In the 1946-1956 period, the rise[2] was 5.5 mm per yr. H. A. Marmar[3] lists approximately 8 mm, annually, from 1930 to 1948, for the entire eastern seaboard of North America. Yet, an appreciable component in this sea-level rise, only approximately 20% is "eustatic" (overall sea level) is possibly due to a secular decelaration of the Gulf Stream. This decreases pressure differences between its right- and left-hand side caused by the Coriolis force. Such pressure fluctuation is noted in the annual changes in the difference of sea level between stations on opposite sides of the Gulf Stream; for example, between Miami and the West Bahamas. This difference is normally of the order of approximately 0.6 m (2 ft). The seasonal variation in water temperature (density of water) is clearly indicated in records along the Southeast coast of Florida where the maximum sea level occurs in September and October, apparently caused by the summer heating of the water, as explained by H. Stommel.[4]

Fig. 3 shows some water-level records for Florida coast cities based on data furnished by the Coast and Geodetic Survey, (United States Department of Commerce (USC&GS). In Fig. 3 the yearly value for each station represents the average height of sea level as determined by averaging the readings of the height of the sea at that station at the beginning of each hour throughout the year. In other words, each value is the average of nearly 9,000 hourly readings. At each station, the hourly heights of the sea were referred to a tide staff, the elevation of which could be kept constant by frequent leveling to a number of adequate bench marks. From this figure it can be seen that the rise in sea level in the 1930-1950 period has been up to 10 mm (0.03ft), on the east coast of Florida [Figs. 3(a) and 3(b)]. On the Gulf Coast [Figs. 3(c) and

3 "Is the Atlantic Sinking? The Evidence from the Tide," by H. A. Marmar, Geological Review, Vol. 38, 1948, p. 652.

4 "The Gulf Stream," by Henry Stommel, Univ. of California Press, Los Angeles, Calif. 1960.

3(d)], it has been up to 8 mm (0.025 ft) per year. Considering the 1930 to 1960 period, it can be seen that the rise of sea level has slowed down somewhat the last decade but may be increasing again. For the entire 1930-1960 period, rise has been 3 mm to 5 mm annually on the east coast of Florida and 3 mm to 4 mm annually on the Gulf side.

The consequences of the rises in sea level on shoreline movements are mentioned in a following section on "Theory of Erosion by Rise of Sea Level."

DIMENSION OF THE NEARSHORE LITTORAL DRIFT ZONE

The dimension perpendicular to the shore of the nearshore littoral drift zone depends on many factors including material, slope, wave, and current

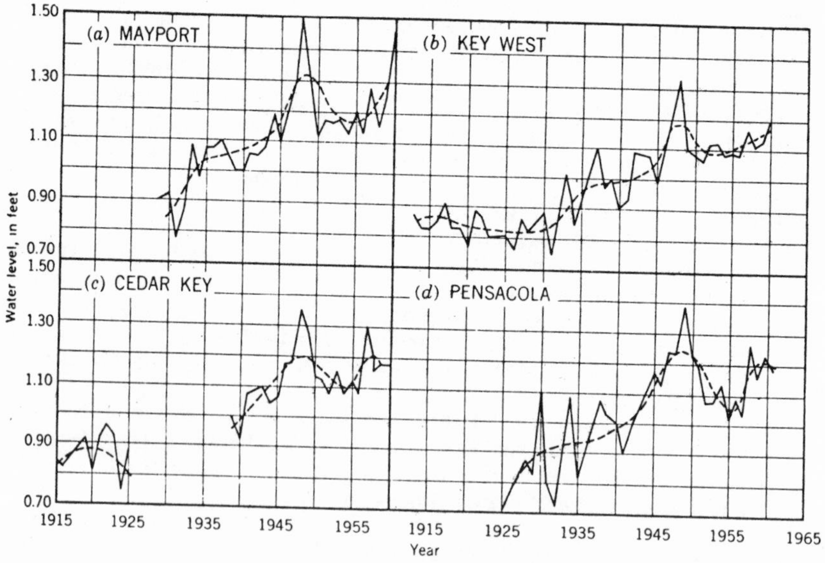

FIG. 3.—VARIATIONS IN AVERAGE YEARLY SEA LEVELS AT FLORIDA TIDE GAGES

characteristics. It may be determined by comparison of shore sediments and offshore bottom sediments, but it is apparent, beforehand, that there will be no clear definition of this zone. Rather, the shore-sediment characteristics on the offshore bottom will gradually taper off with increasing depth and probably be replaced by "offshore sediments" of great variety. This leads to an examination of the so-called "continental shelf."

The continental shelf is a region where continental and marine influences have acted alternately during successive emergences and submergences. Continental shelves have various origins. According to A. Guilcher,[5] the continental shelf is usually wide off low-lying continents and narrow or absent off

[5] "Coastal and Submarine Morphology," by André Guilcher, John Wiley & Sons, Inc., New York, N. Y., 1958.

mountainous regions. The shelf is wide in northwestern Europe, in the Arctic off the USSR and the east and Gulf coasts of the United States. It is narrow in front of the Pacific coast of the United States. The average width of the shelf is approximately 40 miles. The continental shelf meets the continental slope at a slight angle, but this break of slope is not always at the same depth and is often difficult to locate precisely. It is usually found between 120 m and 180 m (70 to 100 fathoms).

The continental shelf has originated by various means. Shelves of limited width may consist of two parts: An inner abrasion platform, and an outer constructional form built up of material eroded during the formation of the inner part. The best example of a constructional form (with subsidence of the sediments) is the shelf off the east coast of the United States, that, to a high degree, was built up of eroded material from the Appalachian rocks. The coastal shelf of the Gulf of Mexico is the same type and has been formed by a great mass of material brought down by the Mississippi River and other streams.

If sea level remained constant for an extended period and if waves were the only important factor in sedimentation, the caliber of the sediments should decrease away from land. In reality, the distribution of sediment on the continental shelf is complex, and many factors seem to be involved in its configuration and composition. Pebbles have been found at great depths in a variety of places. Off Southern California, they have been dredged up from all depths between 18 m and 900 m (60 ft and 3,000 ft). These pebbles were probably in most cases deposited by currents.

Submarine currents may form ripple-marks on sandy bottoms. They have been photographed on George Banks, New England, at a depth of 225 m (750 ft) but they are known from depths up to 2,000 m (approximately 7,000 ft) in restricted straits in the Canaries and East Indies. This indicates that currents with eroding velocities (greater than approximately 0.3 m per sec or 1 ft per sec) exist.

The movement of material from the shore for depositing in the offshore area is probably a slow process by which various kinds of currents—including longshore currents, rip currents, and density currents—are active. With respect to the exchange of material between the shore and the offshore bottom, little is known about the rate of this process because many factors are involved, including the character of shore material and the characteristics of wave and current activity. Most likely, a distinction will have to be made between a short-range process of "fluctuation nature" and a long-range geological adjustment process. With respect to littoral-drift material, the observations by Parker Trask[6] are of particular interest.

Trask explains that present data clearly indicate that sand does move around the rocky California promontories. Meanwhile it seems to be clear that a little sediment is transported beyond a depth of 18 m (60 ft).

D. L. Inman[7] reports that bottom surveys at Scripps Institution of Oceanography, La Jolla, Calif., indicate that most seasonal, offshore-onshore interchanges of sand occur in depths of less than 9 m (30 ft) but that some seasonal effects may extend to greater depths. Inman describes the areal distribution pattern (for this particular location), that shows a pronounced alinement of

6 "Movement of Sand around Southern California Promontories," by Parker Trask, Technical Memorandum No. 52, Corps of Engrs., Beach Erosion Board, 1954.

7 "Areal and Seasonal Variations in Beach and Nearshore Sediments at La Jolla," by D. L. Inman, Technical Memorandum No. 39, Corps of Engrs., Beach Erosion Board, 1953.

sediment properties generally parallel to the beach. There are numerous possible interpretations of the alinement and banding of sediment attributes. Surf beats may be one explanation. Another may lie in the seaward transportation of sediment by diffusion, resulting from a horizontal gradient in concentration of suspended material from the surf zone where concentrations are high, to offshore areas where they are relatively low. In addition to seaward transportation by diffusion, it is well known that a net, onshore transportation of sediment occurs along the bottom because of the differential between onshore and offshore velocities associated with the orbital motion of nearshore waves.

The writer has described[8,9] deep-water erosion on the Danish North Sea Coast of up to 20 m (70 ft) depth.

From the previously noted studies, a reasonable assumption seems to be that with sandy shores of exposed Pacific or Atlantic type, the 18-m (60-ft) depth contour forms some kind of limit between "nearshore" and "deep-sea" littoral drift phenomena that, in this respect, means that short-term exchange of shore material and offshore bottom takes place inside (although not always up to) this depth. It should not be forgotten that the slope of the offshore bottom must be of significant importance. A very gentle slope will undoubtedly slow down transversal migration of material by giving rise to a considerable phase-difference between "action" (rise of sea level) and "reaction" (shore erosion). On the other hand, a very steep, offshore bottom will have the opposite effect manifested in a relatively quick response (in the form of erosion) to rise of water table. In as much as the slope and width of a littoral drift zone are closely connected, it could be expected that a wide "shelf" would demonstrate considerable phase displacement and higher stability than a narrow shelf. A narrow shelf may develop an "equilibrium profile" to a considerable depth indicating displacement of material from the shore to the offshore bottom. If the offshore bottom, such as the southeast coast of Florida at about 18-m (60-ft) depth, turns over to a steeper slope, the tendency to transfer of material to deep water by the assistance of gravity forces may be detectable in the shore stability as an increased shoreline recession.

THEORY OF EROSION BY RISE OF SEA LEVEL

Reference is made to Fig. 4. Consider an equilibrium profile.[8,9] If the water table rises a millimeter, the quantity of material needed to re-establish the same bottom depth over a width of shelf, b, is b times a.

Consider a shoreline that is in longshore quantitative equilibrium, which means that the same quantity of material that is passing in from the updrift side is also passing out downdrift. The quantity b a must be derived from erosion of the shore. This will give rise to a shoreline recession, x. If the elevation of the shore is e, the quantity eroded above sea level is x e. Meanwhile, in order to re-establish the original equilibrium bottom-profile, the entire profile must be moved shoreward by the same distance, x, up to depth, d, at distance, b, from the shoreline. The balance between eroded and deposited

---

8 "Coast Stability," by Per Bruun Danish Tech. Press, Copenhagen, Denmark, 1954.

9 "Coast Erosion and Development of Beach Profiles," by Per Bruun, Technical Memorandum No. 44, Corps of Engrs., Beach Erosion Board, 1955.

quantities by the two independent movements is expressed by

$$x\,e = a\,(b-x)\,d \qquad (1a)$$

or

$$x(e + d) = a\,b \qquad (1b)$$

To test the validity of this concept on a short-term basis, it will be necessary to look for a coastal area at which the phase-difference between rise of sea level and its influence on erosion is relatively small. This, as previously mentioned, will be true for an area with a steep, offshore bottom. Another assumption is that the edge of the continental shelf is no nearer the shore than at approximately the 18-m (60-ft) depth on the exposed Pacific or Atlantic shores.

Such a situation exists along the Southeast coast of Florida between Palm Beach and Miami. Yet part of the shore has rock reefs; and in the Hallandale-Miami Beach Area, a rocky, gently sloping "platform" exists between 12-ft and 20-ft depths. The distance of the 18-m (60-ft) contour from the shoreline

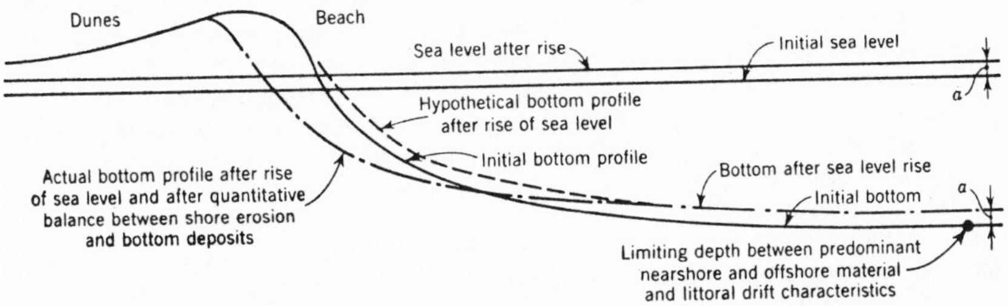

FIG. 4.—INFLUENCE OF SEA-LEVEL RISE ON THE DEVELOPMENT OF BEACH AND OFFSHORE PROFILE

is approximately 2,000 m. Introducing the following figures in Eq. 1b: x denotes the shoreline recession per year; a = the sea level rise per year b = 2, 000 m; e describes about 3 m (10 ft); and d = 18 m (60 ft); yields

$$x(3 + 18) = 2{,}000\ a$$

Referring to the preceding section on rise of the sea table, a rise of 1.2 mm per yr gives x = approximately 11 cm = 1/3 ft per yr. This figure is not in agreement with recent experience. Most likely the 2,000-m shelf is not wide enough to reflect very small but tough, long-term rises. In long-term periods, material is probably "tipping over" the edge of the shelf to deep water; thereby increasing erosion. It is, therefore, more likely that the development on this shore reflects the short-term rapid rises of sea level. Using the 6mm average figure for the rise in recent decades gives: x = approximately 57 cm per year = 2 ft per yr. With a 15-m (50-ft) depth on this short-term basis, x = approximately 2.4 ft.

The 2-ft to 2.5-ft shoreline recession per year is a realistic value when the shores not affected by inlets or groins are considered, but it should still

be borne in mind that even on a short-term basis, it is possible that (fine) material disappears in the deep waters past the edge of the shelf and that this will cause an increase of the shoreline recession.

On the upper east coast, the depth at 2,000 m from the shoreline is approximately 15 m (50 ft), whereas the 18-m (60-ft) contour is from 1.5 miles (St. Augustine-Daytona) to 5 or 6 miles (New Smyrna) out. South of Cape Canaveral to Sebastian Inlet, the 15-m (50-ft) contour is approximately 4,000 m out, while the 18-m (60-ft) contour is from 5 to 8 miles out. Between South Fort Pierce Beach and St. Lucie Inlet, the 15-m (50-ft) contour is 8,000 m out and the 18-m (60-ft) contour is 4 to 6 miles out. Bottom slope between 15-m to 18-m (50-ft and 60-ft) depths is gentle in these areas and will considerably retard the transfer of material seaward. It is not likely, therefore, that the effect of short-term, rapid sea-level rises will make themselves felt on the bottom areas between 15-m to 18-m (50-ft and 60-ft) depths. On the other hand, it may be possible to trace the long-term influence. Introducing, in Eq. 1(b) the following figures for the areas north of Cape Canaveral to Daytona Beach and from the Cape south to about St. Lucie Inlet: a = 1.2 mm per yr; b = approximately 5 miles or about 8,000 m (average); e = 4.5 m (15 ft); and d = 18 m (60 ft); yields

$$x(4.5 + 18) = (8{,}000)\,(0.0012)$$

$$x = 0.43 \text{ m or approximately } 1.4 \text{ ft}$$

This figure also seems realistic when an extended period of time is considered. With respect to the development on a short-term basis (from 1945 or 1950 to 1960), the figures a = 6 mm per year; b = 2,000 m (North of Canaveral, New Smyrna and Daytona); e = 4.5 m (15 ft); and d = 15 m (50 ft); yield

$$x(4.5 + 15) = (2{,}000)\,(0.006)$$

$$x = 0.6 \text{ m or } 2 \text{ ft per yr}$$

which is probably a realistic average figure for recent years—perhaps a little on the high side. It should be remembered that the recording of shoreline movements on a short-term basis involves many uncertainties.

Similar considerations for the Gulf shores are only possible for the upper west coast in as much as the lower west coast is penetrated with inlets and passes to such an extent that they dominate the erosion situation. For the upper Gulf Coast (Santa Rosa Island), the 18-m (60-ft) contour is located on the average 3,000 m from the shore, and beyond this contour the bottom slope is gentle, although no abrupt change in slope occurs at the 18-m (60-ft) depth. Introducing in Eq. 1b the following figures: a = 1.2 mm per yr; b = 3,000 m; e = about 4 m (13 ft); and d = 18 m (60 ft); yield

$$x(4 + 18) = (3{,}000)\,(0.0012)$$

$$x = 0.16 \sim 1/2 \text{ ft per year}$$

which probably is a realistic figure in as much as these shores are rather stable although slight erosion is visible on some of the dunes. It may be asked why a higher figure corresponding to the more rapid rise in sea level in recent years is not used as on the lower east coast? The most logical answer to this question seems to be that said shore, as already mentioned, is stable and does not reflect any response to the rapid rises that, as demonstrated by

Fig. 3, is not either as rapid on the Gulf as it is on the east coast. Consequently, it must be assumed that the development because of the modest wave action is tough and does not respond to quick (but still modest-size) changes.

Model experiments described by George Watts, M. ASCE,[10] on the effect of tidal action on wave-formed beach profiles give certain information on the behavior of profiles with a fluctuating water table. These results should not be transferred, uncritically, to field (prototype) conditions. Although they do not interfere with the previously noted approach, they are of a qualitative nature, only.

An analytical approach based in part on the results from the southeast coast of Florida will be made.

Assuming an equilibrium profile, as indicated in Fig. 4, following the equation

$$y^{3/2} = p\,x \quad \text{. . . . .} \quad (2)$$

in which y is the depth at distance, x, from the shore[8,9] and using the results from the southeast coast of Florida where a rapid rise in sea level of "a" millimeters causes a shoreline recession of "100 a," the intersection point between the old and the new profile corresponding to the rise "a" is found by means of Eq. 2 and

$$(y + a)^{3/2} = p(x + 100\,a) \quad \text{. . . . .} \quad (3)$$

The mathematical expression for the intersection point is, unfortunately, complex and it is easier to find the point using a numerical method.

For the steep profiles on the southeast coast of Florida, $p \sim$ approximately 0.04 (x and y in metric system). With a rise of 1 m, that is 6 mm per yr in 167 yr, or 1.2 mm per yr in 830 yr, the theoretical shoreline recession will be approximately 100 m (335 ft). The intersection point between the old profile without rise in water level and the new profile corresponding to 1 m rise in water level is at a distance of approximately 135 m (450 ft) from the original shoreline and at a depth of approximately 2 m (7 ft) in the original profile (235 m from shore in the new profile at a depth of 3 m).

A 0.3-m (1-ft) rise of the sea level, that may come in 50 yr to 100 yr, may cause shoreline recessions of more than 100 ft on the southeast coast where many beaches at this time (1961) are too narrow to meet that kind of development. This unfortunate situation can only be adequately handled by means of artificial nourishment with suitable sand material.

HOW MUCH MATERIAL IS NEEDED TO MAINTAIN THE FLORIDA SHORES?

It is generally known that erosion is partly a long-range geological process initiated and maintained by nature itself, and is partly caused by man's interference with nature not least in the form of improved inlets—whether these inlets are only dredged or are also jetty-protected and by inadequately designed coastal structures including groins and sea walls. The improved in-

10 "Laboratory Study of Effect of Tidal Action on Wave-Formed Beach Profiles," by George Watts, Technical Memorandum No. 52, Corps of Engrs., Beach Erosion Board, 1954.

let is often responsible for considerable shoal-formations in the sea as well as in the bay or lagoon. In this way, material needed for maintenance of the downdrift-side shores is "wasted to no purpose," and the usual result is an often heavy increase of erosion on the downdrift side (for example, at Palm Beach, Fla).

The total annual erosion along the approximately 1,000 miles of sandy shore in Florida has been estimated[11] to be 10 to 20 million cubic yards. With reference to the computed figure for erosion based on rise of sea level, the following quantitative estimate on natural erosion may be made:

East Coast: (an average of 1.5 ft natural recession per year) approximately 700,000 yd (23) $\left(\frac{1.5}{3}\right)$ = 8,000,000 cu yd

Upper Gulf Coast: (an average of $\frac{1}{2}$ ft natural erosion per year) approximately 400,000 yd (23) $\left(\frac{1}{6}\right.$ = 1,500,000 cu yd

Lower Gulf Coast: estimate based on shoreline recessions = 1,500,000 cu yd

TOTAL: 11,000,000 cu yd

The man-made erosion was estimated to be one-third of the natural erosion[11] and is predominant only in some few areas of limited size particularly at the improved inlets. This adds about 4,000,000 cu yd, making a total of 15,000,000 cu yd or close to the estimate based on actual observations of shoreline movements counting on a short-term basis on an equal landward movement of the depth contour down to 25 ft. This manner of computing erosion quantities assumes that the sea level is constant for a limited period of time. Its assumption is based on surveys of beach and bottom profiles in recent years.

PREDICTION OF SHORE DEVELOPMENT FOR THE IMMEDIATE FUTURE

With respect to the development in sea-level rise during the next few hundred years, it seems to be a general agreement between scientists in the meteorological, climatological, geophysical, and geological fields that (concerning the grand-scale development) the earth is on its way into another glacial period. By comparison with the duration of past interglacial and glacial episodes in the earth's history, it is believed that the new glacial period may be only 10,000 yr to 15,000 yr away. Such a prospect, with its accompanying ice sheets devastating northern lands and settlements is an unfortunate one to contemplate in terms of physical, economic, or political consequences. The countries to be hurt are Canada, the Scandinavian countries, the USA and the USSR. We are, at present, apparently in a short-term, general, world-wide warming trend. In the United States, the rise in mean annual temperature, since 1920, has been approximately 2°C. according to E. Dorf,[12] and the rise

11 "Florida Coastal Problems," by P. Bruun, F. Gerritsen, and W. H. Morgan, Coastal Engineering No. VI, Council on Wave Research, Calif., 1958.

12 "Climatic Changes of the Past and Present," by Erling Dorf, American Scientist, Vol. 48, 1960, p. 341.

in winter temperatures has been approximately twice as much as that in summer temperatures. The rises are unequally distributed.

In recent years it became evident that the Greenland and Alaskan glaciers retreated. At the same time, the codfish from the Atlantic, because of the warming up, replaced the seals in the waters along the coast of Greenland changing the Greenlanders to fishermen (instead of hunters).

With respect to Antarctica the situation may be different. H. Wexler[13] examined eight different ice budgets for the Antarctic Ice Sheet and found that five of these budgets call for rates of increase of the ice that are in good agreement. The observed rise in sea level of the world's ocean would appear to contradict the removal of water required to nourish the Antarctic Ice Sheets. The thermal expansion of ocean waters caused by an increased absorption by the ocean of solar radiation has been invoked to resolve this contradiction but due to the lack of adequate data on long-period thermal changes at all depths of the world's oceans, it still seems unreliable at this time to state that the Antarctic Ice Sheet is either increasing or decreasing. Evidence exists that the glaciers may stop their retreat in the not too distant future if they have not already stopped retreating.

The conclusions from this data seems to be that, for a relatively short period of time, a rise in sea level of the total order indicated by the small fluctuations in Fig. 2 (1 m or 3 to 4 ft) may be expected. Even a rise of only 0.3 m (1 ft), that, as previously noted, may come in 50 yr to 100 yr, will have serious consequences for the erosion situation along the Florida shores, because it may give rise to shoreline recessions of the order of 100 ft or more on the southwest coast of Florida. It is evident that there is a time lag between the rise of the water table and the reaction in the form of erosion. This lag is more pronounced for the gentle sloping, northeastern Florida shores, than for the steep southeastern Florida shores where reactions to fluctuations come more rapidly. These steep shores present less stability than the "flat" shores and any change from rising to lowering, or neutral position of the sea table seems to be reflected more rapidly in these steep-slope shores. With gentle-slope shores (such as the northwestern and western Florida shores) there may be more phase-lag between rise of sea level and erosion. This also means that the steep-slope shore with lowering of the sea level may stop eroding before the gentle-slope shore slowly turns a tendency to erosion into a tendency to accretion. This is all under the assumption that the shores being studied are equilibrium-profile shores with maturity of configurative development.

SUMMARY AND CONCLUSION

1. Eustatic changes are overall changes of the sea level that are independent of land movements and that have many different causes. During the Quaternary, two major and several minor effects are noticeable:[2]

(a) Climatically controlled glacio-eustasy, involving vertical oscillations of a few meters up to 100 m or 125 m (300 ft to 400 ft) in periods ranging from 550 yr to 90,000 yr; and

---

13 "Ice Budgets for Antarctica and Changes in Sea Level," by H. Wexler, Journal of Glaciology, Vol. 3, No. 29, 1961.

(b) geodetic change, associated with either the shape of major ocean basins or with the shape of the geoid in respect to the spheroid, perhaps associated with a polar shift. Several minor geophysical effects are to be expected from glacial loading and unbalancing effects on the globe but their roles have not yet been analyzed.

Analysis in detail of the eustatic effects and their timing over the last 15,000 yr shows that a close correlation is observable between minor oscillations of sea level and climatic events. Every recorded glacial advance of the past 5,000 yr is matched by a eustatic lowering of the order of 3 m to 7 m (10 ft to 22 ft). Pollen analysis from non-glaciated areas confirms that these are climatically cool phases. In the arid American West (New Mexico, and so forth), each of the younger cool phases corresponds to a drought; in temperate belts they are marked by pluvial events.

2. The earth is, at present, in an interglacial stage, but heading toward another glacial stage, perhaps 10,000 yr to 15,000 yr hence.

3. Though marked by minor alternating colder and warmer cycles; the present short-term, general trend of increasing warmth should continue for at least two to three hundred years over most of the lowland regions of the northern hemisphere.

4. Sea level has been relatively stable during the past 5,000 yr, but minor fluctuations up to 3 m to 7 m (10 ft to 20 ft) have occurred. Since the Roman Era, 1,500 to 2,100 years ago, sea level has risen approximately 2 m (7 ft) or an approximate average of 1.2 mm per yr.

5. During the past hundred years, the lowest point of sea level occurred in 1890; the mean annual rise from 1900-1950 was 1.2 mm. The fastest decade was 1946-1956 with 5.5 mm, but patterns vary somewhat if plotted ocean by ocean. The entire eastern seaboard of North America shows an anomalously high, apparent rise of sea level. An appreciable component in the sea-level rise (approximately 20% of which is eustatic) is possibly due to a secular deceleration of the Gulf Stream. For the 1890-1960 period, there is on the East Coast a progressive rise of sea level that is at least 50% greater than any other large departure from other parts of the world. The very rapid rise between 1930 and 1950 has slowed down in the 1950 to 1960 period (as seen from Fig. 3).

6. The effect of sea-level rise seems apparent in the development of erosion along the Florida shore:

Based on the assumption (reasoned in part by the work done by marine geologists) that the 18-m (60-ft) depth contour seems to be the outer limit for the nearshore littoral drift and exchange zone of littoral material between the shore and the offshore bottom area, it is assumed that the offshore bottom for any rise in sea level will undergo a gradual adjustment process tending to keep its "equilibrium form." By this process, the bottom may be raised together with the sea level until it is covered by the same depth of water at the same distance from the (new) shoreline as it was before the rise. The material needed to raise the bottom is assumed to come from the corresponding shore area by movement of material by transversal (rip) currents and by diffusion currents.

7. The rate of the development described above will probably depend to a large extent on the slope of the offshore bottom. Steep profiles are probably more sensitive to short-term rises in sea level than to long-term rises.

Gently sloping profiles may respond to long-term changes only and demonstrate a pronounced phase-lag between rise of sea level and effect on erosion. If the same bottom profile has a nearshore steep part as well as an offshore flat part, the steep part may respond to the short-term fluctuations whereas the profile as a whole, including its flat, offshore part, may respond to the long-term rise of sea level.

8. The validity of the previously noted concept was tested on shoreline recessions on the Florida shores. It appears that the steep profiles along the southeast coast follow the short-term rises of the sea water table (6 mm, annually, in recent years), while the more gently sloping profiles on the northeast and upper Gulf coasts follow the long-term rises of the sea water table (1.2 mm, annually, 1900-1950). With respect to the influence on erosion of the long-term rises of the sea water table on the southeast coast, the width of the 0-m to 18-m (0 to 60-ft) bottom area (up to about the edge of the shelf) seems to be too small to reflect long-term rises. During more extended periods of time, material is probably lost to deeper water outside the 18-m (60-ft) depth contour.

9. The seriousness of the rise of sea level with respect to erosion is demonstrated by the computed results. Even a rise of only 0.3 m (1 ft), that may come in 50 yr to 100 yr, may cause shoreline recessions of more than 35 m (100 ft) with the possibility of much higher recessions in marsh and other low shore areas.

10. Quantitative estimates of the erosion along the Florida sand shores based on computed shoreline recessions caused by rise of the sea water table give an erosion of 11,000,000 cu yd per yr. This quantity does not include the man-made (inlet) erosion that is estimated to be approximately 4,000,000 cu yd per yr. The total is approximately 15,000,000 cu yd. The quantity of accretion is unknown; but, in as much as few shores accrete, it will hardly amount to more than a small fraction of the erosion. Material deposited on shoals is disregarded.

11. It is desirable to test the validity of the assumptions on transfer of material perpendicular to the shoreline. It seems reasonable to assume that such tests can be accomplished, indirectly, by analyses of bottom-material characteristics as compared with beach-material characteristics. A direct method would be tracing of material eroded on steep shores during extreme storms using radio-active or luminescent tracers. Such a method would, needless to say, include some uncertainties and will probably, because of the time limit, indicate more narrow widths than the actual "exchange widths" as referred to herein. It is self explanatory that the previously noted theory should be tested at other places where field data are available for comparison.

12. The only way in which the problem of sea-level rise can be handled is by artificial nourishment to replenish the material eroded and by the construction of dykes or sea walls of proper elevation. Such large scale operations are already badly needed at many places in Florida to cope with storm tides. Most likely it will be necessary to secure part of this material from the offshore bottom simply because no source of material is available on land within reasonable distance. New type dredging equipment may have to be developed to accomplish such a task.[14]

---

14 Discussion by Per Bruun, of "Shark River Inlet By-Passing Project," by W. Mack Angas, Proceedings, ASCE, Vol. 87, No. WW 2, May, 1961.

4

Reprinted, courtesy of the Geological Society of America, from *Geol. Soc. Am. Bull.* **74**:971–990 (1963)

Wave-Base, Marine Profile of Equilibrium, and Wave-Built Terraces: A Critical Appraisal

ROBERT S. DIETZ *U. S. Navy Electronics Laboratory, San Diego, Calif.*[1]

Abstract: The concept of wave-base and its corollaries, the marine profile of equilibrium and the wave-built terrace, are largely erroneous. They do not dominantly control the development of the continental shelf and slope. The treatment of these concepts in geologic textbooks and sourcebooks needs revision.

Wave-base, although a useful concept if applied generally to the surficial zone of the ocean of high-ambient agitation, does not control shelf deposition as the entire shelf is above any wave-base in this sense. Instead, the shelf is a drowned and relict surface, developed by oscillation of sea level and prograding and regressing paralic sediments—and here surf plays the dominant role. Sedimentation on the continental terrace is not controlled by wave-base but chiefly by topography; the continental slope is undergoing erosion rather than prograding, and ultimate deposition occurs on the continental rise. The wave-cut terrace is not cut at wave-base but at surf-base.

Wave-built terraces are nonexistent both on a small scale and on a grand scale as an explanation of the continental slope. A marine profile of equilibrium is developed in the nearshore zone and is associated with a migrating lens of sand, but the outer-shelf profile is not a profile of equilibrium.

CONTENTS

Introduction and acknowledgments 972
Wave-base 973
Marine profile of equilibrium 977
Wave-built terrace 979
Delta terrace 980
Wave-cut (surf-cut) terraces 980
Discussion 984
Concluding remarks 988
References cited 989

Figure

1. Basic environments of deposition showing the development of the continental shelf and slope according to Rich (1951) 972
2. Two similar early views on continental-terrace development 973
3. Various views concerning the origin of the continental shelf, all of which imply a controlling effect of wave-base 974
4. Continental-terrace development 975
5. Wave-cut terraces and wave-built terraces . . 976
6. Various views on the origin of the continental shelf as presented by Shepard (1948) . . 977
7. Profiles off the coast of Madagascar presented by Douglas Johnson (1919) 979
8. Cross section of a wave-built terrace according to Gilbert (1890) 979
9. Schematic diagram of the continental terrace off the eastern United States according to Stetson (1949) 981
10. Interpretation of termination of monoclinally dipping beds (Cretaceous to Recent) extending to the continental slope off the eastern United States according to Heezen and others (1959) 982
11. The delta terrace redrawn from Dunbar and Rodgers (1957) 982
12. Profile of the beach and nearshore region modified after Inman (1962) 983
13. Detailed topographic form of the inner shelf off Panama City, Florida 984
14. Cross section of the continental terrace . . . 985
15. Gulf Coast geosyncline 986
16. Isostasy and the continental embankment . . 987
17. Paralic wedge and the formation of the continental terrace 988

Plate — Facing

1. Wave-cut terraces, San Clemente Island, California 982
2. Wave-cut terraces, Middleton Island, Alaska . 983

[1] Present address: U. S. Coast and Geodetic Survey, Washington, D. C.

INTRODUCTION AND ACKNOWLEDGMENTS

The concept of wave-base with its "corollaries," the marine profile of equilibrium and the wave-built terrace, as applied to the continental shelf are geologic keystones of sedimentology, stratigraphy, and geomorphology. Treatment of these concepts in nearly all geologic textbooks is proof enough of this contention; explained in these terms are the accumulation of the geosynclinal sedimentary prisms. Yet these concepts, as taught and commonly

In contrast Shepard (1948, p. 190–191), for example, has crusaded for many years against the view of the outer continental terrace being a wave-built terrace. Others, too, have concurred in this (*e.g.*, Dietz and Menard, 1951; Dietz, 1952), and so do nearly all modern marine geologists, although Kuenen (1950, p. 302–306) may constitute an exception[2]. But these writings have had little effect in displacing the firmly entrenched classical view made popular by Johnson (1919) and endorsed by Stetson (1949).

Marine geological field evidence shows that

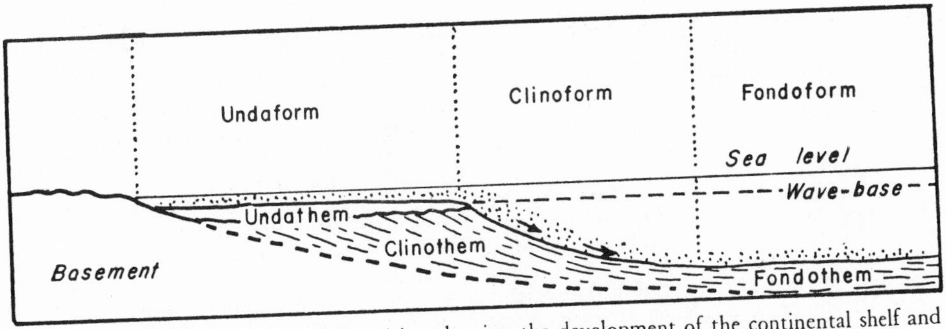

Figure 1. Basic environments of deposition showing the development of the continental shelf and slope according to Rich (1951). His concept follows closely the classical view and he shows the continental terrace to be like a submerged delta with deposition being controlled by wave-base.

accepted, seem to be somewhat confused and partly fallacious.

For example, a modern version of the classical treatment of sedimentary environments in terms of wave-base, etc., is that of Rich (1951). Defining *wave-base* as the greatest depth to which the bottom is stirred by waves during storms, he recognized *undaform* deposits above wave-base, *clinoform* foreset deposits of the wave-built terrace, and *fondoform* bottomset deposits covering the basin floor (Fig. 1). Rich equated his environment with the continental shelf and slope stating that,

> "After a long period of stationary sea level, during which the marine and subaqueous profile of equilibrium had been brought to full maturity, the *undaform* would correspond with the continental shelf, the *clinoform* with the continental slope, and the *fondoform* with the abyssal ocean bottom beyond the base of the slope."

He held wave-base to be the critical plane separating the unda from the clino environment and placed it at a depth of as much as 600 feet for shelves exposed to the full sweep of a stormy ocean. In short, Rich's concept is based in essence on the wave-built terrace.

the continental slope is not composed of the foreset beds of an advancing wave-built terrace (*e.g.*, Heezen and others, 1959, p. 46–47). The persistent tendency to equate the slopes with foreset beds seems to be an attempt to patch up the Johnsonian view and to treat these concepts in a circumspect and vague fashion. Figures 2–6, drawn in most textbooks to explain these concepts, reveal this attempt as they do not give the scale, do not specify if they are including all of the shelf, and, if based on actual field data, resort to Johnson's (1919) few soundings of doubtful veracity taken 4 decades ago.

The field "facts" are never very convincing, as they remain bare empiricisms unless there is some foundation of theoretical or conceptual understanding of why things are the way they are. So it seems useful to broadly review the concepts of wave-base, etc., in a conceptual manner, as these appear to be the root of the

[2] After reading the original manuscript, Kuenen wrote (1962, personal communication) that he is inclined to tentatively accept most of the concepts presented in this paper.

confusion. This essay is attempted, then, in the hope of clarifying these concepts or at least in crystallizing opinion.

The writer wishes to express his appreciation to his many colleagues with whom he has enjoyed fruitful discussions about the concept presented in this paper. Especially to be mentioned are G. Curl, E. L. Hamilton, D. G. Moore, D. Scholl, and R. F. Dill of the Navy Electronics Laboratory; J. R. Curray of Scripps Institution of Oceanography; K. O. Emery of Woods Hole Oceanographic Institution; and Ph. H. Kuenen of the University of Groningen. Don Begnoche assisted closely in the preparation of the manuscript. Sonia Baslee provided clerical assistance and John Holden assisted with the drafting.

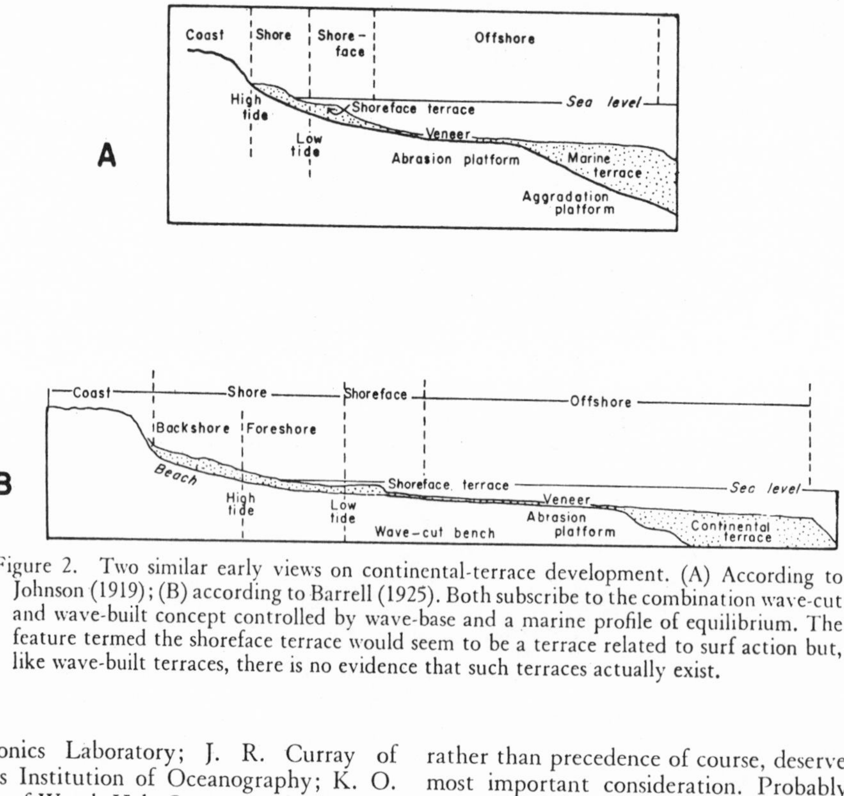

Figure 2. Two similar early views on continental-terrace development. (A) According to Johnson (1919); (B) according to Barrell (1925). Both subscribe to the combination wave-cut and wave-built concept controlled by wave-base and a marine profile of equilibrium. The feature termed the shoreface terrace would seem to be a terrace related to surf action but, like wave-built terraces, there is no evidence that such terraces actually exist.

WAVE-BASE

The term *wave-base* was introduced by Gulliver (1899, p. 176–177) who, without indicating the depth of wave-base, held that it determines the ultimate depth of a platform of marine abrasion. He reasoned that peneplanation is limited to sea level but a platform of marine abrasion is cut to wave-base at a depth considerably below sea level. According to R. D. Salisbury (1899; *see* Howell, 1957) wave-base is "the plane to which waves may degrade the bottom in shallow water." Fenneman (1902) considered it "the depth to which wave action ceases to stir the sediments." Another definition common in the literature is that wave-base is the greatest depth to which the bottom is stirred by storms. Modern usage, rather than precedence of course, deserves the most important consideration. Probably the most generally accepted definition is that wave-base is the depth at which, during times of high swell, deposited sedimentary particles are not again stirred up by oscillatory wave-induced water motion.

This last definition seems satisfactory. Unfortunately the literature does not seem to clearly provide us with parameters about wave-base such as its depth or range of depths, its depth relative to pertinent wave characteristics such as wave length, or the ambient water velocity at wave-base. The assumed depth of wave-base especially is treated in a vague and noncommittal way in the more recent papers.

In contrast, Johnson, in 1919 (p. 80), placed

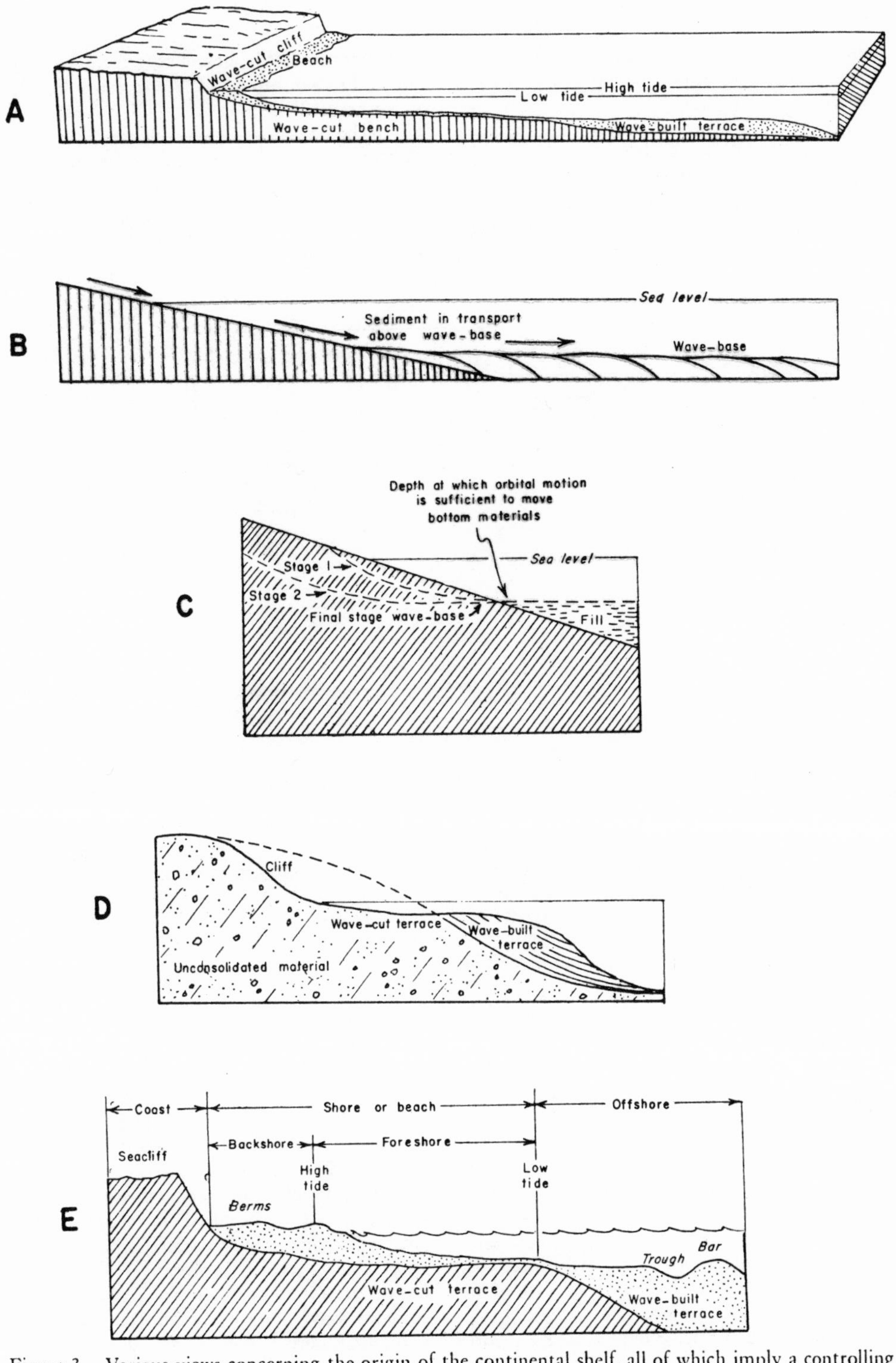

Figure 3. Various views concerning the origin of the continental shelf, all of which imply a controlling effect of wave-base. (A) After Longwell, Knopf, and Flint (1948); (B) after Clark and Stern (1960); (C) after Garrels (1951); (D) after Von Engeln (1942); and (E) after Leet and Judson (1958).

the depth at 100 fathoms or 600 feet, doubtless basing this on the supposed depth of the shelf break. To support this view, he collected observations from fishermen and mariners about the greatest depth at which there seemed to be appreciable water movement. More modern soundings have shown that the shelf break is typically only at 65 fathoms. This, plus the appreciation that sea level has recently risen, has led other writers to propose lesser depths. Thus Daly (1942, p. 9–11, 69) cites 50 fathoms and Barrell (1925, p. 296) 30–40 fathoms.

This vagueness seems to be an admission of uncertainty but the more fundamental objections to wave-base are quite apart from its depth. The author rejects the view that wave-base dominates erosional and sedimentary processes on the continental shelf. And he disagrees with the belief implicit in the wave-built-terrace corollary—that at wave-base, by any definition, there is an abrupt transition from agitated to quiet water. For example, Twenhofel (1950), in his Principles of Sedimentation, supports the wave-base concept in much the common form. He terms wave-base the permanent base-level of erosion and the depth to which marine sediments can build up as well (*e.g.*, Fig. 3B, pt. B). This concept implies a sharp boundary level in the ocean above which the water is in high ambient agitation and below which the water is perfectly still. Also the common diagrams of the continental terrace as a combination wave-cut and wave-built terrace show them to be in flush contact (*see* Figs. 2–6). Hence, wave-base is assumed to be so sharp that the ambient turbulence is sufficient to bevel

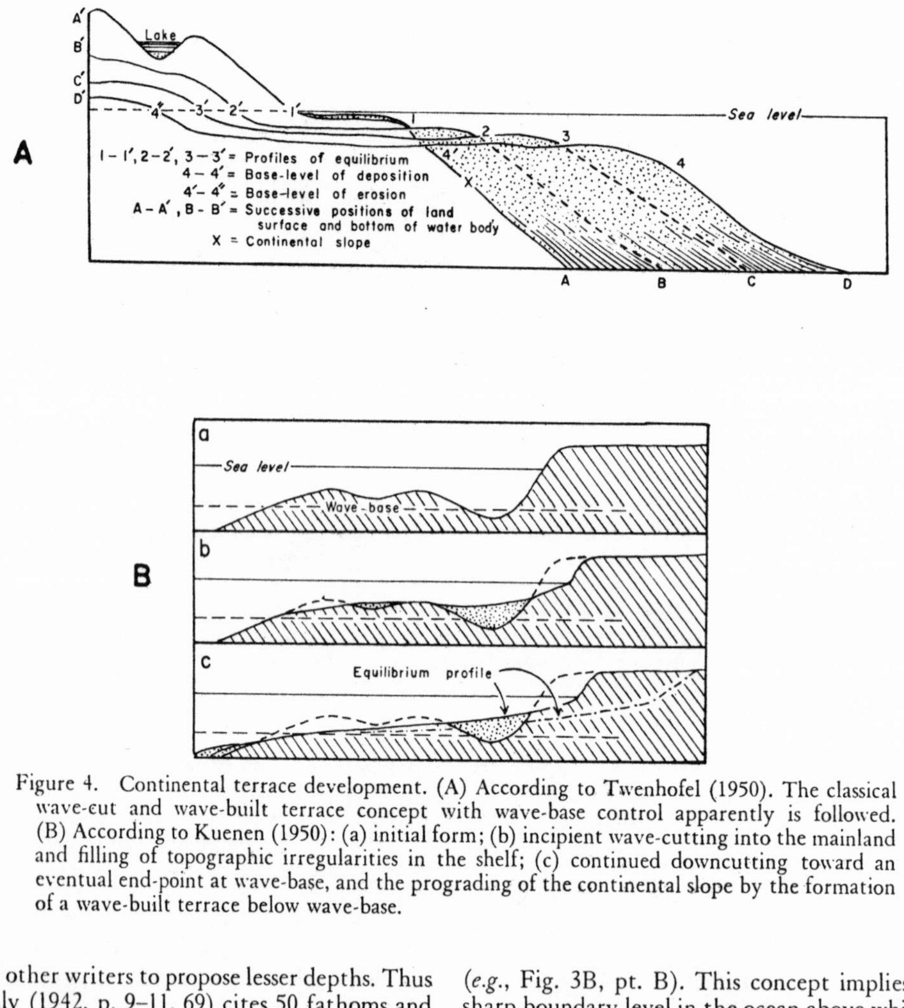

Figure 4. Continental terrace development. (A) According to Twenhofel (1950). The classical wave-cut and wave-built terrace concept with wave-base control apparently is followed. (B) According to Kuenen (1950): (a) initial form; (b) incipient wave-cutting into the mainland and filling of topographic irregularities in the shelf; (c) continued downcutting toward an eventual end-point at wave-base, and the prograding of the continental slope by the formation of a wave-built terrace below wave-base.

bedrock above wave-base, but just below it the water is so still that even the finest floc, once deposited, supposedly is not again moved.

Some investigators (Dietz and Menard, 1951; Fairbridge, 1952; Bradley, 1958) have given 5 fathoms (10m) as the approximate depth of effective wave abrasion. The depth conceivably might be referred to as wave-base. But this would only lead to further confusion, so that "surf-base" or perhaps "surge-base" is a better term for this purpose. This surf-base is not restricted exactly to the surf zone but extends out to the greatest depth where the waves begin to peak up appreciably during storms. It is shoreward of this depth that essentially all bottom loss of wave energy (>95 per cent) occurs (Dietz and Menard, 1951).

Along the present shore line, the most active cutting of rock terraces occurs essentially at sea level; and, in fact, students of sea-level changes commonly equate uplifted terrace levels directly with old sea levels. Under conditions of long sea-level still-stand, however, cutting would probably extend to a depth of about 5 fathoms or possibly 10 fathoms which is the lowest depth of vigorous abrasion. Thus a seamount, for example, would be decapitated to this level but, once waves stopped peaking up over the shoal, it is doubtful that there would be sufficient bottom friction to cause geologically significant planation.

Emery (1961) raises an interesting point casting doubt on the outer shelf being at wave-base. He finds that all shelves which have been well studied contain relict sediments along their outer part. This alone means that these outer shelves are *below* any wave-base of erosion and *above* any wave-base of deposition. Gunnerson and Emery (1962) emphasize the probable importance on the shelf of internal waves (and internal surf from breaking internal waves), especially at the thermocline depth. They suggest that turbulence so generated can cause the seaward transfer of silt and clay over the shelf break.

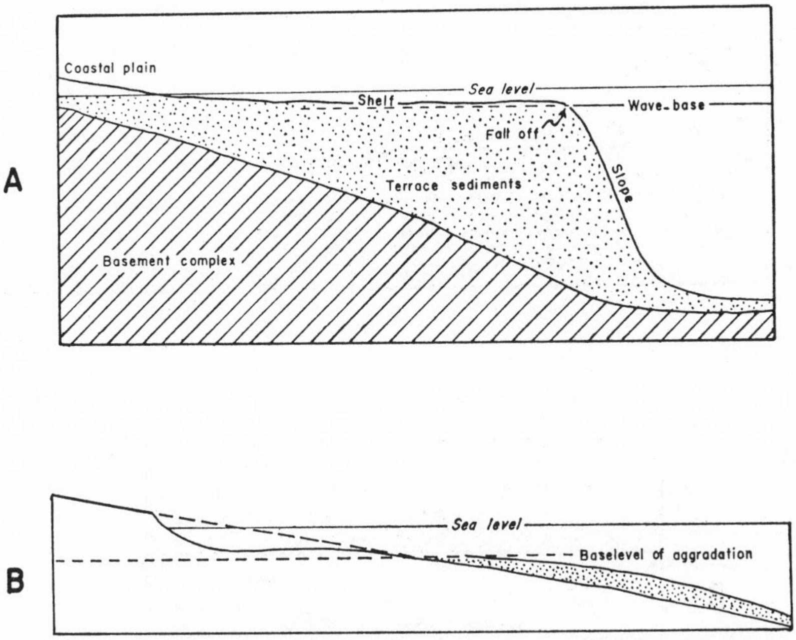

Figure 5. Wave-cut terraces and wave-built terraces. (A) Development of the continental slope off the eastern United States after Stetson (1949); (B) an apparent similarity between base level of aggradation and wave-base after Dunbar and Rodgers (1957).

WAVE-BASE OF STRATIGRAPHY: Stratigraphers commonly refer to ancient marine sedimentary rocks as having been laid down either above or below wave-base (*e.g.*, Pettijohn, 1957, p. 593), speaking of the cratonic platform and/or miogeosynclinal deposits in contrast to the eugeosynclinal graywackes or flysch facies. This seems

to be a meaningful and useful application of the term wave-base when employed in this general way. In any event, this usage is deeply ingrained so it is likely to continue.

Unquestionably the upper levels of the ocean are in a higher state of ambient agitation than the deeper water. Discovery of deep ripple marks and scours does not change this generalization. In this sense, the entire continental shelf, both in extent and depth, is above wave-base; sediments laid down here are shifted about and tend to attain textural and minerological maturity. The above wave-base facies apparently is easily recognized in the geologic column. They are, in general, the miogeosynclinal facies characterized by pure sands, gravel, coquinas, calcarinites, etc. The sand is well sorted, rounded, rippled, and cross-bedded. Diastems are numerous; many are marked by glauconite and phosphorite. We must bear in mind, too, that the continental shelf of the Holocene probably is unusually deep, so that its typical 70-fathom outer depth may be geologically uncommon. The present shelf depth presumably reflects the great Pleistocene swings at sea level terminated by the Holocene rise for which there seems to be growing evidence of as much as 120 meters since 20,000 years B. P. (Curray, 1960; 1961).

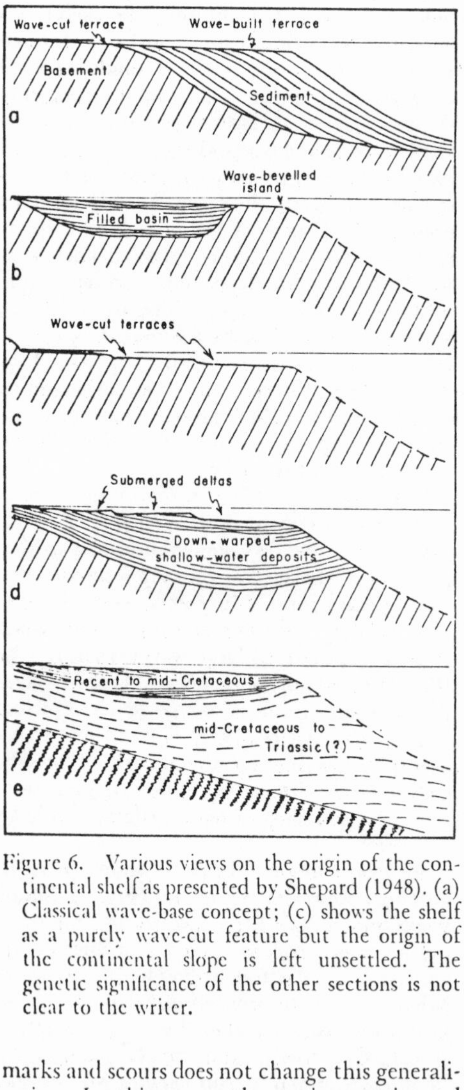

Figure 6. Various views on the origin of the continental shelf as presented by Shepard (1948). (a) Classical wave-base concept; (c) shows the shelf as a purely wave-cut feature but the origin of the continental slope is left unsettled. The genetic significance of the other sections is not clear to the writer.

By way of contrast, many other ancient sediments presumably are laid down in quiet deep water. Most obviously so are the graywackes and flysch facies of eugeosynclines which apparently are turbidites. They commonly are considered to be poured-in dirty sediments and gravels intermixed with clay. Tidal motion and other deep currents apparently do not produce significant reworking of these deposits. But, as will be discussed later, if we deny the existence of wave-built terraces, such flysch-type deposits cannot be continental-slope deposits. Rather it is presumed that they are predominantly continental-rise deposits (Dietz, 1963), so that there is a considerable topographic gap between the above wave-base miogeosynclinal deposits and those of the eugeosyncline.

There are, of course, many other below wave-base deposits in the stratigraphers' sense which are not turbidites, *e.g.*, the Chattanooga Shale. Such deposits could not have been laid down on open exposed shelves but rather in some protected environment. A topographic depression within the continent, even if of small depth, could provide a suitable environment for the deposition of some fine-grained deposits. Even if stirred up, the fine material would not necessarily be flushed out but would simply settle to the bottom again within the depression.

MARINE PROFILE OF EQUILIBRIUM

A valid concept for the profile of equilibrium has been developed for rivers with sea level providing the ultimate base-level: there can be no river erosion significantly deeper than sea level. Such a profile of equilibrium is reached when transporting power and sedimentation are everywhere in balance. In maturity, such a river is concave-upward in profile near its head and becomes almost flat near its mouth.

Similarly, the continental-shelf profile has been generally held to conform to a marine

profile of equilibrium with wave-base serving, like sea level for rivers, as the ultimate base of sediment transport. Most marine geologists do not now accept this view although Stetson and Kuenen have advocated it (Stetson, 1949; Kuenen, 1950, p. 302–306). Accordingly, the shelf is said to be concave-upward at the shore end, passing into a horizontal stretch farther out, and then merging into a convex-downward depositional region beyond wave-base. Supposedly, the slope of the shelf is adjusted to the average particle size of the detritus moved by the waves at each particular point along it. The inclination is supposedly just steep enough to keep material in slow transport out to wave-base, much as a graded stream ultimately conveys its sediment load to sea level. Suffice to say, because wave motion simply falls off asymtopically with depth and because many other motions (especially tidal) pervade the entire ocean, there can be no abrupt and specific depth at which sediment transport is suddenly completely checked. Without a wave-base, in this sense, there can be no profile of equilibrium.

What about field evidence? For this the southern California region is particularly instructive. Acoustic probing (Moore, 1960) and scuba diving have made this shelf quite well known. Unlike the Atlantic coast, it is easy to identify the modern sediment, as this overlies a relict surf-cut rock terrace, usually with angular unconformity. A marine profile of equilibrium *does* exist in the nearshore or paralic region between a depth of about 20 m and the highest reach of the waves on the beach. Here a concave-upward lens of sand, commonly with an underlying basal conglomerate, is developed which shows 12-day tidal and seasonal changes and which tends to decrease in grain size offshore. This lens rests on the surf-cut terrace of bedrock. The lens, however, is not related to wave-base, but rather to wave action in the surf and pre-surf zone. The processes occurring in the surf region, of course, are complex. But perhaps a dominating effect is the pushing of the tractive sand load toward shore, as waves have a fast push followed and balanced by a longer but slower (and so less competent) withdrawal. The sand is "bulldozed" toward shore, building a profile of equilibrium slope that is balanced against gravity and wave energy. This sand lens is an integral and dynamic part of the shore and it tends to move along with any transgressions or regressions of the shore line.

If a profile of equilibrium does not exist sea-

wa
ex
te
pr
up
th
ge
an
b
p
a

penetration records revealing sub-bottom stratification show that this rock platform is covered with scattered lenses and sheets of unconsolidated sediment, probably mainly paralic beds left behind during the recent postglacial transgression. But some of the sand bodies apparently are modern offshore deposits. These tend to fill topographic depressions and lap against submerged rock terraces, forming a lapping apron of detritus, but *not* a sedimentary terrace which could be interpreted as a wave-built terrace. These deposits have a smoothing effect on a continental-shelf profile which contrasts strongly with the irregularity of island shelves where sediment is scarce.

The marine-profile-of-equilibrium concept seems to have stemmed in large part from Johnson's (1938, Fig. 34) presentation of two shelf profiles off Madagascar showing a smooth concave-upward form. More important, the shelf is much deeper off the southeast coast, where there is exposure to large waves, than it is off the protected northwest coast. These Madagascar profiles appear in many texts, of which Gilluly and others (1959, Figs. 15–16) is one example. Considering the wealth of modern data, it is odd that we continue to depend upon a few old soundings from a geographically remote site.

Johnson's original figure is shown here including its few spot soundings (Fig. 7). Simple inspection reveals little justification for drawing a smooth concave-upward profile; on the contrary, the bottom seems highly irregular. Johnson obviously smoothed the sounding data (and arbitrarily so, rather than by any accepted statistical method), assuming the soundings to be in error at depth. But persons who do such surveying know that lead-line soundings are not likely to be wrong at depth although their position might easily be off. In neither profile does Johnson show any soundings which justify the drawing of the continental slope, although one is arbitrarily shown for the upper profile. We can conclude that Johnson's demonstration

of a profile of equilibrium off Madagascar is unsatisfactory.

Study of the best available charts around Madagascar (U.S. Hydrographic Office chart 3826–3829 and French charts: see especially Service Hydrographique de la Marine Charts Nos. 4174, 5976, 5461, 5506, and 5561) reveals that the region of Johnson's deep profile is a typical shelf and the shelf break is actually at

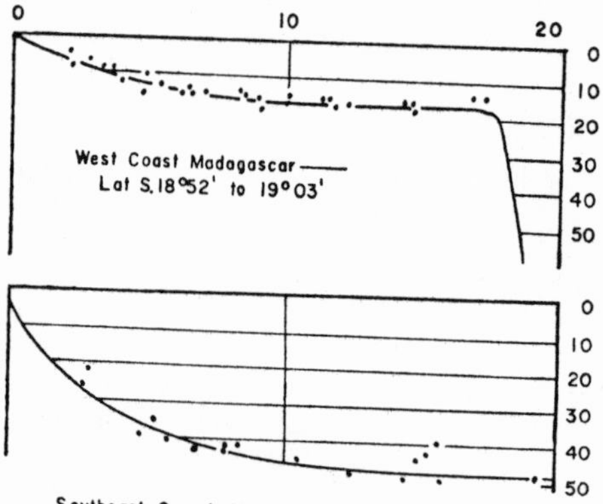

Figure 7. Profiles off the coast of Madagascar presented by Douglas Johnson (1919, p. 212) which purport to show that shelves generally have a concave-upward profile throughout their length and that shelves off exposed coasts are deeper than those off protected coasts. The dots represent spot soundings upon which the smooth profiles are drawn. As explained in the text, neither conclusion is justified. However, Johnson's conclusions are repeated in modern textbooks and these profiles are still used.

about 70 fathoms. The shelf around Johnson's smooth shallow profile is somewhat unusual in being highly irregular and shallow. But the reason is quite evident as this is a coralline region; very probably we are dealing with a reef-infested bottom and quite likely there is a drowned barrier reef at the shelf margin. It would not be useful to critique this "evidence" for a marine profile of equilibrium any farther.

WAVE-BUILT TERRACE

The term *wave-built terrace* was coined by Gilbert (1890) in his classic monograph on Lake Bonneville of which the Great Salt Lake is the remnant. Gilbert's usage of the term deserves scrutiny, not only because of originality, but also because Lake Bonneville is frequently referred to as the type locality, suggesting that wave-built terraces actually exist there.

Gilbert's wave-built terraces are features which everyone will agree are real. But Gilbert knowingly and explicitly described the wave-built terrace as a shore-deposited sand body and not a subaqueous terrace (Fig. 8). His feature is described as being constructed by sediment being drifted along the shore rather than offshore. He described its subaerial surface as being level in gross but uneven in detail, consisting of curved parallel ridges. He states (p. 56–57),

> ". . . each of these is referable to some exceptional storm, the wave of which threw the shore drift to an unusual height. Where the shore drift consists wholly or in large part of sand . . . , the wave-built terrace gives origin to dunes which are apt to make its normal ribbed structure."

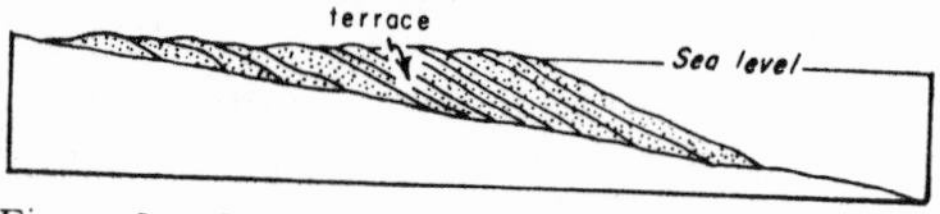

Figure 8. Cross section of a wave-built terrace according to Gilbert (1890) who originally coined the term. Obviously, as shown here, Gilbert was referring to a prograding shore deposit (a mixed subaerial-subaqueous boundary form) and not to a wave-built terrace in the modern usage of this term.

Thus Gilbert's wave-built terrace is generally equivalent to a cuspate bar in modern terminology and has nothing at all to do with wave-built terrace as the term is used now.

In geomorphology it is useful to have a type example or locality, as we do for a monadnock, a louderback, or a meander. As noted, Lake Bonneville fails to provide such a type locality. A brief review of the literature has revealed no examples of allegedly uplifted wave-built terraces, as for example around uplifted islands. Raised wave-cut terraces are among the commonest physiographic forms and, if wave-built terraces are real, some raised wave-cut terraces should grade into wave-built terraces. So far as the writer can discover, there are no examples, and it would seem that they are nonexistent on the small scale.

Turning now to the grand scale, it is evident that Johnson (1919), Daly (1942), and most others had in mind the outer continental terrace generally, with the Atlantic seaboard providing the type example. Among French geologists, equating the outer continental terrace

with a wave-built terrace is evident by their term for the continental slope, that is, *talus continentale*. Geosynclines, especially eugeosynclines, supposedly are examples of wave-built terraces according to some stratigraphers.

In many papers, Shepard (*e.g.*, 1948, p. 158–160) has objected strongly to the belief in wave-built terraces. He has pointed out that the extensive rock areas on outer shelves and their considerable irregularity contradict this concept. From his extensive compilation of shelf-sediment charts, there seems to be no general gradation from coarse to fine sediments as would be required by the wave-built-terrace concept. The depth of the shelf edge on a world-wide basis shows no relation to the exposure to waves; for example some of the largest waves strike Morocco and the Washington-Oregon coasts but the shelf-break depth is normal. Finally, the shelf-edge depth is the same for both the lee and exposed sides of an island, *e.g.*, Catalina Island (Shepard and Wrath, 1937). Thus Shepard has compiled a rather formidible array of field evidence against the outer continental terrace being a wave-built terrace.

Stetson (1949) dredged some of the submarine canyons off the Atlantic coast expecting to find the young foreset beds of the wave-built terrace. Instead, he dredged rocks as old as Cretaceous, forcing him to conclude that the foreset beds (which he still insisted must be present) were implaced farther seaward. (Fig. 9). This reduced the bulk of the supposed foreset beds to only a minor fraction of the entire bulk of the continental terrace, relegating foreset deposition to a minor role, at most, in creating that continental terrace. On the basis of more recent dredging and geomorphic interpretation, *e.g.*, by Heezen and others (1959, p. 43–50), we can dismiss construction of the Atlantic continental slope by foreset prograding because monoclinally thickening beds are found to outcrop in submarine canyons near the slope and in some cases directly on the slope (Fig. 10). As is to be expected, sub-bottom acoustic reflection studies are beginning to show some mantling of this continental slope by dip-slope strata, but the thickness of these beds is very small as compared with the over-all width of the shelf (C. Drake, 1962, personal communication). Such mantling is probably bed-rock supported so that it can not be considered a true and active prograding.

In summary, wave-built terraces, whether large or small, appear nonexistent. Doubtless, the desire to hold on to the concept in the face of the formidable evidence to the contrary stems from the need to account for great sedimentary terrace wedges like that now forming our Atlantic terrace. Also we need an environment of deposition for the ancient eugeosynclinal prisms. We can account adequately for these wedges by explaining the terrace as forming largely in terms of prograded paralic beds and by equating the eugeosynclines with continental-rise deposits (Dietz, 1952; 1963). Admittedly, this is major geologic surgery, but the patient does survive.

DELTA TERRACE

The hypothetical wave-built terrace is supposedly much like the delta terrace with topset (if sea level rises), foreset, and bottomset beds. Doubtless the delta terrace has provided the inspiration for the wave-built terrace. Without question, the delta terrace is a common physiographic form. Drowned delta terraces or those preserved in dry lakes might easily be misinterpreted as wave-built terraces. Let us then consider the nature of the delta terrace to avoid any confusion between it and the wave-built terrace.

When a sediment-laden stream enters a still deep body of water and the stream flow is suddenly checked, a simple delta is formed with topset, foreset, and bottomset beds to create the delta terrace (Fig. 11). Such a delta terrace is a mixed subaerial-subaqueous form, with the water level intersecting essentially at the nick-point of the terrace. The delta terrace now growing into Lake Mead from the Colorado River is an excellent example (Gould, 1960). Such terraces form best in lakes, as here waves are very small; large waves would destroy the text-book perfection of such a terrace, as indeed they do in most marine environments.

In contrast, the hypothetical wave-built terrace is wholly a subaqueous form. The terrace nick-point is supposedly provided by the sudden checking of water agitation at wave-base. But we have already given reasons why sudden checking of water velocity does not occur in the ocean outside of the surf zone.

WAVE-CUT (SURF-CUT) TERRACES

Wave-cut terraces are among the commonest of all geomorphic forms. They tend to be cut essentially at sea level, but with continued cutting during stationary sea level they tend to cut somewhat deeper. The author rejects the conventional notion that they can be cut down to

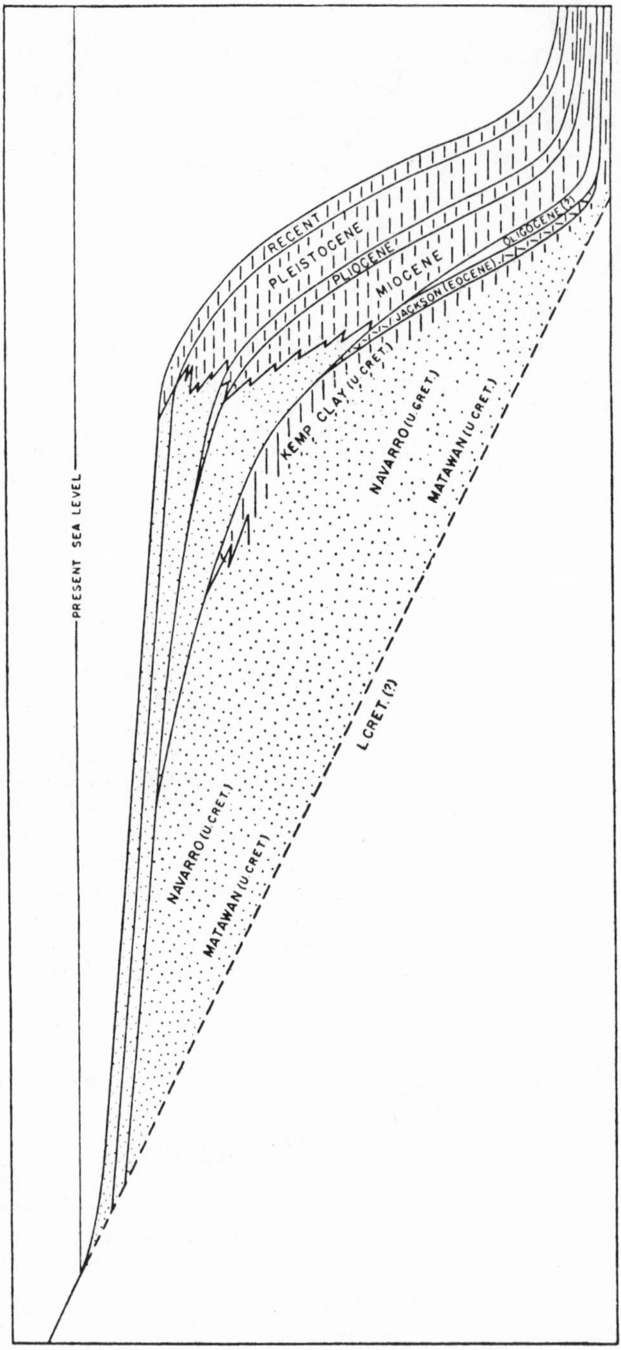

Figure 9. Schematic diagram of the continental terrace off the eastern United States according to Stetson (1949) showing his adherence to the classical wave-built terrace concept

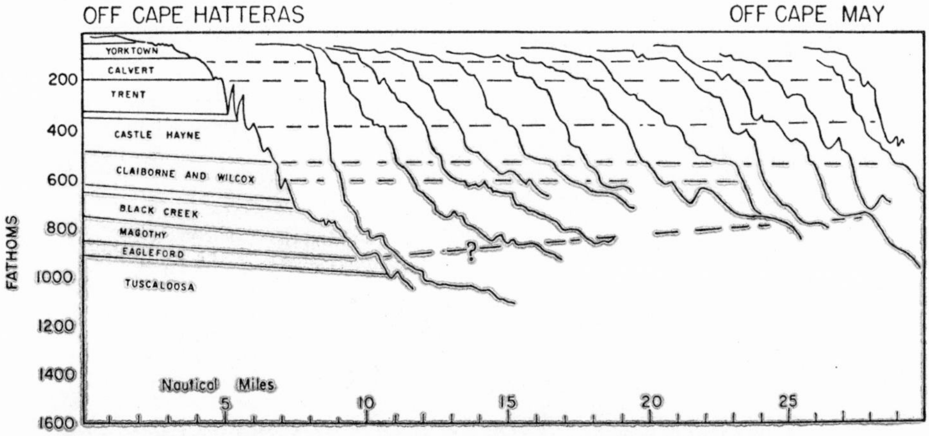

Figure 10. Interpretation of termination of monoclinally dipping beds (Cretaceous to Recent) extending to the continental slope off the eastern United States according to Heezen and others (1959). This diagram shows that terrace levels appearing on continental-slope profiles can be correlated with outcropping formations. These structural benches indicate that the classical continental slope is not built up of foreset beds as advocates of wave-base have assumed. (Soundings by U.S. Coast and Geodetic Survey; 35°30′N. to 38°30′N.)

wave-base. Instead he believes that appreciable rock erosion (and especially rock truncation) can continue down only where waves exert considerable bottom stress as marked by their peaking up and breaking into surf. For geological purposes, this is surf zone; so it is more explicit and precise to speak of surf-cut rather than wave-cut terraces, and this terminology will be used here.

Sub-bottom acoustic profiling (Moore, 1960) and oil-company explorations have demonstrated that the shelf off southern California is an almost-pure surf-cut terrace truncating dipping beds and creating an angular unconformity. This situation is typical of high-relief coasts undergoing diastrophic uplift. Oceanic volcanic islands, too, display surf-cut platforms all the way out to the shelf break. Wave-cut terraces of San Clemente Island off California are shown in Plate 1.

Middleton Island (Pl. 2) off the bight of Alaska displays a remarkable series of raised surf-cut terraces (Miller, 1953). Presumably, the island has undergone a series of steplike diastropic uplifts. A widely used textbook of geology (Gilluly and others, 1959) presents this same photograph as demonstration of the marine profile of equilibrium. It is difficult to understand in what sense a profile of equilibrium is demonstrated. A study of more detailed photographs of this island, kindly furnished by the U.S. Geological Survey, shows

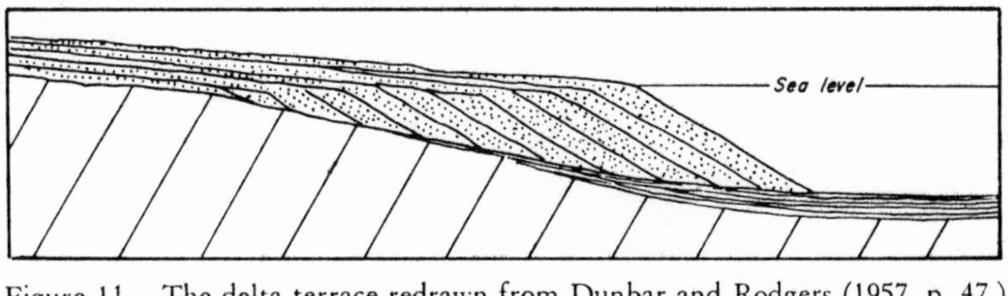

Figure 11. The delta terrace redrawn from Dunbar and Rodgers (1957, p. 47.) Topset, foreset, and bottomset beds are shown to develop when a sediment-laden river enters a quiet body of water. Such delta terraces are common geomorphic forms; they should not be confused with the entirely hypothetical wave-built terrace. Note especially that the nick-point is at water level and not below; wave action would tend to modify or destroy the delta terrace. A rising water level would drown it, producing a feature which erroneously might be interpreted as a wave-built terrace.

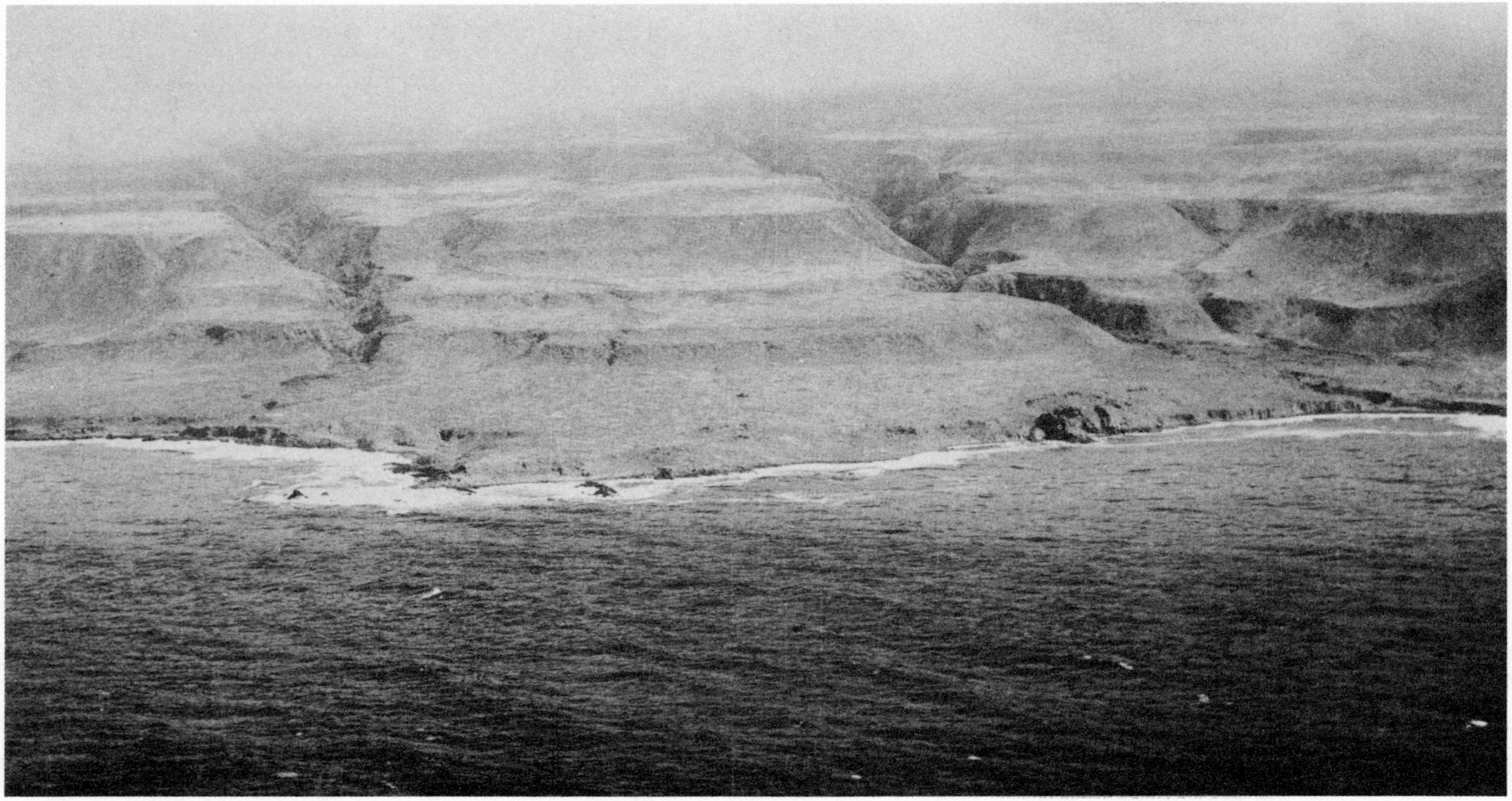

Aerial view shows well-defined recently uplifted terraces at various elevations on this andesitic island. Such terraces appear to have been cut in the surf zone and essentially at sea level, rather than at wave-base. No wave-built terraces are to be found on this island. (U.S. Navy photograph by R. S. Dietz)

WAVE-CUT TERRACES, SAN CLEMENTE ISLAND, CALIFORNIA

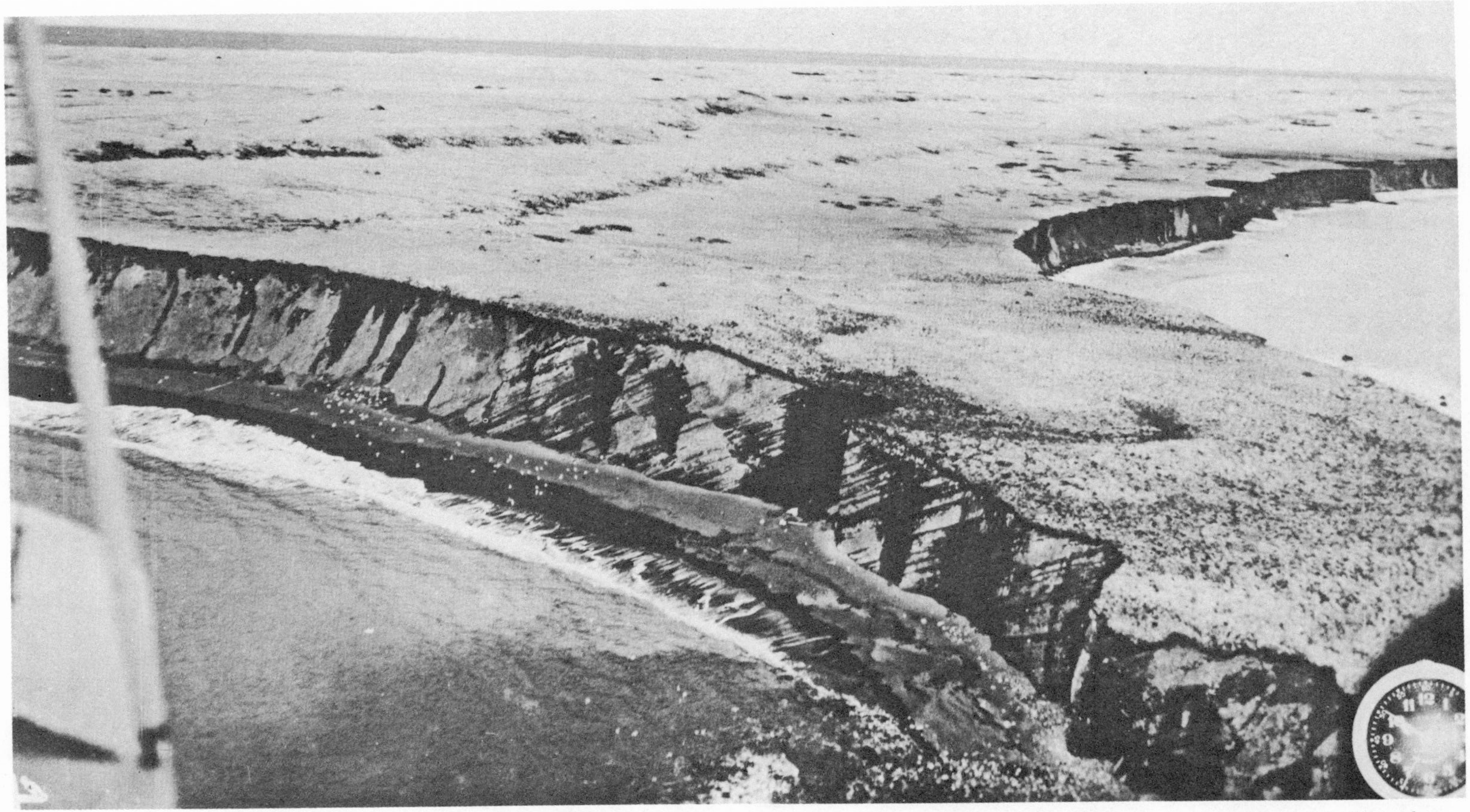

Aerial view shows extensive terraces truncating steeply dipping strata on this recently uplifted isolated island far out on the shelf off southern Alaska. No wave-built terraces are to be seen; nor is there any profile of depositional equilibrium. (U.S. Geological Survey photograph by S. A. Capps)

WAVE-CUT TERRACES, MIDDLETON ISLAND, ALASKA

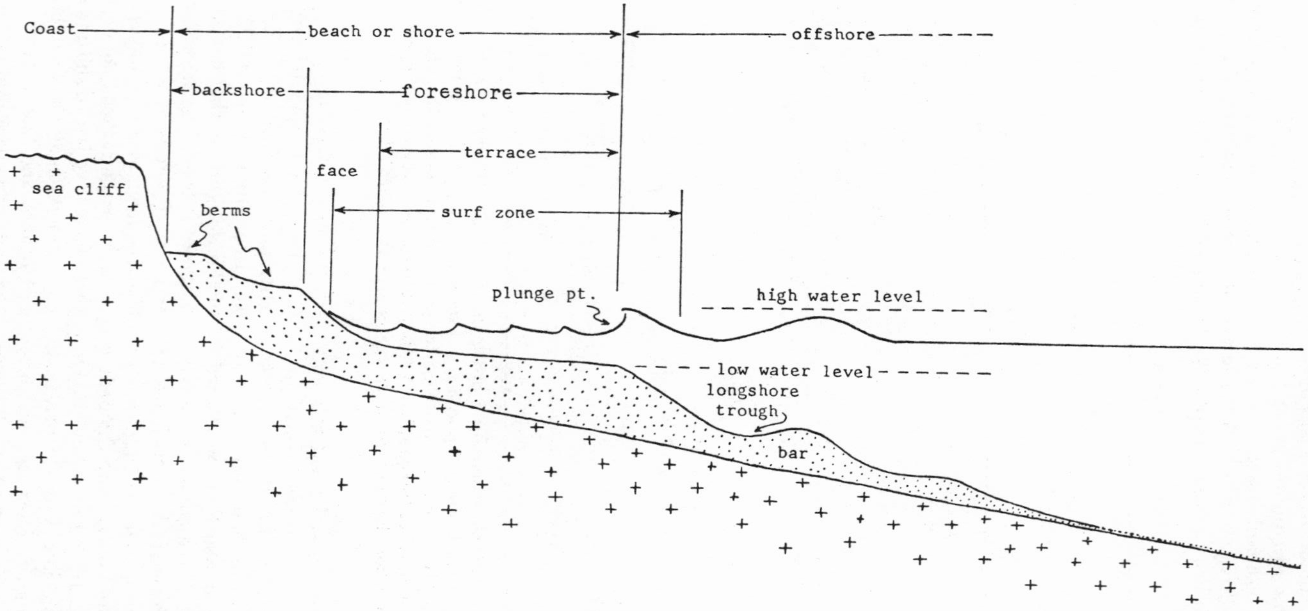

Figure 12. Profile of the beach and nearshore region modified after Inman (1962). The position of the active wedge is especially emphasized; it pinches out offshore into a relict shelf. The principal effect of waves is to push sand against the sea cliff rather than to construct a wavebuilt terrace.

some gravel lenses at the base of the ancient sea cliffs as well as the modern one. These appear to be beach gravels which were laid down in equilibrium with surf conditions, but this does not constitute a profile of equilibrium in the sense that this term is ordinarily used.

Even the inner shelf regions of the Atlantic and Gulf Coast cannot be interpreted as purely surf-cut as they are capped with great sediment wedges. But this does not necessarily require that they be interpreted as wave-built terraces. The author would explain these as built of prograded wedges of "hinter-surf" or paralic sediments (Dietz, 1952). These coasts are heavily alluviated and of low relief. A small change of sea level causes considerable transgression or regression of the shore line, causing the paralic sediments to be over-ridden or stranded. On the average, the shore line is overloaded with sediment so that prograding occurs, and these deposits are geologically preserved by continental downwarping. Along much of the eastern Atlantic Coast, the original surf-cut basement unconformity was formed in the pre-Cretaceous. Since then there has been only disconformable minor surf-cutting and reworking of the paralic and shelf beds.

DISCUSSION

Having discussed the various concepts of nearshore and continental-terrace formation, it seems worthwhile to present diagrams showing the writer's viewpoint which avoids invoking wave-base or the wave-built-terrace concepts. Therefore a diagram of the beach region (Fig. 12) according to Inman (1962, p. 300–306), is modified slightly so that not only the principle parts of the beach region are shown but also the active nearshore sand wedge. This emphasizes that the main action of waves is to move sand onto the beach rather than sweep it across the shelf. Extensive lateral shifting of this sand occurs, straightening the shore line and filling in re-entrants, and finally causing prograding of the shore deposits seaward although this aspect is not shown in the diagram. As an example to show the reality of such a shore lens, an area near Panama City, Florida, contoured in fine

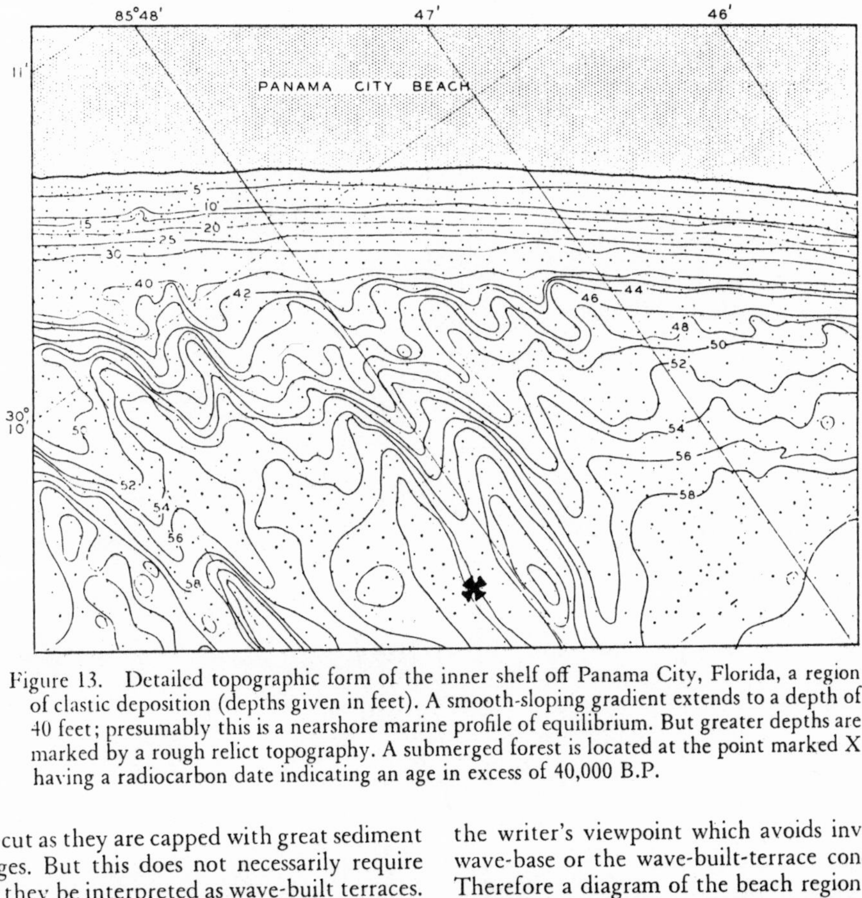

Figure 13. Detailed topographic form of the inner shelf off Panama City, Florida, a region of clastic deposition (depths given in feet). A smooth-sloping gradient extends to a depth of 40 feet; presumably this is a nearshore marine profile of equilibrium. But greater depths are marked by a rough relict topography. A submerged forest is located at the point marked X having a radiocarbon date indicating an age in excess of 40,000 B.P.

detail, is shown. There (Fig. 13), a smooth and gently dipping active shore wedge extends to a depth of 40 feet. Beyond this is a rough relict shelf, as proved by the discovery of a submerged forest of partially fossilized stumps (Shumway and others, in press). The shore wedge seems to be at present migrating seaward so that the buried forest will eventually be covered with a prograded bed.

The writer's preferred view of continental-terrace development is shown in Figure 14. The Texas Gulf Coast section, a continental terrace in old age, is of course quite different, especially in that the continental shelf and slope are now merged.

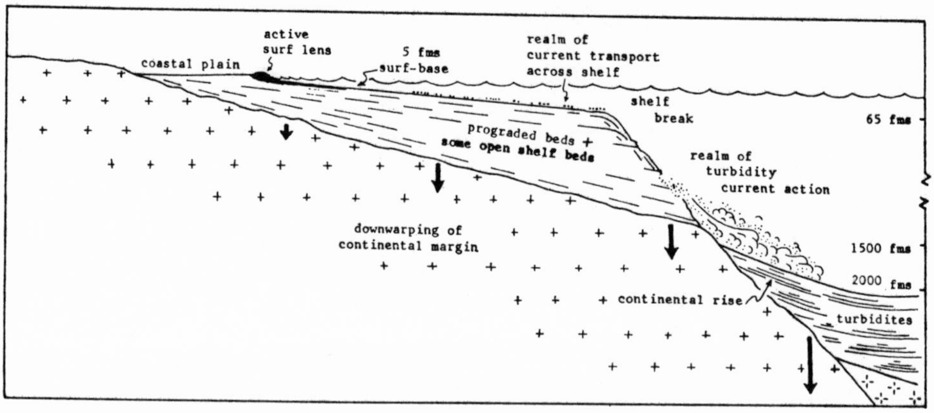

Figure 14. Cross section of the continental terrace. This is a greatly simplified and schematic diagram of the writer's preferred view on the development of a continental terrace in maturity such as that off the eastern coast of the United States (Dietz, 1952). Emphasis is on prograded beds in a purely clastic environment, along with marginal downwarping of the continental block.

Mention has been made above of an initial structural continental slope of considerable declivity, implying that sedimentation is only a modifying process. The origin and evolution of continental slopes will be discussed in a subsequent paper. Suffice to say here that the writer supposes that these slopes are primarily the flanks of orogenically collapsed continental rises which are accreted to the continental block by a continent-ward thrusting of the sea floor. Later continental rifting may also account for some continental slopes, but this process, of course, can be only a secondary structural process.

If wave-built terraces are nonexistent, how can one explain a prograded continental terrace? The view has already been expressed that the Atlantic continental slope is not prograded, even though some draping of dip-slope strata may be present, so that this is a somewhat mantled continental slope. On the other hand, the prograding of the terrace off Louisiana and Texas is, in all probability, large. A possible explanation for such prograding is shown in Figure 15. Accordingly, the initial continental slope, of structural origin, is too steep (about 6°) to permit permanent slope sedimentation. However, a continental rise gradually uplaps the continental slope. Eventually, the initial slope is fully covered so that the continental slope is, in reality, the upper surface of a continental rise with a slope of about only 2°. If the continental rise was laid down by turbidity-current deposition, its surface should be at grade for this process—or, in other words, it would be a profile of equilibrium for turbidity currents. If this is so, the continental terrace could now undergo rapid and effective progradation by delta foreset beds and by any other detritus carried across the shelf. Such a terrace would not be, correctly speaking, a wave-built terrace, as its development was not controlled by wave-base. Instead, material is dumped on the continental slope as delta foreset beds and by other sediments swept across the shelf. At least in part, this terrace of accumulation may be viewed as a delta terrace with the shelf-break being a drowned nick-point. A useful term for such a terrace of accumulation is a *continental embankment*.

Figure 16 gives some idea why the simple classical and purely intuitive view of the continental terrace being made up of an abrasion

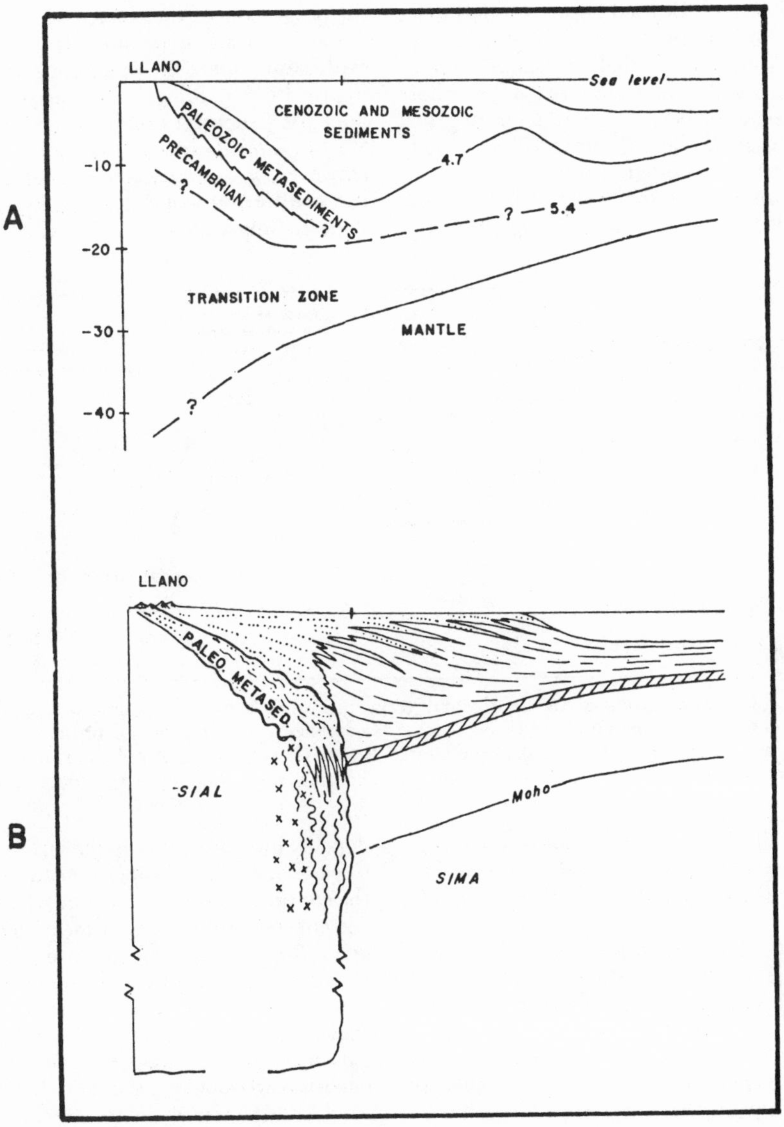

Figure 15. Gulf coast geosyncline. (A) Generalized diagram from the Llano uplift to the central Gulf adapted from Cram (1962); (B) a geologic interpretation in accordance with the concept presented in the paper. The geosyncline is shown as a composite structure consisting of a terrace wedge (miogeosyncline), a continental-rise prism (eugeosyncline), and a continental embankment (mixed facies). The last two are laid down on oceanic sima whereas the terrace wedge is laid down on the margin of the continental block.

The basal ruled line is the Jurassic Louann salt interpreted as a deep-sea salt. The concepts of wave-base and of the wave-built terrace are not needed to explain this development.

platform and a wave-built embankment can not exist in nature. The topographic contrast between the continents and the ocean basins is so large that isostatic downbowing of the unstable margin occurs, owing to the loading of quate data, was based on deduction alone. It called for an inner abrasion platform and an outer embankment of talus. It was a reasonable concept *except* that it failed to take into account the effect of isostasy. In the *absence* of

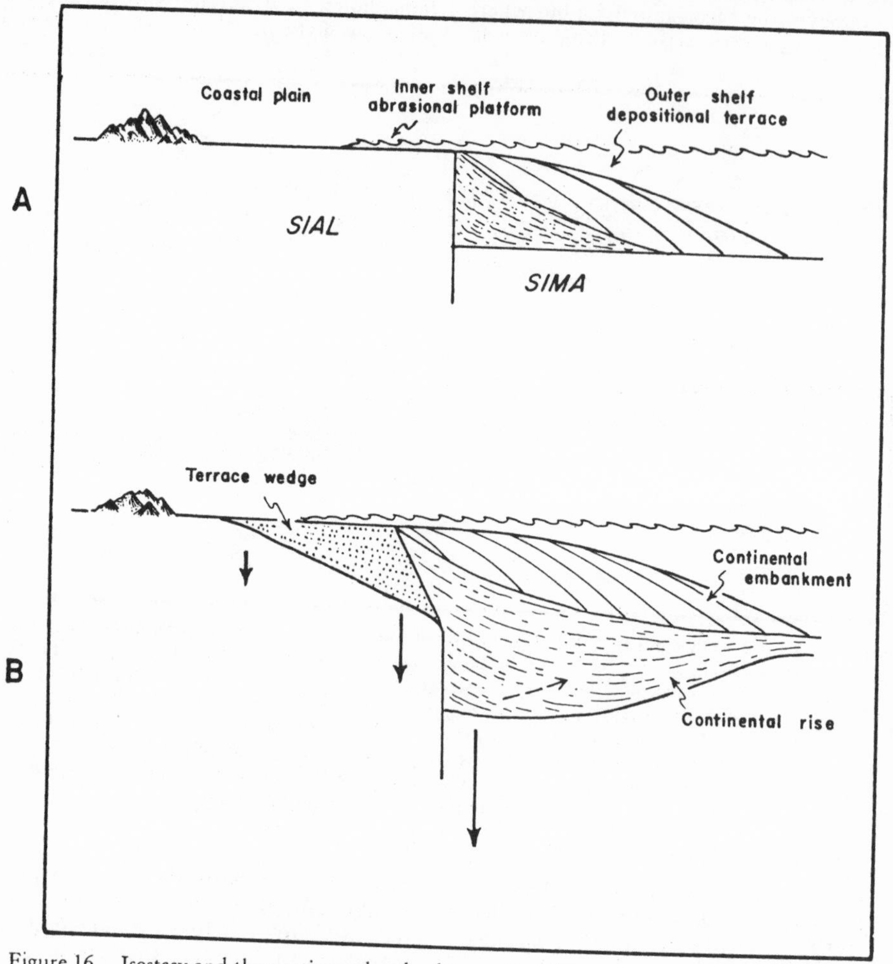

Figure 16. Isostasy and the continental embankment. Simplified diagram to show the build-up of a continental rise uplapping the continental slope. This is followed by the prograding of a continental embankment. Without the effect of isostasy an inner abrasion platform and an outer terrace of accumulation would form (A), but because isostasy does come into play, a terrace wedge develops instead of an abrasion platform (B). Because of the continent–ocean-floor topographic contrast, the rise thickness attains great dimensions before reaching isostatic equilibrium. According to Lawson (1942) such thickness in the Mississippi delta region would be 40,000 feet. As shown by the dashed arrow, the sedimentation axis moves seaward with time.

the sea floor by the growing continental rise. Hence, a continental-terrace wedge composed of prograded paralic beds is invariably developed. The classical theory of continental-margin development, in the absence of any ade- isostasy (Fig. 16) there would be prograding to form a continental embankment, once the continental rise uplapped the initial structural continental slope. However, the slope relief is so grand that isostatic downwarping of the

continental block necessarily occurs. Hence the inner abrasion platform becomes buried under a terrace wedge of sediments.

To clarify the belief that a continental-slope face can be built up by prograded paralic beds, Figure 17 is presented. Once away from the active passes of the Mississippi delta but where longshore processes are actively bringing sediment, paralic strata are being laid down shoreward of the 25-foot isobath (Gould and McFarlan, 1959). Recent sedimentation offshore is nil and Pleistocene beds are commonly exposed. Such paralic deposition may build up a continental-terrace wedge of beds which thickens seaward and an upper continental slope as well.

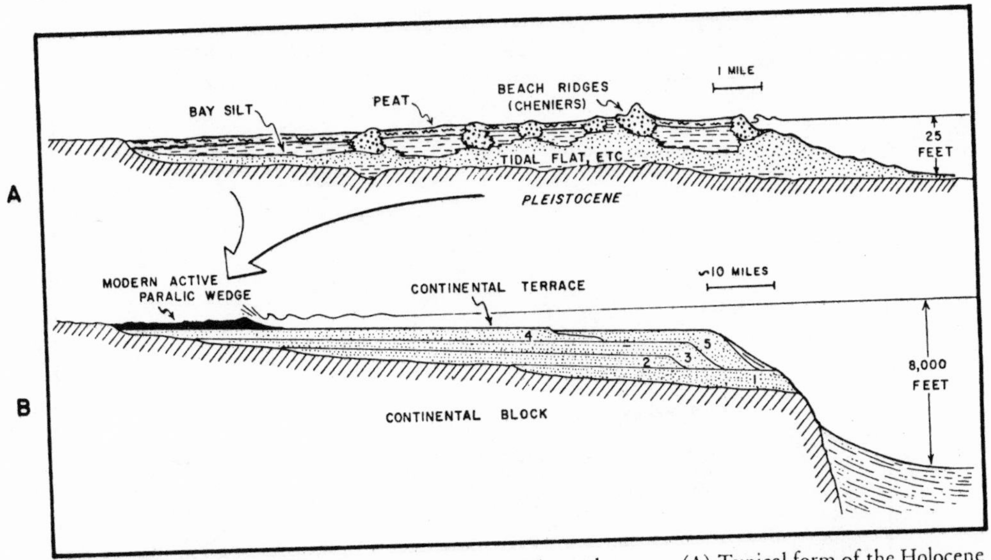

Figure 17. Paralic wedge and the formation of the continental terrace. (A) Typical form of the Holocene transgression of paralic sediments (along the Louisiana coast to the east of the Mississippi Delta) over the Pleistocene Prairie Formation. The rising Pleistocene sea level reached its present still-stand about 4000 years ago. Note that a relatively steep face extends outward from the shore line to the 25-foot isobath beyond which no Holocene sedimentation exists. (Modified and greatly simplified from Gould and McFarlan, 1959, Pl. 2.) (B) By the successive prograding and overleafing of paralic beds on a subsiding continental margin a terrace wedge can be built. Note that the paralic beds thicken seaward and the fore-face is sufficiently steep to further build up the continental slope, though less steep than the initial continental slope.

CONCLUDING REMARKS

The purpose of this paper is to record a marine geologist's objections to the wave-base concept as it now pervades geologic thinking. The title may be rankling, but this will serve a useful purpose if geologists will review and clarify their own ideas on this subject.

In the interest of clarity and more precise definition and to make this critique more constructive, it seems useful to make some recommendations. These are:

(1) The concept of the *wave-built terrace* should be discarded as this seems to be wholly fictitious.

(2) The concept of a *marine profile of equilibrium* should be used only with respect to surf processes of the nearshore or paralic zone. The gradient of the outer shelf is not a profile of equilibrium relative to any sort of wave-base. The shelf surface is drowned and out of equilibrium with currents now affecting it. The continental rise, abutting the base of the continental slope, does have a concave-upward form and this apparently is a profile of equilibrium relative to turbidity currents.

(3) As those marine-rock terraces now termed wave-cut terraces are apparently surf-cut, they are related to surf-base rather than to wave-base. In the interest of being explicit and precise, we should employ the term *surf-cut terrace* in place of *wave-cut terrace*. And they are cut to surf-base rather than to wave-base.

(4) The term *wave-base*, as now used, is both vague and confusing. Certainly wave-base is

not the dominant process which controls continental-terrace development and should not be treated as such in our textbooks; surf action, for example, is much more important. And wave-base can be neither a cause of rock terracing nor can it form a wave-built terrace. The term does retain stratigraphic usefulness when applied to the miogeosynclinal sediments, for example where sorting, etc., indicate deposition in agitated water. Much of this reworking presumably would have been done by surf, and depth-wise, the entire continental shelf would be included in such a wave-base zone.

REFERENCES CITED

Barrell, J., 1925, Marine and terrestrial conglomerates: Geol. Soc. America Bull., v. 36, p. 279–341

Bradley, W., 1958, Submarine abrasion and wave-cut platforms: Geol. Soc. America Bull., v. 69, p. 967–974

Clark, T., and Stearn, C., 1960, Geological evolution of North America: New York, Ronald Press Co., 434 p.

Cram, Ira H., Jr., 1962, Crustal structure of Texas coastal plain region: Am. Assoc. Petroleum Geologists Bull., v. 46, p. 1721–1727

Curray, J. R., 1960, Sediments and history of Holocene transgression, continental shelf, northwest Gulf of Mexico, p. 221–266 *in* Shepard and others, *Editors*, Recent sediments, N. W. Gulf of Mexico: Tulsa, Am. Assoc. Petroleum Geologists

—— 1961, Late Quaternary sea level: A discussion: Geol. Soc. America Bull., v. 72, p. 1707–1712

Daly, R., 1942, The floor of the ocean: Univ. North Carolina Press, 177 p.

Dietz, R. S., 1952, Geomorphic evolution of continental terrace (continental shelf and slope): Am. Assoc. Petroleum Geologists Bull., v. 36, p. 1802–1820

—— 1963, Collapsing continental rises: an actualistic concept of geosynclines and mountain building: Jour. Geology, v. 71, p. 314–333

Dietz, R., and Menard, H., 1951, Origin of abrupt change in slope at continental-shelf margin: Am. Assoc. Petroleum Geologists Bull., v. 35, p. 1994-2016

Dunbar, C., and Rodgers, J., 1957, Principles of stratigraphy: New York, John Wiley and Sons, Inc., 356 p.

Emery, K. O., 1961, Submerged marine terraces and their sediments: Zeitsch. fur Geomorphologie Supp. 3, p. 2–29

Fairbridge, R., 1952, Marine erosion: 7th Pacific Sci. Cong., Pacific Sci. Assoc. Proc., v. 3, p. 347–358

Fenneman, N. M., 1902, Development of the profile of equilibrium of the subaqueous shore terrace: Jour. Geology, v. 10, p. 31

Garrels, R., 1951, Textbook of geology: New York, Harper and Bros., 511 p.

Gilbert, G. K., 1890, Lake Bonneville: U.S. Geol. Survey Mon. No. 1, 438 p.

Gilluly, J., Waters, A., and Woodford, A., 1959, Principles of geology (2d ed.): New York, W. H. Freeman and Co., 534 p.

Gould, H. R., 1960, Character of the accumulated sediment, p. 149–186 *in* Smith, W. O., and others, Comprehensive survey of sedimentation in Lake Mead, 1948–1949: U. S. Geol. Survey Prof. Paper 295, 254 p.

Gould, H. R., and McFarlan, E. M., 1959, Geologic history of the chenier plain, southwestern Louisiana: Gulf Coast Assoc. Geol. Socs. Trans., v. 9, p. 1–10

Gulliver, F., 1899, Shoreline topography: Am. Acad. Arts and Sciences Proc., v. 34, p. 151–258

Gunnerson, G. G., and Emery, K. O., 1962, Suspended sediment and plankton over San Pedro basin, California: Limnology and Oceanography, v. 7, p. 14–20

Heezen, B., Tharp, M., and Ewing, M., 1959, The floors of the oceans. I. The North Atlantic: Geol. Soc. America Special Paper 65, 122 p.

Howell, J. V., *Chairman*, Glossary of geology and related sciences: Am. Geol. Inst., 325 p.

Inman, D., 1962, Beach processes, p. 299–306 *in* McGraw-Hill encyclopedia of science and technology: New York, McGraw-Hill Book Co., Inc.

Johnson, D., 1919 (1938, 2d ed.), Shore processes and shoreline development: New York, John Wiley and Sons, Inc., 584 p.

Kuenen, Ph., 1950, Marine geology: New York, John Wiley and Sons, Inc., 568 p.

Lawson, Andrew C., 1942, Mississippi delta—a study in isostasy: Geol. Soc. America Bull., v. 53, p. 1231–1254

Leet, L. D., and Judson, S., 1958, Physical geology (2d ed.): Englewood Cliffs, New Jersey, Prentice Hall, Inc., 502 p.

Longwell, C., Knopf, A. K., and Flint, R., 1948, Physical geology: New York, John Wiley and Sons, Inc., 543 p.

Miller, D. J., 1953, Late Cenozoic marine glacial sediments and marine terraces of Middleton Island, Alaska: Jour. Geology, v. 61, p. 17–40

Moore, D., 1960, Acoustic-reflection studies of the continental shelf and slope off Southern California: Geol. Soc. America Bull., v. 71, p. 1121–1136

Pettijohn, F., 1957, Sedimentary rocks (2d ed.): New York, Harper and Bros., 718 p.

Rich, J. L., 1951, Three critical environments of deposition and criteria for recognition of rocks deposited in each of them: Geol. Soc. America Bull., v. 62, p. 1–20

Shepard, F., 1948, Submarine geology: New York, Harper and Bros., 348 p.

Shepard, F., and Wrath, W., 1937, Marine sediments around Catalina Island: Jour. Sed. Petrology, v 7, p. 41–50

Shumway, Ç., Dowling, G. B., Salsman, G., and Payne, R. H., 1961, Submerged forest of mid-Wisconsin age on the continental shelf off Panama City, Florida: Geol. Soc. America Special Paper 68, 271 p.

Stetson, H. C., 1949, The sediments and stratigraphy of the East Coast continental margin: Georges Bank to Norfolk Canyon: Woods Hole Oceanographic Inst., Papers in Phys. Oceanography and Meteorology, v. 11, no. 2, p. 1–60

Twenhofel, W., 1950, Principles of sedimentation (2d ed.): New York, McGraw-Hill Book Co., Inc., 673 p.

Von Engeln, O., 1942, Geomorphology: New York, MacMillan and Co., 665 p.

5

Reprinted, courtesy of Geological Society of America, from *Geol. Soc. Am. Bull.* **75**:1267–1273 (1964)

WAVE-BASE, MARINE PROFILE OF EQUILIBRIUM, AND WAVE-BUILT TERRACES: DISCUSSION

DAVID G. MOORE
JOSEPH R. CURRAY

Abstract: Some concepts presented in a recent paper by Dietz (1963) can be enlarged upon with an attempt to emphasize what are believed to be shortcomings. Other concepts can be negated by newer field evidence.

The old concept of *wave-base* should not be completely discarded but should be modified in accordance with present knowledge of the transportation and deposition of sediments. Where modern rates of supply are high enough, sediment is transported to and deposited on the inner and central parts of continental shelves, generally with decrease of grain size offshore. The gradient in particle size cannot be predicted merely by considering instantaneous conditions but may theoretically be predicted over the *net cyclic period of the region*, perhaps in the range between 1 and 100 years. The effects of variations in energy and supply and the occurrence of extreme flood or storm conditions are thus averaged out. The morphological *marine profile of equilibrium*, a direct corollary, has not yet been attained on most continental shelves because of the effects of rapid, wide Quaternary fluctuations of sea level.

Use of the terms *delta*, *topset*, *foreset*, and *bottomset* should be restricted to known marine or lacustrine deltas. In particular, continental slope deposits should not be termed foresets unless they are of known deltaic origin.

Dietz's model of the structure of the continental terrace is incorrect for the surveyed portions off the coasts of North America. The basic sedimentary framework of most of these continental terraces appears to be upbuilding by shelf and paralic facies and outbuilding by slope deposition. This basic structure is locally modified by salt-dome tectonics, folding, faulting, and mass slumping. The continental slope is believed to be predominantly a realm of deposition rather than of turbidity-current degradation.

Introduction

In a recent paper, Dietz (1963) attempted to update the concepts and use of the terms *wave-base*, *marine profile of equilibrium*, and *wave-built terrace*. He very convincingly pointed out original definitions, examples of misuse, and a recent opinion of the proper application and status of these terms. We submit that, whereas much of this discussion was apropos and certainly necessary to clarify these concepts, Dietz did not sufficiently consider newer field data. As a result, he has held to some very questionable concepts, particularly regarding the structure of the continental terrace, and he has failed to reconcile these concepts to the present state of our knowledge.

Wave-Base

The concept of wave-base, popular during the early part of this century as an explanation of the origin of the continental shelf, has fallen into disrepute. (See, for example, Dietz and Menard, 1951; Dietz, 1952; Fairbridge, 1952; Bradley, 1958; Dietz, 1963, Shepard, 1963.) The old concept of wave-base implied an abrupt base level below which sedimentary particles were not stirred by wave action and above which particles could not come permanently to rest because of wave action. We know that this oversimplified concept is not valid, but in re-evaluating it, we should not swing the pendulum too far and disregard what we know about the relationship between orbital velocity associated with waves and depth of water.

Dietz (1963), following Dietz and Menard (1951), rejected "wave-base" in favor of "surf-base" or "surge-base," defined as the approximate depth of effective wave abrasion. He cites evidence that this effective depth of wave abrasion is a maximum of approximately 5 fathoms or 10 m, although wave-cut terraces on bedrock are formed essentially at sea level. We concur in this concept but believe it is important to emphasize that this depth of effective wave abrasion is dependent upon the type of material to be eroded and upon the length of time available for erosion. For example, "surf-base" would be essentially sea level for a short period of time on a resistant bedrock base; "surf-base" would be much

deeper for a longer given period of time or over soft, unconsolidated sediment.

Confusion surrounding the wave-base concept has stemmed partly from two factors. First, some earlier investigators spoke of wave-base in regard to erosion of pre-existing material; others used wave-base in referring to a base of deposition of sediment; and still others equated these two depths. The processes of deposition of sediment and erosion of rock obviously should be considered separately. Secondly, earlier investigators attempted to equate wave-base, in connection with both deposition and erosion, with the present edge of the continental shelf. The depth of the edge of the continental shelf is now, however, generally regarded by most marine geologists as relict from conditions of lower stands of sea level during the Pleistocene. (*See*, for example, Shepard, 1948; 1963; Dietz and Menard, 1951; Dietz, 1952.) Also relict from lower stands of sea level are many of the coarse sediments now found on continental shelves where not yet covered with sediments in equilibrium with present conditions. As pointed out by Emery (1952) much confusion has resulted from failure to distinguish these relict sediments from modern detrital sediments.

For any simple wave train acting in deep water, the orbital velocity decreases exponentially with depth of water. As the waves approach shallow water, this exponential decrease with water depth is modified and diminished but can still be approximated theoretically. Good data are now available on the threshold water velocities necessary to pick up particles from the bottom and on the energy conditions necessary to maintain them in suspension (Hjulström, 1939; Sundborg, 1956; Inman, 1963; Inman and Bagnold, 1963). If we assume the simplest possible conditions at any instant of time, that is, a wave train of single wave height and period and a given depth of water, we can approximate the largest size of the bed load particles. Thus, over a range of depths, we would have an effective *zone* of wave-base corresponding to a range of particle sizes. This effective, instantaneous "wave-base" would be "shallower" for gravel and coarse sand than for medium to fine sand. If these simple wave conditions persisted for a period of time, silt- and clay-size particles would be maintained in suspension until they reached their effective wave-base on the continental shelf where they would be deposited. Once the silt- and clay-size particles were deposited and remained on the bottom for a significant period of time, higher velocities would be required to resuspend them because of cohesion between the particles. Further complications in nature are the extreme variability from day to day in wave conditions, the presence of and periodic variations in tidal currents, and the presence of permanent currents and other wind-induced currents.

It is futile, therefore, in dealing with an accumulated column of sediment to speak of instantaneous wave-base or wave conditions; we must consider instead the net effect over what we might define as the *natural cyclic period* of a region, (Curray and Moore, 1964). This *natural cyclic period* of a region, perhaps in the range of 1–100 years, is that period of time over which we must integrate sedimentary processes in order to average out variations in wave and other current conditions and variations in rate of sediment supply. The variations in wave and current conditions include normal waves, extreme storm or hurricane waves, and coincidence of extreme storm conditions on periods of spring tidal range. These extreme, although rare, conditions may affect the nature of deposition more than do mild, normal conditions.

If we now consider this *natural cyclic period* and the net rate of sediment supply over this period of time, we can arrive at depositional equilibrium for a given region. Where the sediments involved in this equilibrium are thick enough to cover underlying older sediments, they normally show a gradation to relatively finer sediment farther from shore or source.

The process of transportation is discontinuous within the basin of deposition throughout the *natural cyclic period*. It consists of many temporary periods of deposition and "erosion" or penecontemperaneous resuspension before sedimentary particles come to an ultimate resting place and burial. For any given particle size, there will be an equilibrium depth at which the net rate of deposition will equal the net rate of removal, the point of depositional equilibrium. Below this depth zone, net deposition will occur out to the range limit of the transporting processes (Curray, 1964). Beyond the range limit of the transporting processes the present shelf surface may be covered with sediments relict from previous equilibrium conditions. Above the point of depositional equilibrium, only net transportation of this given particle size will result. In actual fact, this situation has

been attained in our modern seas only on narrow shelves or on the inner parts of some wide shelves because of the recency of sea level changes associated with the Pleistocene. Even where attained in either modern or ancient seas, a clean separation of particle sizes is not produced because silt- and clay-size particles are not resuspended easily. Nevertheless, there is a gradation from coarse to fine in most modern continental shelf sediments where they are in equilibrium with present conditions.

The old concept of wave-base then is grossly oversimplified but should be modified rather than rejected. We submit that over the natural cyclical period of a region, an equilibrium occurs between deposition and removal for each particle size. This *depositional equilibrium* is a function of wave and current conditions and the rate or supply of sediments.

Marine Profile of Equilibrium

The marine profile of equilibrium, a direct corollary to the wave-base concept, has been shown by Dietz to apply to modern seas in the nearshore region. Beyond this nearshore region, many continental shelves are relict both in surface sediment distribution and in morphology. The effects of wide, rapid Pleistocene fluctuations of sea level are far reaching. The present configuration of the continental shelf is a result of the transgressions and regressions accompanying glacial advance and retreat. Some shelves show a truncation of bedrock produced by a migration of the surf zone back and forth across the shelf. Other shelves, in predominantly subsiding regions or in regions of more abundant sediment supply, show the effect of deposition of paralic, nearshore, and open marine shelf sediments as the shore line and the river sources of the sediments have moved back and forth across the shelf. In any event, the configuration of the shelf is primarily a result of these fluctuations in sea level. There has been insufficient time since sea level has reached its present position for the attainment of any kind of profile of equilibrium except in localized areas of rapid deposition. These localized areas primarily include nearshore areas or the shore face of the beach, as pointed out by Dietz (1963) and as discussed by Inman and Bagnold (1963, p. 530). Beyond this nearshore area, the relict topography is in places covered with a thin veneer of modern shelf-facies muds and in no sense can be called a profile of equilibrium. Elsewhere modern shelf sediments are known to attain thicknesses of many tens of feet (Moore, 1960), and the shelf may be starting to approach a profile of equilibrium.

Confusion perhaps has resulted from the fact that some investigators refer to topography or morphology in speaking of a profile of equilibrium, whereas other investigators have referred to the nature of the bottom sediments. Beyond the nearshore profile of equilibrium, discussed by Dietz, many shelf surfaces of relict morphology are indeed covered with a veneer of modern shelf-facies sediments in equilibrium with the present condition (Curray, 1960). The sediments are thus in equilibrium, although a *morphological* profile of equilibrium has not yet been attained.

Wave-Built Terrace

Dietz (1963) reviewed the origin of the term *wave-built terrace* (Gilbert, 1890). Gilbert clearly referred to shore line and nearshore deposits of sand formed by wave action and longshore drift. Regardless of the internal sedimentary structure of the continental terrace (coastal plain, continental shelf, and continental slope), it should not be called a wave-built terrace, in view of Gilbert's original definition. Furthermore, today we have better terminology for shore line and nearshore features. We urge then, that the term wave-built terrace be abandoned and that the continental terrace be referred to simply as the continental terrace or its component parts. In suggesting the discarding of this term, we concur with Dietz; however, our reasons for discarding the term are different. Dietz proposes abandoning the term because of a misconception about the internal structure of the continental terrace. By his reasoning, if a continental terrace structure consisting of both shelf and slope deposits existed and was formed by transgressions and regressions of the sea, he would apply the term wave-built terrace. We will show that this is indeed the internal structure of some continental terraces, but we see no necessity for applying the term wave-built terrace, in contradiction to the original definition by Gilbert.

Delta Terrace

Dietz's (1963) discussion of the delta terrace correctly differentiates processes that form a true river delta and those of the so-called wave-built terrace. We feel that Dietz did not go far enough in his critical appraisal of the use of the terminology *topset*, *foreset*, and *bottomset* beds.

Topset, foreset and bottomset are classical

terms used to describe the fundamental beds of a delta complex. Deltas, by definition, require subaerial, fluvial, and marine or lacustrine deposition in combination. We feel strongly that these terms should be restricted to their original meaning and should not be used to describe purely marine deposition. In particular, the use of foreset beds to describe continental slope deposition is unfortunate and confusing. There is a clear distinction between delta foreset beds and slope deposits, and this distinction should be recognized in the terminology. Off the Costa de Nayarit, Mexico (Curray and Moore, 1964), for example, both kinds of deposits are present, and in fact adjoining (Fig. 1). In this area foreset beds of a Pleistocene delta have nearly prograded across the pre-existing continental shelf. Below the foreset beds and beyond the old shelf break are similarly dipping, but genetically different, deposits of the upper continental slope, which in this case are the bottomset beds of the delta complex. It is true that in cases where deltas have built completely across the shelf and deposited foreset beds on the upper continental slope, there exists the possibility of confusing the two kinds of deposits. However, where deposits are of known deltaic origin, they should retain delta-bed terminology; similarly, where beds are known to be nondeltaic continental slope deposits, they should be termed continental slope deposits, not foreset beds.

In the case of topset beds, the terminology can be similarly analyzed. Sediments on or under and parallel to the present surface of the continental shelf may consist of marine, paralic, or continental facies, but they should not be called topset beds unless known to be deltaic. Likewise, deposits of the continental rise or abyssal depths at the base of the continental slope should not be referred to as bottomset beds. The continental terrace may contain deltaic portions, but it is not, in itself, a giant deltaic wedge.

Continental Terrace Structure

We have thus far attempted to enlarge upon Dietz's (1963) discussion of the concepts of wave-base, marine profile of equilibrium, wave-built terrace, and delta terrace. Our difference with Dietz is more a question of emphasis than disagreement. It is the final section of Dietz's paper, however, which we feel to be most open to criticism. His theory of the development of the continental terrace is essentially unchanged from that of his stimulating paper of 1952, except that he has added the *continental embankment* concept. Briefly, Dietz's concept minimizes permanent significant continental slope deposition and suggests that the slope is a realm of general degradation by sliding and turbidity currents. He further minimizes the importance of open continental shelf deposits and would have the body of the terrace built up of shallow water and other paralic deposits accumulated as a result of subsidence. He explains the subsidence as isostatic adjustment to the load of continental rise deposits accumulating at the base of the initial, structural continental slope as turbidites. In an old-age terrace, he postulates that these turbidites cover the eroded slope face by embankment.

At the time of its initial publication in 1952, Dietz's theory on geomorphic evolution of the continental terrace was new and challenging. It explained some known instances of ancient rock outcrops in deep submarine canyons and was in tune with contemporary ideas of turbidity-current action. Re-affirmation of this same theory in the 1963 paper, however, shows a disregard of data collected during the past few years.

The recent development of continuous reflection profiling techniques (*see*, for example, Hersey, 1963) has made possible the observation of the internal structure of the continental terrace in two of the key regions discussed by Dietz. In the northwest Gulf of Mexico (Moore and Curray, 1963a), reflection records show conclusively that, at least as far back as the Late Tertiary, the structure is formed by simple sedimentary upbuilding and outbuilding with accompanying subsidence. No evidence of extensive erosion has been found on the upper slope, although complex slope topography has developed as a result of salt-dome tectonics. Uniformity and continuity of bedding suggest that open-shelf and slope deposits are the predominant facies. No structural evidence of a facies change from prograded paralic beds to turbidites of an embanked continental rise has been found.

Present data on the stability of submarine sedimentary slopes (Moore, 1961), while admittedly scant, are not in accord with one of Dietz's (1963, p. 985) basic presumptions; *i.e.*, that a 6-degree continental slope is unstable and subject to degradation by slumping and turbidity currents, whereas a 2-degree slope can be prograded to form a vast sedimentary embankment. This arbitrary division seems unrealistic when the complex factors controlling

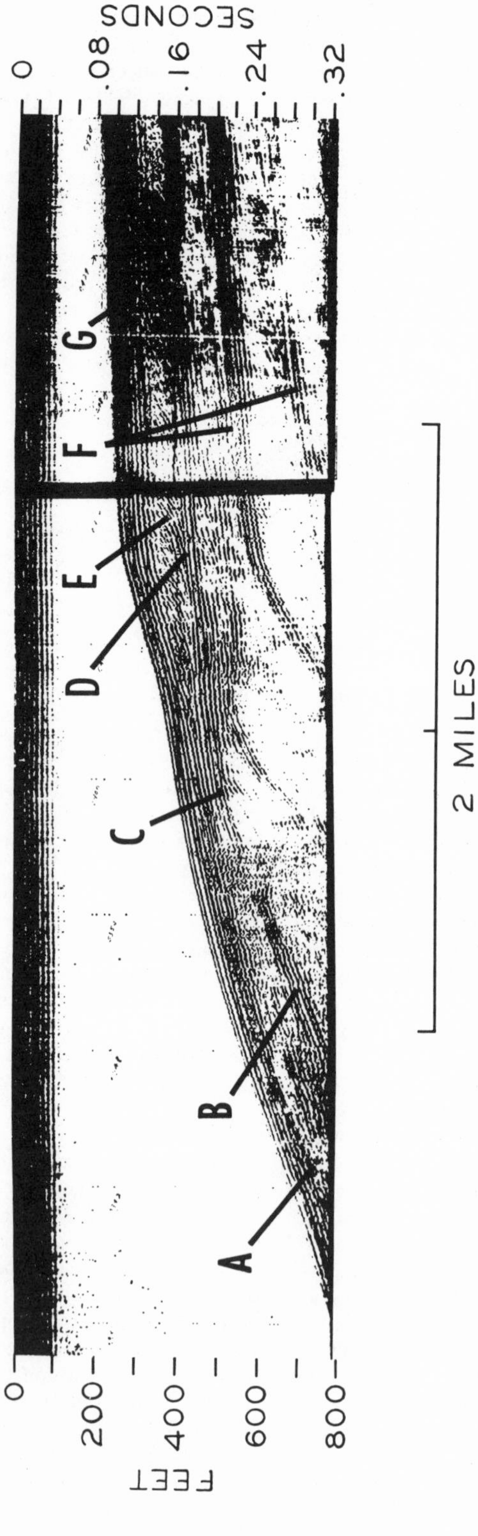

Figure 1. Continuous reflection record (using 2000 Joule spark sound source) from outer continental terrace off west-central Mexico (Curray and Moore, 1964), showing sub-bottom strata. (A) Continental slope deposits, the bottomset beds of the delta shown at E, (B) Former surface of the continental slope, predating the delta, (C) Former shelf break, predating the delta, (D) Former surface of the continental shelf, predating the delta, (E) Foreset beds of the Pleistocene, lowered sea level delta, (F) Multiple reflections of the sea floor and sub-bottom layers, (G) Continental shelf deposits overlying the delta

slope stability are considered. Slope stability is a function not only of declevity but of physical properties of the sediments and rates of deposition. These factors control cohesion, angle of internal friction, and degree of consolidation of the slope deposits. They can result in stable deposits on 20-degree slopes, or unstable deposits on slopes of less than 1°, such as those that occur off the modern Mississippi Delta (Shepard, 1955; Moore, 1961).

In introducing the continental embankment idea, Dietz (1963) adapted his 1952 theory to comply with the known extensive deposits on the continental slope of the Gulf of Mexico. This adaptation involves some conflicting concepts and unacceptable models of the postulated embankment structure. In describing the development of the "continental embankment," Dietz suggests (1963, p. 985) that "... its development was not controlled by wave-base. Instead, material is dumped on the continental slope as delta foreset beds and by other sediments swept across the shelf." What is this if not wave-base? Furthermore, the schematic diagram by which Dietz represents the structure resulting from this sweeping of material across the shelf to be dumped on the slope (Fig. 16B), closely resembles his Figures 3B, 6A, and 9 which were intended as examples of a structure that would result by control of wave-base.

Further evidence contradicting Dietz's "mature" terrace structure (Dietz, 1963, Fig. 14) is found in the continuous reflection profiler records collected from the continental terrace off Norfolk, Virginia (Moore and Curray, 1963b). This is the type section of Dietz's terrace of prograded paralic beds cropping out on an erosional slope. Figure 2 is a line drawing of reflecting horizons from continuous reflection profiling records. The structure in no way resembles that of Dietz's model. Instead, the fundamental structure again suggests outbuilding by slope deposition and upbuilding by open shelf deposition accompanied by subsidence. This record shows structures indicating accumulation of slope deposits to a thickness of at least 5000 feet (assuming an average velocity of 6700 feet/sec.).

Outbuilding and upbuilding by slope and shelf deposition is by no means unique to these two areas. Detailed reflection profiling records (Fig. 1) from the Costa de Nayarit off west-central Mexico (Curray and Moore, 1964) show extensive terrace progradation by slope deposition, and unpublished records from northern California show the same structure. We do not suggest this to be the only method of continental terrace development, but it is, at least, a common type. Furthermore, this fundamental sedimentary structure is known to be locally modified by massive slumping on the upper slope. We do not believe, however, that beds thus exposed, by slumping, or faulting, constitute proof that the slope is an erosional realm. Continental terraces, like mountain ranges, must be regionally different, depending on the balance between such factors as rates of supply of sediment, intensity of oceanographic processes, rates of subsidence or uplift, and influence by tectonics (van Andel and Curray, 1960).

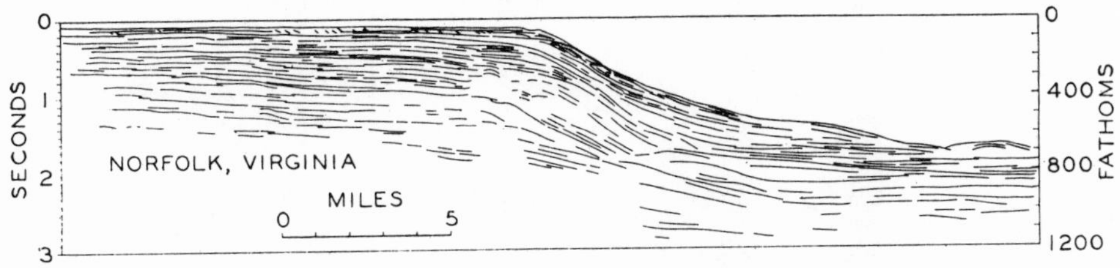

Figure 2. Line drawing of continuous reflection record (Electro-Sonic Profiler, Arcer) across continental shelf and slope off Norfolk, Virginia. Note the thick sections of sediments underlying both the shelf and slope and the general conformity of bedding with the present surface.

Acknowledgments

The origin of ideas is difficult to trace; some are most certainly generated in casual discussion with colleagues. Many marine geologists have basically the same ideas on matters they consider "concepts," but most have never taken the time to write them down. When they do, they differ in principal emphasis. We gratefully acknowledge the assistance of many of our colleagues for discussion of "concepts" before and after preparation of this note and for critical review of the manuscript, especially R. S. Dietz, F. P. Shepard, Tj. H. van Andel, E. L. Hamilton, E. C. Buffington, G. H. Curl, and D. L. Inman.

References Cited

Bradley, W., 1958, Submarine abrasion and wavecut platforms: Geol. Soc. America Bull., v. 69, p. 967–974

Curray, J. R., 1960, Sediments and history of Holocene transgression, continental shelf, Northwest Gulf of Mexico, p. 221–266 *in* Shepard and others, *Editors*, Recent Sediments, N. W. Gulf of Mexico: Am. Assoc. Petroleum Geologist, 394 p.

—— 1964, Transgressions and regressions, p. 175–203 *in* Miller, R. L., *Editor*, Papers in Marine Geology: Shepard Commemorative Volume, New York, Macmillan, 531 p.

Curray, J. R., and Moore, D. G., 1964, Pleistocene deltaic progradation of continental terrace, Costa de Nayarit, Mexico, *in* van Andel and Shor, *Editors*, Marine Geology of the Gulf of California: Am. Assoc. Petroleum Geologists, p. 193–215

Dietz, R. S., 1952, Geomorphic evolution of continental terrace (continental shelf and slope): Am. Assoc. Petroleum Geologists Bull., v. 36, p. 1802–1820

—— 1963, Wave-base, marine profile of equilibrium, and wave-built terraces: A critical appraisal: Geol. Soc. America Bull., v. 74, p. 971–990

Dietz, R. S., and Menard, H., 1951, Origin of abrupt change in slope at continental-shelf margin: Am. Assoc. Petroleum Geologists Bull., v. 35, p. 1994–2016

Emery, K. O., 1952, Continental shelf sediments of Southern California: Geol. Soc. America Bull., v. 63, p. 1105–1108

Fairbridge, R., 1952, Marine erosion: 7th Pacific Sci. Cong., Pacific Sci. Assoc. Proc., v. 3, p. 347–358

Gilbert, G. K., 1890, Lake Bonneville: U. S. Geol. Survey Mon. No. 1, 438 p.

Hersey, J. B., 1963, Continuous reflection profiling, Chap. 4, p. 47–72 *in* Hill, M. N., *Editor*, The Sea, v. 3: New York, Interscience Publishing Co. Inc., Div., John Wiley and Sons, 963 p.

Hjulström, F., 1939, Transportation of detritus by moving water, p. 5–31 *in* Trask, P. D., *Editor*, Recent Marine Sediments: Am. Assoc. Petroleum Geologists, 736 p.

Inman, D. L., 1963, Sediments: Physical properties and mechanics of sedimentation, Chap. V, p. 101–147, *in* Shepard, F. P., Submarine Geology (2d ed.): New York, Harper and Row, 557 p.

Inman, D. L., and Bagnold, R. A., 1963, Beach and nearshore processes, Part II, Littoral Processes, Chap. 21, p. 529–553, *in* Hill, M. N., *Editor*, The Sea, V. 3: New York Interscience Publishing Co. Inc., Div., John Wiley and Sons, 963 p.

Moore, D. G., 1960, Acoustic reflection studies of the continental shelf and slope off southern California: Geol. Soc. America Bull., v. 71, p. 1121–1136

—— 1961, Submarine slumps: Jour. Sed. Petrology, v. 31, p. 343–357

Moore, D. G., and Curray, J. R., 1963a, Structural framework of the continental terrace, northwest Gulf of Mexico: Jour. Geophys. Research, v. 68, p. 1725–1747

—— 1963b, Sedimentary framework of the continental terrace off Norfolk, Virginia, and Newport, Rhode Island: Am. Assoc. Petroleum Geologists Bull., v. 47, p. 2051–2054

Shepard, F. P., 1948, Submarine Geology: New York, Harper Bros., 348 p.

—— 1955, Delta-front valleys bordering the Mississippi distributaries: Geol. Soc. America Bull., v. 66, p. 1489–1498

—— 1963, Submarine Geology (2d ed.): New York, Harper and Row, 557 p.

Sundborg, Å., 1956, The river Klarälven—a study of fluvial processes: Geografisk Ann., Stockholm, v. 38, p. 127–316

Van Andel, Tj. H., and Curray, J. R., 1960, Regional aspects of modern sedimentation in northern Gulf of Mexico and similar basins and paleogeographic significance, p. 345–364 *in* Shepard and others, *Editors*, Recent Sediments, Northwestern Gulf of Mexico: Am. Assoc. Petroleum Geologists, 394 p.

6

Reprinted, courtesy of the Geological Society of America, from *Geol. Soc. Am. Bull.* **75**:1275–1281 (1964)

WAVE-BASE, MARINE PROFILE OF EQUILIBRIUM, AND WAVE-BUILT TERRACES: REPLY

ROBERT S. DIETZ

Abstract: This paper is a reply to the critique by David G. Moore and Joseph R. Curray (1964) of my paper "Wave-base, marine profile of equilibrium, and wave-built terraces: A critical appraisal." Our points of difference seem only to involve minor aspects and emphasis.

In contrast to Moore and Curray, I doubt the importance of continental terrace prograding. Even for the example they cite off Norfolk, Virginia, this extension of the terrace is shown to be only slight. Although I do not deny the existence of some open-shelf and continental-slope deposition, I consider their volume to be greatly less than either paralic or continental-rise deposition.

I re-emphasize that wave-base can only be of minor and subtle geologic importance and does not control the building of the continental terrace. In contrast, surf-base exerts a strong influence. In common with earlier writers, Moore and Curray fail to specify the depth of wave-base or its exact position on the self profile; therefore their concept of wave-base remains vague. I suggest placing wave-base at the shelf break, regardless of its depth. Actually the realm of sea-water agitation extends deeper than shelf depths everywhere, hence shelf deposits tend to be sorted or "above wave-base" as this term is used by stratigraphers. Beyond the shelf break the gradient of the continental slope probably exerts a stronger influence on sediment facies than does the degree of water agitation. In this context, wave-base is controlled by topography and not topography by wave-base.

Introduction

I appreciate this opportunity to reply to Moore and Curray's (1964) discussion of my recent paper (1963b). We appear to enjoy a broad area of common agreement about the fundamental geologic concepts of wave-base, marine profile of equilibrium, and wave-built terraces. Their critique, however, also raises questions of differing emphasis and some points of basic disagreement. I hope that this exchange of views will further clarify these concepts, bringing us closer together on certain points, and at the same time clarifying any persisting differences of viewpoint.

I agree with the critique that my treatment was oversimplified. However, it was not directed to the specialist in marine geology but to the student and the textbook writer. I attempted to show in simplest terms that there is a satisfactory alternate to the Johnsonian concept of the continental terrace which depends upon wave-base as its fulcrum. The wave-base concept, in its classical form (which I consider erroneous), still strongly pervades geology. It is treated in nearly all textbooks although the two recent editions of Longwell and Flint (1955; 1963) are notable exceptions.

Deposition and Erosion on Continental Terraces

Much of the critique pertains to the Gulf Coast (Moore and Curray, 1963a). They correctly point out that I have attempted to enlarge upon my earlier treatment of the geomorphic evolution of the continental terraces (Dietz, 1952) by adding the concept of a continental embankment, for it is rather evident that the Gulf Coast slope has undergone a very extensive prograding. Since slope-erosion processes are gravity controlled, there must be some gradient, for a particular slope at a given time, below which slope sedimentation dominates over slope erosion (2° is suggested as an average). Of course, rate and fluctuation of sedimentation are also important factors which I glossed over and which must be considered in any complete treatment. To a first approximation, the Gulf continental embankment may be likened to a huge delta terrace developed by the Mississippi River and its ancestral equivalents which provided a point-source of sediment. This migrated northeastward during the Cenozoic from Mexico to the present "depo-center" off Louisiana. Rates of sedimentation have varied greatly. Quaternary rates may be three times the Miocene rate and 35 times the Cretaceous rate (Hardin, 1962). Unlike the delta terrace (an air-water interface terrace) which develops from a point-source of sediment, the hypothetical wave-built terrace (a wholly subaqueous terrace) develops from a line-source of sediment.

The sub-bottom acoustic profile obtained off Norfolk and presented in their critique is an example of a continental-slope terrace in "late maturity" (Dietz, 1952), extensively modified by sedimentation. The continental rise off Norfolk has nearly uplapped the continental slope. The contact between the slope proper and the rise now is at about 300 fathoms if one uses the line of maximum flexure to determine this juncture (Fig. 1). Judging from one sub-bottom reflecting horizon, which seems to show an earlier shelf break, one could suggest that a prograding of the shelf edge of nearly 1 mile seems established. This is a rather small amount considering that the entire terrace is 150 miles from the Fall Line to the shelf edge. The volume of slope-facies sediment is small compared to the stratiform or supra-shelf facies and the volume of the continental rise (Fig. 1). I would suppose that the stratiform shelf beds would be composed of mixed open-shelf and paralic facies with the latter greatly predominating. Much hinges on one's interpretation of what constitutes an important amount of prograding of the continental slope. If the diagrams of Johnson (1919; 1938) or Rich (1951) are used as a model, a prograding of about 75 miles would be expected!

Extensive sub-bottom profiles of the continental terrace have been obtained along the northern portion of the eastern United States by K. O. Emery (personal communication). He finds that, by sub-bottom echo, the continental-slope profile invariably extends beneath the continental-rise sedimentary prism. This suggests that the continental rise is a separate depositional realm which uplaps the slope rather than being simply an extension of the continental-slope regime.

I should re-emphasize that I do not deny the existence of some open-shelf and slope deposition, but in contrast to the classical view, I consider their volume to be of lesser importance than either the paralic or the continental-rise realms. In general continental slopes must retrograde more than they prograde. Prograding is only locally important where deltas are present and where the continental slope has been uplapped by the continental rise, or nearly so. This scheme pertains only to the clastic regimen and not to the carbonate environment.

Moore and Curray (1963b) cite their new evidence as inconsistent with my interpretation. In turn, they tend to ignore or consider discredited older data. If their section off Norfolk, Virginia, is typical for the Atlantic terrace, how can we understand the dredging of Mesozoic and Tertiary rock from the slope face and from submarine canyon walls (Stetson, 1936; 1949; Northrop and Heezen, 1951; J. Ewing and others, 1963)? What about the deep terraces on the slope face which seem to be related to the outcropping of gently dipping terrace-top strata (Heezen and others, 1959)? The Norfolk sub-bottom reflection section may be located at the ancestral seaward limit of the Susquehanna-Chesapeake River and Jamestown River systems. The dip-slope beds may be ancient deltaic deposits.

The Moore and Curray sections may not be typical of modern shelves generally since the Atlantic Coast and Gulf Coast shelves from which their results are obtained are underlain by thick, nonfolded, wedge-shaped, and stratiform deposits. These are commonly termed paraliageosynclines (Kay, 1951) or modern miogeosynclines (Drake and others, 1961; Dietz, 1963a). Fairbridge's (1957) world survey of nonfolded geosynclines and sedimentary basins revealed that only 30 per cent of the shelf areas are covered by such stratiform sheets. Although lacking sub-bottom acoustic penetration surveys in their comprehensive study of the Pacific continental slope between San Francisco and central Baja California, Uchupi and Emery (1963) found no evidence of prograding, and they generally concurred in the mode of slope evolution of Dietz (1952). Even for the Atlantic terrace, J. Ewing and others (1960; 1963) find only minor slope mantling in the Hudson Canyon area. Drake (personal communication) has found at most only minor mantling of the slope between New Jersey and Newfoundland.

A growing number of geologists now accept continental drift. South America and Africa especially seem to have once been contiguous prior to mid-Mesozoic. If drift *is* the answer to the excellent fit between these two continents, this would seem to rule out any extensive prograding of these particular continental margins in the last 200 million years, excepting the Niger delta.

One point of basic disagreement concerns the role of turbidity currents in continental-slope development. My evaluation of the over-all evidence, especially as adduced from the writings of Kuenen and Heezen, remains that turbidity currents *do* play the dominant role in slope erosion although they are not the sole process. However, all the processes involved in slope degradation are gravity propelled and,

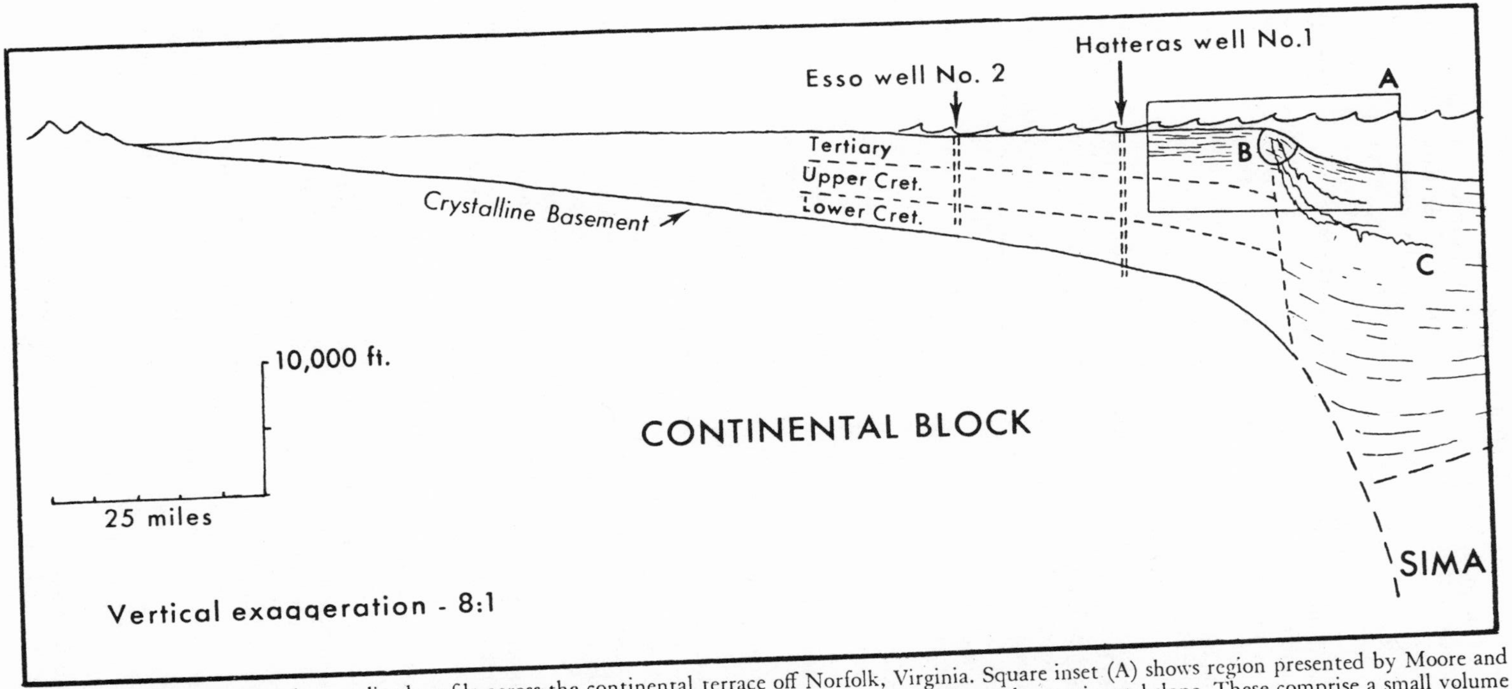

Figure 1. Diagrammatic and generalized profile across the continental terrace off Norfolk, Virginia. Square inset (A) shows region presented by Moore and Curray (1964, Fig. 2). The solid semicircle (B) covers the section where they found dip-slope beds on the continental slope. These comprise a small volume relative to the entire shelf beds in this diagram. Three continental-slope profiles (C) (from Heezen and others, 1959) illustrate that the profile off Norfolk is considerably gentler than typical continental slopes. Two wells, Esso No. 2 and Cape Hatteras No. 1, projected onto the section, are taken from Heezen and others (1959, p. 4). Data on the presumed thickness of the continental rise taken from M. Ewing (1963). Width of the shelf was adapted from Murray (1961)

therefore, act in a downslope direction. Accordingly, there must be a gradient below which these processess are no longer effective and deposition takes over. Of course, many parameters are involved (*i.e.* rate of supply, frequency of seisms, grain size, *etc.*) making it impossible to identify precisely the grade at which deposition will exceed erosion. If one uses canyon development as the criterion for an erosional regime, however, a 2° declivity may be a useful round number for separating slopes which are mainly aggradational from those which are dominantly erosional.

Continental slopes which are furrowed by numerous submarine canyons must have undergone erosion. Since canyon axes are gentler than the declivity of the continental slope into which they are cut, the slopes too must be undergoing degradation. This is of course, simply a guideline and not a hard and fast rule. A slope may be depositional during a time of rapid sedimentation when a delta terrace is forming, but after its cessation, the front may become erosional and furrowed by canyons. Nevertheless, the suggestion that canyons are erosional because they represent a special environment, whereas the interfluves are simultaneously depositional, seems to me unlikely. This, however, does seem likely to Moore and Curray (personal communication). Slopes lacking submarine canyons and having low declivity may be reasonably interpreted as depositional.

The critique seems to object to the identification of any slope angle marking a usual transition between a depositional continental slope and an erosional one. Mention is made of sedimentary slopes of 20° which are stable. Although probably true in rare instances, it is not Moore's and Curray's (personal communication) intention to imply that the continental slopes can prograde at this angle. Such steep continental slopes can at most be only thinly mantled by sediment which is in turn supported by underlying bedrock. Continental slopes of such precipitousness must be structural and not sedimentary (Dietz, 1964). In their discussion, Moore and Curray (1963b) themselves suggest mass degradation of the continental slope off Rhode Island which has an inclination of only 4°.

Wave-Base and its Relation to Terrace Development

It would not be useful to return to Gilbert's definition of wave-built terrace. His usage is archaic and is an example where priority should be laid aside. Modern usage will and should prevail in spite of any opinions to the contrary. The concept of a *net cyclic period* as advanced by Moore and Curray which has a range between 1 and 100 years is doubtless valid and useful for some purposes. Exceptional long-period waves of rare occurrence may well play some part in determining sediment patterns of the outer shelf. Another long-term cyclic period is of even more importance. The over-all morphology of the outer modern continental terrace appears to be dominated mostly by past eustatic oscillations of sea level. These strong oscillations encompass the entire Quaternary, or a period of 1 m.y., and it is the minimum levels which play the greatest role rather than the average position of sea level during the Quaternary.

In common with nearly all writers of the past, Moore and Curray failed to specify a depth, or a range of depths, for wave-base even for any specified locality. They consider it to be at some depth less than that of the shelf edge which typically lies at about 65 fathoms (120 m). If wave-base is considered to be exactly at the depth of the shelf edge, it must be construed as fortuitous or arbitrary, for it seems unlikely that the buildup of the shelf could have kept pace with the rapid rise of sea level from about —400 feet since 19,000 B.P. (Curray, 1961). If wave-base is *on* the shelf and capable of building a wave-built terrace, many incipient examples should exist on modern shelves. A still-stand of sea level is a prerequisite, but most modern students of sea-level change (Curray included) agree that sea level has remained essentially fixed for the last 3000–6000 years, subsequent to the rapid postglacial rise from 20,000 years B.P. However modern wave-built terraces on the outer shelf have not as yet been reported. (Moore and Curray [personal communication] agree that outer shelf wave-built terraces appear to be nonexistent, but they believe that sheet deposits related to wave-base are present.)

As applied to the continental shelf, the profile-of-equilibrium viewpoint usually holds that the outward slope of the shelf represents some sort of equilibrium profile between the shoreline and wave-base. In my opinion, this is not true. The comparison drawn between the profile of equilibrium of a river bed is neither valid nor appropriate, and shelves do not have a concave profile as often supposed. The concept does have validity within the paralic zone alone (0–15 or 20 m) where a concave profile does in-

deed develop. In simplest terms, this profile seems to represent a balance between gravity forces tending to level the bottom and the onshore push of waves. Waves have a quick shoreward stroke followed by a slow return stroke which favors shoreward transport of the tractive load. (Moore and Curray [personal communication] suggest that modern shelves may be grossly out of adjustment, but if given sufficient time, a shelf would develop some type of profile of equilibrium.)

I agree with the critique that there is some evidence for a decrease in grain size of clastic sediment across the shelf although thoughtful work is required to establish this. Emery (1952), for example, has classified shelf sediments off parts of California into authigenic, organic, residual, relict, and modern detrital types. The modern detrital sediments, taken alone, show a slight gradation from coarse to fine seaward. However, Shepard (1963, p. 258) describes this supposed gradation across the shelf as an entrenched geologic dogma; he does agree that a few exceptional areas may exist. In short, there is some evidence for a decrease in grain size with depth across some shelves when modern detrital sediment is taken alone. Thus wave-base may have some residual significance in controlling sediment distribution but is not a dominating process in shelf development. Gradation is readily apparent in the paralic zone, but this is related to surf-base.

Although the term *wave-base* should be retained in some redefined sense as meaning the depth at which wind waves no longer appreciably stir the bottom it should be recognized that wave-base is only of minor geologic importance as compared to *surf-base*, the depth at which waves peak up and then break into surf. Less than 5 per cent of the energy of a wave is spent prior to breaking; therefore it is in the surf zone where important corrasion, *etc.* occurs. *Surf-base* pertains to the paralic zone whereas *wave-base* pertains to the outer shelf where it has some effect in controlling patterns of sedimentation. Doubtless, a greater ambient water energy above wave-base contributes to sediment sweeping and reworking on the outer shelf, but wave-base is quite incapable of building a wave-built terrace anywhere and cannot account for the nick-point or shelf-break at the shelf edge. (Moore and Curray [personal communication] concur in this.) In short, the morphologic development of the continental terrace is not controlled by wave-base. Wave-base should be greatly downgraded in importance as a geologic process but not wholly discarded.

We need to use both terms—surf-base and wave-base. We should not confuse the issue by bringing wave-base into the paralic zone, recognizing the greater importance of surf effects. We still need the term wave-base to apply to deep-water effects. Considering wind waves and swell alone, these decay asymptotically with depth; hence the decay is rapid although not abrupt enough to create a wave-built terrace. Also, and quite importantly, water motions from other processes come into play. However, it is certainly true that shallow water (*i.e.* all water at normal shelf depths) is much more highly agitated than abyssal water. Where wave-base is eventually established will be largely a matter of one's definition of this term. Curray and Moore (1964) suggest wave-base should be placed somewhere on the shelf and correlated in some way with shelf sheet deposits. I suggest that it *arbitrarily* be placed at the shelf break, regardless of the depth of this break, as this would be of greatest usefulness in stratigraphy. I reason that actually the realm of moderate sea-water agitation extends below shelf depths everywhere; hence, shelf deposits tend to be sorted, winnowed, *etc.* rather than "poured in." However, beyond the shelf break, the gradient of the slope is likely to exert a strong and perhaps dominating influence tending to result in a change of sediment facies. It is well known too that the topographic form of the shelf edge causes a speeding up of any currents which approach from the deep sea. In this context, wave-base is controlled by topography and not topography by wave-base.

Concluding Remarks

The most fundamental questions to be asked of wave-base are: (1) Does it dominate continental-terrace development? (2) Is it capable of constructing a wave-built terrace? (3) Is continental-shelf profile a profile of equilibrium somehow related to wave-base? We agree that the answer to all three questions is no. Moore and Curray (1964) also suggest that it plays a role in sweeping detritus onto the continental slope, and in this respect I agree with them.

Very likely, our most basic difference aside from emphasis is one of philosophic approach. I have supposed that a classification and a geomorphic evolution of continental terraces can be achieved based on a few dominating processes and especially without invoking the concept of wave-base and its corollaries—the wave-built

terrace and the marine profile of equilibrium. My interpretation seems to me valid but by no means all inclusive or definitive. In particular, much additional treatment should be accorded the effect of deltaic processes and the delta terrace. As is usual in geology, "textbook examples" are few, and actual continental terraces can be regarded as departures from idealized examples. We *do* need to identify and develop an understanding of the dominating processes operating. To this end, this exchange of views serves a useful purpose.

In the preparation of this reply, I have profited by discussions with Francis P. Shepard, Lewis Butler, John Holden, K. O. Emery, B. C. Heezen, Charles Drake, Richard Malloy, and E. L. Hamilton. It should not be construed, however, that any of these persons necessarily share the writer's views.

References Cited

Curray, J. R., 1961, Late Quaternary sea level: A discussion: Geol. Soc. America Bull., v. 72, p. 1707–1712

Dietz, R. S., 1952, Geomorphic evolution of the continental terrace (continental shelf and slope): Am. Assoc. Petroleum Geologists Bull., v. 36, p. 1802–1820

—— 1963a, Collapsing continental rises: an actualistic concept of geosynclines and mountain building: Jour. Geology, v. 71, p. 314–333

—— 1963b, Wave-base, marine profile of equilibrium, and wave-built terraces: A critical appraisal: Geol. Soc. America Bull., v. 74, p. 971–990

—— 1964, Origin of continental slopes: American Scientist, v. 52, p. 50–69

Drake, C., Ewing, M., and Sutton, G., 1961, Continental margins and geosynclines: The east coast of North America north of Cape Hatteras, p. 110–198 *in* Aherns, L. H., Rankama, K., Press, F., and Runcorn, S. K., *Editors*, Physics and Chemistry of the Earth, V. 3: New York, Pergamon Press, 464 p.

Emery, K. O., 1952, Continental shelf sediments off California: Geol. Soc. America Bull., v. 63, p. 1105–1108

Ewing, J., Le Pichon, X., and Ewing, M., 1963, Upper stratification on the Hudson Apron region: Jour. Geophys. Research, v. 68, p. 6303–6316

Ewing, J., Luskin, A., Roberts, A., and Hirshman, J., 1960, Sub-bottom reflection measurements on the continental shelf, Bermuda banks, West Indies arc, and in the west Atlantic basins: Jour. Geophys. Research, v. 63, p. 2849–2859

Ewing, M., 1963, Sediments of ocean basins, p. 41–59 *in* Higginbotham, *Editor*, Man, Science, Learning and Education: Houston, Rice Univ. Press

Fairbridge, R. W., 1957, Statistics of non-folded basins: Pub. Bur. Central Seimologique International, ser. A, fasc. 20, p. 419–440

Hardin, G. C., 1962, Notes on Cenozoic sedimentation in the Gulf Coast geosyncline, p. 1–15 *in* Rainwater, E. H., and Zingula, R. R., *Editors*, Geology of the Gulf Coast and central Texas: Houston Geol. Soc., 391 p.

Heezen, B. C., Tharp, M., and Ewing, M., 1959, The floors of the oceans. I. The North Atlantic: Geol. Soc. America Special Paper 65, 122 p.

Johnson, D., 1938, Shore processes and shoreline development, 2d ed.: New York, John Wiley and Sons, Inc., 584 p.

Kay, M., 1951, North American geosynclines: Geol. Soc. America Memoir 48, 143 p.

Longwell, C. R., and Flint, R. F., 1955, Introduction to physical geology: New York, John Wiley and Sons, Inc., 432 p.

—— 1962, Introduction to physical geology, 2nd ed.: New York, John Wiley and Sons, Inc., 504 p.

Moore, D. G., and Curray, J. R., 1963a, Structural framework of the continental terrace, northwest Gulf of Mexico: Jour. Geophys. Research, v. 68, p. 1725–1747

—— 1963b, Sedimentary framework of the continental terrace off Norfolk, Virginia, and Newport, Rhode Island: Am. Assoc. Petroleum Geologists Bull., v. 47, p. 2051–2054

—— 1964, Wave-base, marine profile of equilibrium, and wave-built terraces: Discussion: Geol. Soc. America Bull., v. 75, p. 1267–1274

Murray, G., 1961, Geology of the Atlantic and Gulf Coastal province of North America: New York, Harper and Bro., 692 p.

Northrop, J., and Heezen, B. C., 1951, An outcrop of Eocene sediment on the continental slope: Jour. Geology, v. 59, p. 396–399

Rich, J. L., 1951, Three critical environments of deposition and criteria for recognition of rocks deposited in each of them: Geol. Soc. America Bull., v. 62, p. 1–20

Shepard, F. P., 1963, Submarine geology, 2d ed.: New York, Harper and Row, 557 p.

Stetson, H. C., 1936, Geology and paleontology of Georges Bank canyons: Geol. Soc. America Bull., v. 47, p. 339–366

—— 1949, The sediments and stratigraphy of the east coast continental margin; Georges Bank to Norfolk Canyon: Woods Hole Oceanographic Inst. Papers in Phys. Oceanography and Meteorology, v. 11, no. 2, p. 1–60

Uchupi, E., and Emery, K. O., 1963, The continental slope between San Francisco, California and Cedros Island, Mexico: Deep-Sea Research, v. 10, p. 397–447

7

Reprinted from *Science* **176**:282–284 (1972)

RIVER DELTA MORPHOLOGY: WAVE CLIMATE AND THE ROLE OF THE SUBAQUEOUS PROFILE

L. D. Wright and J. M. Coleman
Coastal Studies Institute, Louisiana State University

Shoreline geometries and dominant wave regimes strive to attain mutual adjustment along coasts that are molded primarily by the action of ocean waves (*1*). The coastline strives to develop in such a way as to minimize or eliminate longshore gradients in wave forces; in turn, the sediment budget is balanced by redistribution of longshore and onshore wave forces. The resultant subaqueous and subaerial morphology tends to attain quasi-equilibrium with the long-term wave climate. The reciprocation between this morphology and the wave regime is considerable; depositional patterns depend on the spatiotemporal distribution of wave forces; the latter is a function of subaqueous and shoreline topography. Waves are modified by the bottom through refraction, shoaling (*2*), and frictional attenuation (*3, 4*). To date, in most studies of the relationships between wave forces and landforms the effects of bottom friction on waves have not been considered quantitatively; however, in many instances these effects may exert the dominant control on the magnitude and distribution of wave forces in the nearshore zone.

Near the mouths of large rivers the equilibrium between wave regime and depositional topography is upset by the continued introduction of sediments from a source (river channels) external to the littoral transport system. Resultant forms depend not only on the magnitude and distribution of wave forces but also on the ability of the river to supply sediments. Hence river deltas have a spectrum of configurations and landform combinations ranging from those which have been produced solely by the debouchment of the river, without significant interference from wave activity (for example, digitate prograded distributaries), to those which reflect complete dominance by the waves in distributing sediments and straightening the coastline according to their own design (for example, deltas impounded by barrier islands or spits). Whether forms reflect fluvial or wave dominance depends largely on the ability of the river to supply sediments relative to the ability of waves to rework and redistribute them. For the purpose of obtaining a first-order estimate, the sediment-transporting ability of the river is assumed to be proportional to river discharge (*5*), whereas the ability of waves to rework and redeposit sediments is considered proportional to wave power. In a general sense, then, the greater the river discharge the greater is the wave power required to rework the sediments into a wave-dominated configuration.

As part of a continuing research program aimed at explaining the variability of the major river deltas of the world in terms of their process environments (*6*), we have developed a procedure for analyzing the discharge and wave-power climates of delta coasts from published data, maps, and aerial photographs. Sets of deepwater wave characteristics (specific combinations of direction, height, and period) are obtained from published records or by hindcasting from wind data (*7*) where wave data are unavailable. The average relative frequency of each set of wave characteristics occurring during each month is recorded. These values serve as input to a comprehensive computer program developed to compute the effects of refraction, shoaling, and frictional attenuation incrementally along the wave orthogonal by a finite-difference solution to the equations of Bretschneider (*3*) and Bretschneider and Reid (*4*). Wave-power values are calculated from the changing wave characteristics at each 3-m depth interval from deep water to the 3-m contour. From this depth to the shoreline, calculations are made at 0.3-m intervals. We weight these values according to the relative frequencies of the initial wave conditions and calculate weighted mean power values for each month along different sectors of the coast. The ratio of the average discharge per unit channel width to the

Table 1. Characteristic delta morphologies.

Coastline and river mouth configuration	Delta shoreline landforms	Delta plain landforms
	Mississippi	
Highly indented coastline, multiple extended digitate distributaries—"bird foot"	Indented marsh coastline, sand beaches scarce and poorly developed	Marsh, open and closed bays
	Danube	
Slightly indented with protruding river mouths	Marsh coastline with sand beaches adjacent to river mouths	Marsh, lakes, and abandoned beach ridges
	Ebro	
Smooth shoreline with single protruding river channel	Low sand beaches and extensive spits with some eolian dunes	Salt marsh with a few beach ridges
	Niger	
Smooth, arcuate shoreline, multiple river mouths slightly protruding	Sand beaches nearly continuous along shoreline	Marsh, mangrove swamp, and beach ridges
	Nile	
Gently arcuate, smooth shoreline with two slightly protruding distributary mouths	Broad, high sand beaches and barrier formation with eolian dunes, beach ridges at distributary mouths	Floodplain with abandoned channels and a few beach ridges, hypersaline flats and barrier lagoons near present shoreline
	São Francisco	
Straight, sandy shoreline with single slightly constricted river mouth	High, broad sand beaches with large eolian dunes	Stranded beach ridges and dunes
	Senegal	
Straight coastline with extensive barrier deflecting river mouth, no protrusion	High, broad sand beaches with large eolian dunes	Large linear beach ridges and swales, eolian dunes

average nearshore power per unit crest width yields an index of the sediment-transporting ability of the river relative to the reworking ability of the waves. We refer to this as the discharge effectiveness index. Although this ratio is not dimensionless, relative values and ordering between deltas remain independent of units. However, it should be noted, that absolute values have no physical meaning.

We have applied our wave climate analyses to seven of the world's major deltas: Mississippi (United States), Danube (Romania), Ebro (Spain), Niger (Nigeria), Nile (Egypt), São Francisco (Brazil), and Senegal (Senegal). These deltas possess unique morphologies reflecting differing relative degrees of riverine and wave control. In the order listed above they range in configuration from types that are almost totally river produced, through those reflecting approximately equal contributions from riverine and wave forces, to types that are completely wave dominated. Distinguishing morphological characteristics of each delta are listed in Table 1; corresponding mean annual values for deepwater and nearshore wave power, mean annual discharges, discharge effectiveness indexes, and attenuation ratios are presented in Table 2. Table 1 indicates a progressive decrease in river-built features—such as digital distributaries, open bays, broad marshes, and crevasses—and an ascendancy of features —such as beach ridges, dunes, spits, and barrier formations—which are responses to high wave energy. These geomorphic tendencies are not explicable in terms of either deepwater wave power or river discharge (Table 2); however, there is a close correlation with the discharge effectiveness index and, to a slightly lesser extent, with the nearshore wave power.

Notably, the Mississippi has higher deepwater wave power than the Danube, Ebro, and Niger, and the value for the São Francisco is more than twice that of the wave-dominated Senegal. The fact that morphology and nearshore wave power in no way parallel deepwater wave power is of primary significance. This reflects the fundamentally important role of frictional attenuation and the consequently varying degrees of power loss owing to different subaqueous slopes. Attenuation increases with decreasing offshore slope. Figure 1 shows the average offshore slope configuration for each of the seven deltas. Slopes are flat and convex seaward of the low-energy deltas; they increase in steepness and concavity toward the high-energy end of the delta spectrum. Similarly, between sectors of individual deltas steep, concave slopes are contingent with zones of high nearshore wave power. This is consistent with the general tendency for wave-built profiles of equilibrium to be concave (*8*).

The extent to which power is lost through friction between deep water and the shore is indexed by an attenuation ratio:

$$P_0 r_s^2 / P_s$$

where P_0 and P_s are the deepwater and shore power values, respectively, and r_s is the refraction coefficient nearest the shore. A value of 1 for this quantity indicates complete conservation of power; values greater than 1 are proportional to the degree to which power is lost. If it is assumed that the bottom roughness is constant, frictional attenuation rates increase as water depth decreases; and, for the most commonly occurring wavelengths, the attenuation ratio depends largely on the bottom slope from a depth of 10 m to the shoreline and to some extent on the dominant wave characteristics and the obliquity of wave approach. Table 2 indicates the general increase in attenuation ratio that corresponds to the decrease in offshore slope, as illustrated in Fig. 1. The high attenuation ratio for the Ebro (relative to offshore slope) is attributable to the high obliquity of wave incidence around most of the delta front. The subaqueous profile fronting a delta can therefore be regarded as a control that determines the power of waves that reach the shore. For any particular regional continental shelf slope, the subaqueous slope will depend on the ability of the river to discharge sediments faster than they can be removed by waves.

We conclude from the analyses so far completed that the morphologies of river deltas are, to a considerable degree, functions of river discharge and the strength of wave forces near the delta shoreline. The latter cannot be assumed to be proportional to the wave power in deep water but depends primarily on the subaqueous profile. Hence, before a river can effectively oppose the sea to develop a river-dominated deltaic configuration it must first build a flat, shallow offshore profile to attenuate incoming waves. The actual subaqueous profile which accumulates in front of a particular delta depends partially on the regional slope

Table 2. Mean annual discharge and wave-power climate indexes. Wave-power values are given per centimeter of wave crest. The discharge effectiveness index is normalized to maximum discharge.

River	Wave power (erg/sec)		Discharge rate $\times 10^3$ (m^3/sec)	Discharge effectiveness index	Attenuation ratio
	Deepwater	Nearshore			
Mississippi	1.06×10^8	1.34×10^4	17.69	1.00	7913.3
Danube	2.30×10^7	1.40×10^4	6.29	2.14×10^{-1}	2585.0
Ebro	7.28×10^7	5.09×10^4	0.55	4.87×10^{-2}	1299.5
Niger	6.76×10^7	6.59×10^5	10.90	8.03×10^{-4}	102.8
Nile	1.36×10^8	3.21×10^6	1.47	5.86×10^{-4}	42.46
São Francisco	3.71×10^8	9.97×10^6	3.12	2.37×10^{-4}	37.2
Senegal	1.56×10^8	3.77×10^7	0.77	4.75×10^{-5}	4.16

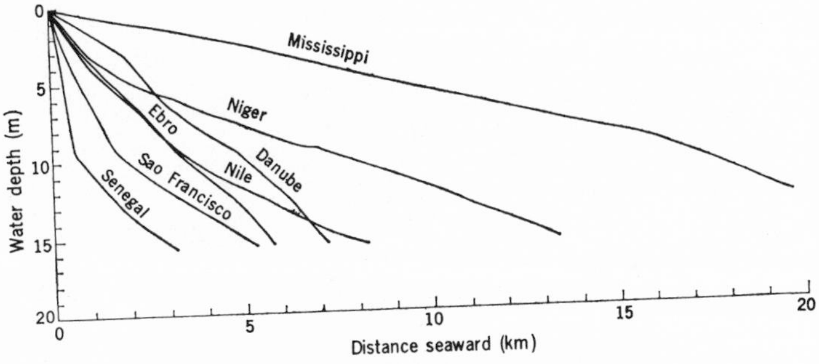

Fig. 1. Average subaqueous profiles of seven deltas.

of the continental shelf but largely on the rate at which the river can supply sediments to the nearshore zone. In summary, wave forces are the primary mechanism whereby the sea reworks and molds deltaic sediments; the construction of flat offshore profiles is the basic mechanism by which the river is able to overcome wave effects.

References and Notes

1. W. V. Lewis, *Proc. Geol. Ass.* **49**, 107 (1938); J. L. Davies, *Geogr. Stud.* **5**, 51 (1958); J. N. Jennings, *Aust. Geogr.* **5**, 36 (1955); V. P. Zenkovich, *Processes of Coastal Development,* J. A. Steers, Ed. (Wiley, New York, 1967), pp. 383–433; J. Larras, *Plages et Côtes de Sable* (Eyrolles, Paris, 1957).
2. R. L. Wiegel, *Oceanographical Engineering* (Prentice-Hall, Englewood Cliffs, N.J., 1964), pp. 150–179.
3. C. L. Bretschneider, *U.S. Army Corps of Engineers Beach Erosion Board Technical Memorandum 46* (U.S. Army Corps of Engineers, Baltimore, Md., 1954).
4. ——— and R. O. Reid, *U.S. Army Corps of Engineers Beach Erosion Board Technical Memorandum 45* (U.S. Army Corps of Engineers, Baltimore, Md., 1954).
5. Actual sediment transport data are unavailable for most of the rivers examined.
6. J. M. Coleman and L. D. Wright, *Coastal Studies Institute Technical Report 95* (Coastal Studies Institute, Louisiana State University, Baton Rouge, 1971).
7. C. L. Bretschneider, in *Estuary and Coastline Hydrodynamics,* A. T. Ippen, Ed. (McGraw-Hill, New York, 1966), pp. 133–196.
8. V. P. Zenkovich, *Processes of Coastal Development,* J. A. Steers, Ed. (Wiley, New York, 1967), pp. 101–119; J. W. Johnson and P. S. Eagleson, in *Estuary and Coastline Hydrodynamics,* A. T. Ippen, Ed. (McGraw-Hill, New York, 1966), pp. 449–462.
9. Supported by Geography Programs, Office of Naval Research, under contract N00014-69-A-0211-0003, project NR 388 002, with the Coastal Studies Institute, Louisiana State University.

Part II

COASTAL DEPOSITS

Editors' Comments on Papers 8 Through 12

8 **CURRAY**
Transgressions and Regressions

9 **PILKEY and FRANKENBERG**
The Relict-Recent Sediment Boundary on the Georgia Continental Shelf

10 **CURRAY, EMMEL, and CRAMPTON**
Holocene History of a Strand Plain, Lagoonal Coast, Nayarit, Mexico

11 **SWIFT el al.**
Textural Differentiation on the Shore Face During Erosional Retreat of an Unconsolidated Coast, Cape Henry to Cape Hatteras, Western North Atlantic Shelf

12 **SHERIDAN, DILL, and KRAFT**
Holocene Sedimentary Environment of the Atlantic Inner Shelf off Delaware

The papers in this section implicitly or explicitly acknowledge the equilibrium profile concept, but their main concern is with the implications of the theory for stratigraphy and the distribution of surficial sediments.

Paper 8 constitutes a true benchmark paper in the classical "mainstream" of papers on coastal sedimentation. There are two new elements. First, the focus is now on the uniformitarian application of insights from modern stratigraphic examples to stratigraphic problems in the rock record. Second, modern oceanographic techniques, appearing first indirectly in Fischer's Paper 2 in Part 1, are now brought to bear on many of the areas being studied, as a consequence of Curray's involvement in API Project 51, a large-scale sedimentological study of the northwest Gulf of Mexico coast and shelf. However, while this investigation was interdisciplinary in the sense that both sedimentary deposits and fauna were studied, fluid and sediment dynamics were not.

Although Paper 8 does not mention the equilibrium profile by name, it nevertheless constitutes a critical step forward in the understanding of coastal dynamics in that it systematically applies the insights of classical stratigraphy to problems of modern coastal sedimentation and in particular to the interrelationship of process variables. The key generalization concerns the rate and sense (landward or seaward) of shoreline migration as a function of rate of sedimentation and rate of sealevel displacement. It is first presented by Curray in Paper 8, Figure 1, applied to specific cases in Paper 8, Figures 2 through 7, and is presented in its final, fully developed form in Paper 8, Figure 10.

In Paper 9, we feel the impact of one of the major conceptual models for shelf sedimentation—that is, Emery's (1952) classification of detrital shelf sediments as modern (recently introduced into and in textural equilibrium with the depositional environment in which it is found) or as relict (from an earlier, different depositional environment). The Pilkey-Frankenberg paper occupies an interesting position in the evolution of this idea. Kuhn (1970), in his study of scientific revolutions, has shown that the progress of science is not linear, but step-like. A scientific revolution is not the result of new data; it occurs when a perceptive person produces a great, unifying hypothesis that reconciles and orders apparently contradictory or random facts that have been long available. The new synthesis holds for a period while those who conduct "normal science" busily exploit the new concept by working it out to its logical conclusions, but the success of the revolutionary concept is its own undoing. It leads workers to yet new facts, not all of which respond to the concept's explanatory powers, and eventually the undigested residue becomes so great that the old conceptual scaffolding yields to a new one that incorporates the anomalous data yet goes beyond; for example, as the Copernican universe followed the Ptolemaic or as Einstein's physics followed Newton's.

In the Pilkey-Frankenberg paper we are close to such a turning point. The sediment types fit the relict-recent model as closely as one could wish. The distinction between "modern" sands landward of the six-fathom isobath and "relict" sediments seaward of it is clear-cut, but how is the pattern explained? None of the classical explanations seem to fit. Perhaps, the authors speculate, the grain size pattern is a function of local energy conditions rather than sediment source. We shall see that the sharp "relict-recent" boundary occurs on most retreating coasts and is generally easier to explain in terms of differing hydraulic environments than it is in terms of the relict-recent model.

Paper 10 is another Curray classic. The field work for Paper 10 clearly informed the generalizations of Paper 8, but the paper nevertheless deserves to be included in unabridged fashion. It is the first and still the most carefully documented study of the stratigraphy of a prograding coast. We learn that if the rate of sediment input is high relative to the rate of sea-level rise, the coastal profile does not translate landward and upward, but instead seaward and upward. Shoreface progradation is not smooth and continuous as implied by Bruun (Paper 3) and Fischer (Paper 2) for shoreface retreat. It is cyclic, through the repeated formation of beach ridges (Paper 10, Figure 7). Here is new light on the old issue of barrier formation. The beach ridges are not barrier islands; they are genetically different features of much smaller scale. The entire Nayarit Strand Plain is in a sense a barrier grown immensely wide by beach ridge accretion.

In Paper 11, we return to the problems of sedimentation on wave-dominated retreating coasts. Alerted by Pilkey and Frankenberg's study, the authors have examined the distribution of grain size facies on the North Carolina coast in systematic detail and have supplemented their observations with information about North Carolina coastal stratigraphy and about coastal hydrodynamics gleaned from the literature.

The observations are strikingly similar to those of Pilkey and Frankenberg. Narrow belts of fine, medium, or coarse sand lie on the beach foreshore, in the long shore trough, or on the breakpoint bar; they are beyond the resolution of the sample net. An apron of fine, seaward-fining sand extends down the shoreface. It terminates abruptly at ten to twelve meters (Paper 11, Figures 3, 11, 13). The seaward-fining gradient of the upper shoreface fine sand apron is quite uniform, in marked contrast to the highly variable and generally coarser sands of the lower shoreface. At twelve meters on the facies boundary, bimodal sands occur. The "near-shore modern" facies is much narrower than on the Georgia coast, but it extends to exactly the same depth. There are no rivers to blame, and littoral drift diverges from the center of the study area towards both ends so the fine sands cannot be conveniently supplied by this method. The upper shoreface blanket of fine seaward-fining sand is more widespread on retreating coasts than we realized at the time, as can be found by carefully reading Van Straaten's benchmark study of the Dutch coast (in Schwartz, 1973).

The pattern is widespread on retreating coasts, and it cannot be explained by bringing the fine sand in from somewhere else. Therefore the sand has to originate in place. We had the benefit of

Cook's careful field observations of sand transport on the California coast (see Cook and Gorsline, Part IV), and we could suggest *how* it originates in place: We inferred that it is a fallout-blanket of sand suspended in the surf and moved seaward by rip currents. If the narrow fine sand belt of the Carolina coast originated in this way, then does the broad fine sand belt of the Georgia coast perhaps form from the fallout from the ebb tidal jets of those closely spaced estuary mouths? The direct observations that would serve to test the fine sand fallout hypothesis are yet to be made. In the meantime, Paper 11 offers an alternative to the relict-recent model for coastal sedimentation. The key distinction is not which sands were transported here now and which were transported here before; it is instead: How is the coastal hydraulic regime fractionating the debris of shoreface erosion into various grades of sand and where is it depositing these sand types?

Paper 12 represents the triumph of technology in the study of coastal deposits. The model presented in papers by Bruun (Paper 3), Fischer (Paper 2), Moody (1964), and Swift and others (Paper 11), for erosional shoreface retreat are tested and found to be valid, through application of the three basic tools of the coastal marine stratigrapher: seismic profiling, vibracoring, and radiocarbon dating. All three techniques had been available for over a decade but had lagged in application.

With these tools, Sheridan and colleagues (Paper 12) are able to demonstrate that erosional shoreface retreat beveled the lagoonal deposits of the leading edge of the Holocene transgression, and the retreating shoreface has left a veneer of recent marine sand over the lagoonal mud. See Stahl and others (1974) for a similar study of the New Jersey coast.

While some of us were drawing broad conclusions on relatively narrow data bases, others were reducing the spatial scale of their observations of coastal deposits and were producing far more systematic and detailed descriptions of the variation of the substrate with depth and distance from shore. While space does not permit the inclusion of these papers, this collection would not be complete without reference to the school of coastal studies exemplified by Clifton and others (1971), Reineck and Singh (1971), Howard and Reineck (1972), and Kumar and Sanders (1976).

REFERENCES

Clifton, H. E., Hunter, R. E., and Phillips, R. L. 1971. Depositional structures and processes in the non-barred, high energy nearshore. *J. Sed. Pet.* **4**: 651–70.

Emery, K. O. 1952. Continental shelf sediments of Southern California. *Geol. Soc. Am. Bull.* **63**:1105–08.

Howard, J. P., and Reineck, H. E. 1972. Georgia Coastal Region, Sapelo Island, USA: sedimentology and biology. IV. Physical and biogenic sedimentary structures of the nearshore shelf. *Senckenbergiana Maritima* **4**:81–123.

Kuhn, T. S. 1970. *The Structure of Scientific Revolutions.* Chicago: University of Chicago Press, 2nd ed.

Kumar, N., and Sanders, J. E. 1976. Characteristics of shoreface storm deposits: Modern and ancient examples. *J. Sed. Pet.* **46**:145–62.

Moody, D. W. 1964. Coastal morphology and processes in relation to the development of submarine sand ridges off Bethany Beach, Delaware. Ph.D. dissertation, Johns Hopkins Univ., 167 pp.

Reineck, H. E., and Singh, I. B. 1971. Der Golf Von Gaeta (Tyrrhenisches Meer). III. Die Gefuge Von Vorstrand und schelf sedimentation. *Senckenbergiana Maritima* **3**:135–83.

Schwartz, M. L. 1973. *Barrier Islands.* Stroudsburg, Pa.: Dowden, Hutchinson & Ross, 451 pp.

Stahl, L., Koczan, J. Jr., and Swift, D. J. P. 1974. Anatomy of a shoreface-connected and ridge on the New Jersey shelf: Implications for the genesis of the shelf surficial and sheet. *Geol.* **2**: 117–20.

Van Straaten, L. M. J. U. 1965. Coastal barrier deposits in south and north Holland—in particular the area around Scheveningen and Ijmuden. *Meded. Geol. Sticht.* **NS17**:41–75.

8

Reprinted from pp. 175–203 of *Papers in Marine Geology,* R. L. Miller, ed., Macmillan, 1964, 531 pp.

Transgressions and Regressions

by JOSEPH R. CURRAY
SCRIPPS INSTITUTION OF OCEANOGRAPHY, LA JOLLA, CALIFORNIA*

Introduction

MARINE TRANSGRESSION and regression is a fundamental concept in geology because many ancient sediments represent environments of deposition near the shoreline. The migrations of the shoreline through a section are of considerable importance in understanding the paleogeography, sources of supply, environments, and deposition mechanisms of sediments. Some of the more difficult determinations in the interpretation of transgressions and regressions in ancient sediments are lengths of time involved, rates of supply, rates of dispersal, and rates of deposition of sediments relative to rates of eustatic sea level change or tectonic movement, and cause of shoreline migration. In most studies, these questions cannot be answered with any degree of certainty because of the indirect and speculative nature of our knowledge of the process of lateral shoreline migrations.

For several years the writer has had the opportunity to participate in the detailed study of a transgressive sequence and small parts of the preceding regressive sequence in the Holocene sediments of the northwestern Gulf of Mexico. Consideration of the lithologic and biologic attributes of the sediments, the oceanographic conditions, the known sources and approximate rates of supply of sediments, as well as the known time interval have made it possible to recreate some the details of the events that took place. Subsequent study of other shelf and coastal plain sediments and perusal of the literature

*Contribution from the Scripps Institution of Oceanography. This work has been partly supported by the American Petroleum Institute Project 51, and partly by the National Science Foundation. Acknowledgement is gratefully made to the following for assistance in the formulation of ideas expressed and for review of the manuscript: Robert H. Parker, Tj. H. van Andel, Fred B. Phleger, Robert S. Dietz, David G. Moore, and Robert L. Miller.

suggest the general applicability of these principles to a variety of depositional situations. Holocene eustatic transgressive and regressive deposits clearly represent only a few of the many combinations of factors, but the principles involved in these studies may aid in the interpretation of ancient equivalent sediments.

Definitions

Definitions of the terms overlap, onlap, offlap, etc., as they apply to the process of transgression or regression of the shoreline, have been the subject of considerable controversy in the geological literature (Grabau, 1906; Melton, 1947; Lovely, 1948; Swain, 1949; Lahee, 1949). Most stratigraphers appear to prefer the terms "onlap" or "transgressive overlap" in referring to a transgressive sequence, and "offlap" or "regressive overlap" in referring to a regressive sequence, since they are primarily concerned with the positions of the younger beds relative to the older or partially underlying beds. The interpretation that transgression or regression has occurred is based largely on the overlapping relationships of shallower or deeper water facies. In this paper the terms transgression and regression will be used, since our main concern is with the process of migration of the shoreline rather than with the overlapping relationships of sedimentary facies. Some combinations of the factors affecting the nature of transgressions and regressions produce sequences of sediments in which the overlapping relationships are not simple textbook examples. The objectives of this paper are not to engage in semantics, but rather to attempt to contribute some understanding of an important geologic concept from the study of Quaternary sediments.

The term transgression is, therefore, used in this paper to mean the process of migration of the shoreline of a water body in a landward direction. Regression is the reverse, or a migration in a seaward direction. Nothing is implied concerning progressive pinching out of the sediment body toward or away from the margins of the depositional basin nor about textural changes up and down in the section.

Factors Influencing Transgressions and Regressions

The products of transgressions and regressions depend upon many factors, most of which are interdependent and cannot necessarily be evaluated separately even in examples of Holocene sediments. The factors include rate and

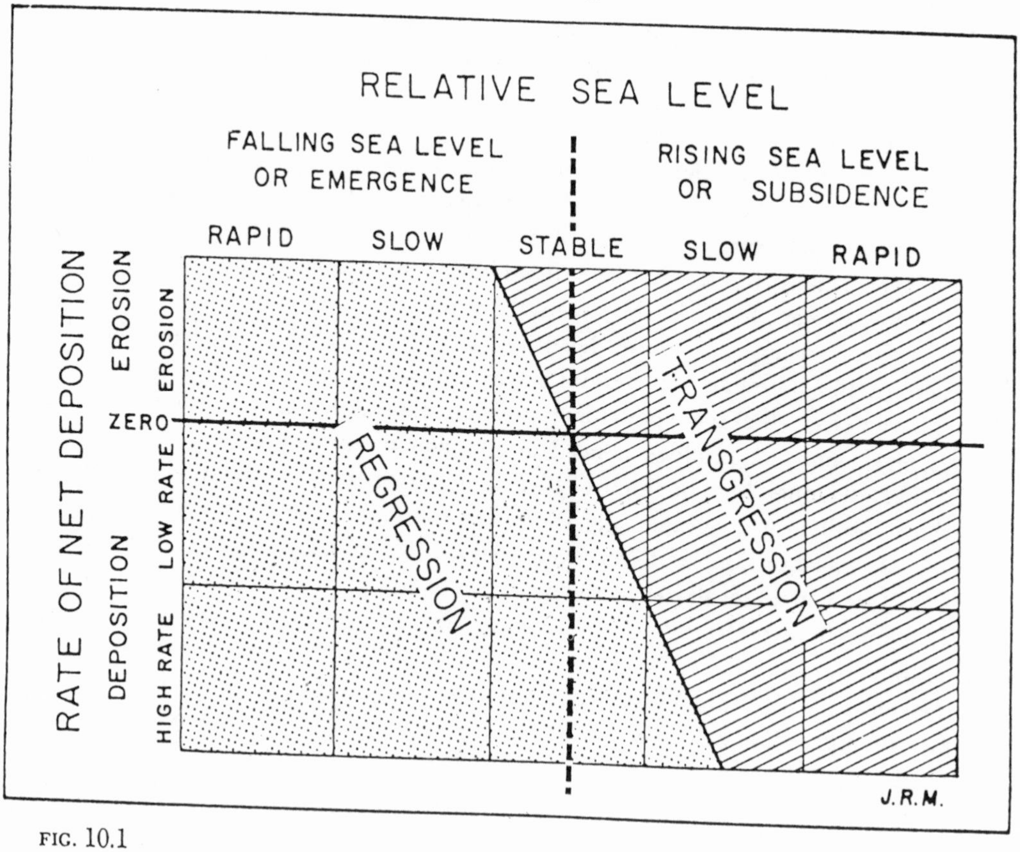

FIG. 10.1

Diagram of the effects of the rate of change of relative sea level and of the local net rate of deposition on lateral migration of the shoreline in a direction perpendicular to the shoreline. Transgression is defined as a landward migration of the shoreline, and regression is defined as a seaward migration. With no net deposition or erosion and with stable sea level, the shoreline remains geographically stationary. Rising relative sea level usually results in transgression, but a high rate of deposition can offset this tendency and cause progradation of the shoreline, or regression. Similarly, falling relative sea level usually results in regression, but excess of erosion over deposition can result in recession of the shoreline under conditions of significant falling of sea level.

nature of supply of sediments, intensity of the oceanographic processes which disperse, sort, cause, and prevent deposition of the sediments, width and configuration of the "shelf" or underlying surfaces, and rate and direction of relative sea level change. Ultimately, most of these factors can be attributed to tectonics, regional geology, climate, and eustatic sea level changes, which in turn are also interdependent.

To provide a basis for discussion, these factors can be simplified by grouping to the two most important; rate of deposition, and rate and direction of relative sea level change. These variables are plotted diagrammatically in Figure

10-1 to show their effects on lateral migration of the shoreline. Both variables are continuous functions, but are arbitrarily shown as divided into discrete classes.

Figure 10-1 diagrammatically shows the effects of change of relative sea level and net rate of deposition on lateral shoreline migration in a direction perpendicular to the shoreline. Rising relative sea level usually results in transgression, or landward migration of the shoreline. Conversely, falling relative sea level usually results in seaward migration of the shoreline, or regression. This process can, however, be reversed by the effects of deposition or erosion. With slowly rising relative sea level and a high enough rate of deposition, for example, the shoreline can prograde seaward as in a delta. Rapid erosion can similarly offset the tendency for regression caused by a very slowly falling sea level, and can produce a net transgression of the shoreline.

The diagonal line of Figure 10-1 thus shows the combination of factors which result in a geographically stationary shoreline. The regions of the graph labeled "transgression" and "regression" show diagrammatically the combinations of factors which result respectively in landward or seaward migration of the shoreline. Rate of transgression or regression will, in general, be proportional to the distance from this diagonal line. Maximum rate of transgression will then occur with rapid erosion and rapidly rising relative sea level. Maximum rate of regression will occur with rapidly falling relative sea level and rapid deposition.

Rate of deposition must be considered here as the local, instantaneous net rate at the shoreline, compounding the effects of rate of introduction of sediment into the basin, oceanographic conditions affecting the dispersion and deposition of the sediments, and erosion of previously deposited sediments. The amount of deposition is also dependent upon the rate of lateral migration of the shoreline. For example, with conditions of rapidly rising sea level and hence rapid transgression, the local instantaneous rate of deposition may be high but may result in little sediment deposition, because the environments of deposition are constantly and rapidly moving up slope. Under conditions of low rate of supply of sediment and moderate-to-intense oceanographic conditions, the net local result can be erosion rather than deposition. For this reason, then, the effects of rate of supply of sediment and intensity of oceanographic conditions can be compounded into one variable for the purposes of the following discussion.

The other variable, rate and direction of change of relative sea level, is fairly obvious: rising sea level or subsidence = transgression, falling sea level or emergence = regression. Here again, it is necessary to impose some restrictions, since it is desirable to combine several factors into one variable. It may not ordinarily be possible to distinguish between eustatic and tectonic causes in relative sea level change in ancient sedimentary records. The rather extreme

eustatic fluctuations of sea level associated with a glacial period are considered rare in the geologic record, and we know very little concerning eustatic changes of sea level due to other causes. Relative sea level change will suffice in most discussions, although locally – when something is known concerning the tectonism of a period – we may surmise that the relative sea level changes are caused by the uplift or subsidence of the land. Finally, local compaction of underlying sediments may cause subsidence of the coastline under consideration.

If the rate and direction of local change of sea level and the instantaneous local rate of deposition are known, they can be represented by a point on the graph (Figures 10-1 and 10-10). Another location on the same shoreline a few miles away may at the same time have quite a different local rate of subsidence and deposition, plotted on a different location on the graph. Similarly, if secular changes occur at any given place, the position of the point on the graph will also change. The history of transgression, regression, deposition, and erosion along the shoreline can thus be described by tracing the position of the point on the graph representing the changing conditions. This device will be used in the next section to describe some Holocene and Pleistocene sediments of the northwest Gulf of Mexico and Pacific Coast of California and Mexico.

Quaternary Examples of Transgression and Regression

Northwestern Gulf of Mexico

Excellent examples of various combinations of factors affecting transgressions and regressions can be drawn from the continental shelf and coastal plain of the northwest Gulf of Mexico because of the wide range of rates of deposition and because of the great concentration of information from this one depositional basin.

The chronology of the late Wisconsin regression and the Holocene transgression (Figure 10-2) is rather complex, because sea level fluctuated rather rapidly during parts of this period (Curray, 1960 and 1961; Shepard, 1960a). Sea level dropped from somewhere near its present level 25,000 to 30,000 years B.P. (Curray, 1961; Frye and Willman, 1961) to a maximum late-Wisconsin lowering of about 65 fathoms. The Holocene transgression from the maximum lowering started at about 20,000 B.P. as an overall transgression interrupted by several brief regressions. These brief regressions will not be considered in this discussion, though they had a profound effect on the details of the bathymetry, sediment, and faunal distributions. The general rate of transgression was rapid from about 20,000 to about 7000 B.P. (period I in the following discussion) and was slow or beginning to stabilize after that time (7,000

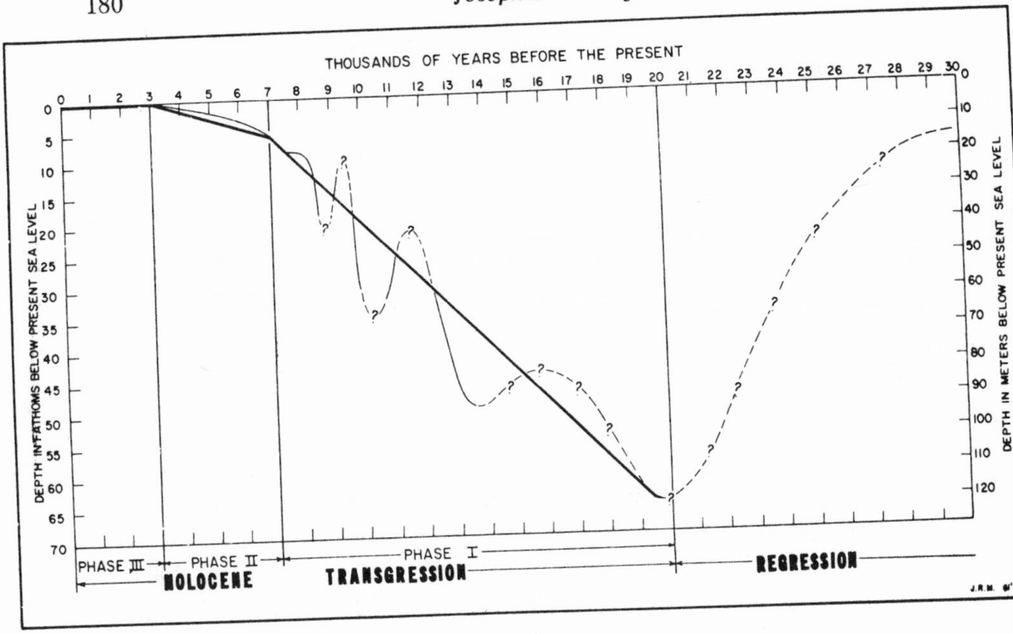

FIG. 10.2
Postulated curve of eustatic changes of sea level for the past 30,000 years. The dashed curve (Curray, 1961) shows possible fluctuations of sea level during the Holocene period of general transgression. For purposes of the present discussion, the rise of the last 20,000 years is generalized into three phases representing rapid rise, slow rise, and relative stability.

to 3,000 B.P., period II). Recent opinion is that sea level has been comparatively stable at its present level for the last 3000 years (period III) (Gould and McFarlan, 1959; Byrne, LeRoy, and Riley, 1959; and Jelgersma and Pannekoek, 1960), although this is not without controversy (Fairbridge, 1958, 1960, 1961; Shepard 1960a). Little is known of the rates of ancient transgressions, but it seems probable from the extremely long periods of time geologists attribute to the deposition of ancient sediments that these late Quaternary fluctuations must be extremely rapid. From this Holocene record, however, we can derive information on the results of "rapid" transgression, "slow" transgression, and relative stability by separate consideration of the products of these three periods of time (Figure 10-2): period I, 20,000 to 7000 B.P.; Period II, 7000 to 3000 B.P.; and Period III, 3000 B.P. to present.

Continental Shelf off Rockport, Texas

Rate of deposition has ranged from moderate to very low or zero on different parts of the continental shelf off Texas. Off Rockport, Texas (Figure 10-3), the rate of supply to the shelf has probably changed from moderate to low

during the last 20,000 years as the Pleistocene river valleys were drowned to become estuaries, which trap some of the sediment from the rivers. The section probably consists of basal transgressive sands overlying pretransgressive alluvium or bay deposits (Curray, 1960, p. 251). These basal transgressive sands, deposited as littoral, beach, dune, and nearshore sands, are in turn overlain by shelf facies muds deposited seaward of the nearshore sands. Bay deposits lie landward of the present barrier island (Figure 10-3). Although this section has been deposited continuously through the entire Holocene, it is important to identify at least by tentative interpretation the sediments deposited during each stage of the period from 20,000 B.P. to present. From about 20,000 to 7000 B.P., the conditions of deposition at the shoreline would plot at about point I in the graph of Figure 10-3. Rate of deposition was intermediate and rate of change of sea level rapid, resulting probably in a thin veneer of basal sand and a thin deposit of shelf muds seaward of the sands. The thicknesses are largely speculative because little information exists on sediment thicknesses on this shelf area (Curray, 1960, p. 251). Since the shelf

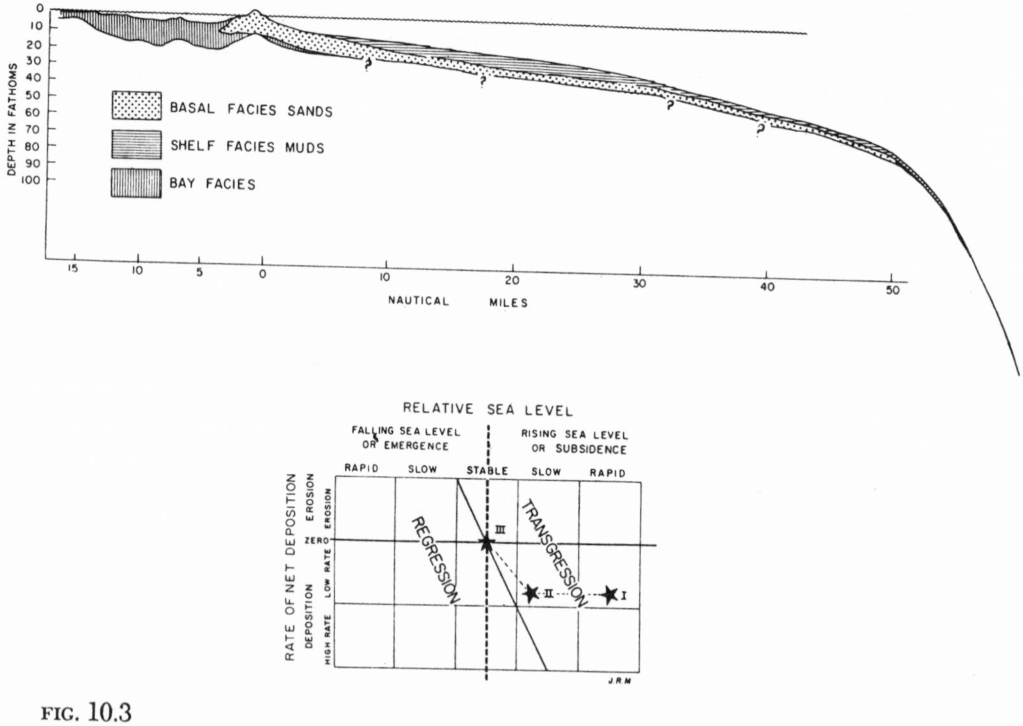

FIG. 10.3

Generalized section of late Quarternary sediments in a line perpendicular to the coast near Rockport, Texas, showing basal transgressive sandy sediments, shelf facies muds, and various bay sediments grouped as a bay facies. Sediment thicknesses on the shelf were deposited largely during transgression phases I and II, as were probably the bulk of the shelf facies muds. The barrier island, bay facies, and the inner shelf muds were probably deposited during phases II and III.

off central Texas is wide, the rise of sea level caused very rapid lateral migration of the shoreline. Early in this stage the marine muds were deposited on the upper continental slope as well as on the shelf.

Between about 7000 and 3000 B.P. (period II), when the rate of rise of sea level slowed, the conditions might be indicated by point II (Figure 10-3) with almost the same rate of deposition. During this period the barrier island was built upward (Shepard, 1956a; Parker, 1959; Shepard and Moore, 1955, 1960). The present shoreline (point III of Figure 10-3) is virtually stationary (Shepard and Moore, 1960), indicating little or no net deposition or erosion at the shoreline. Slow deposition is occurring farther out on the shelf where the shelf muds are gradually thickening; but no transgression or regression is occurring. If the present conditions continue for any length of time, a slow depositional regression will probably occur after the bays and estuaries have filled and more of the river sediment is carried directly into the gulf.

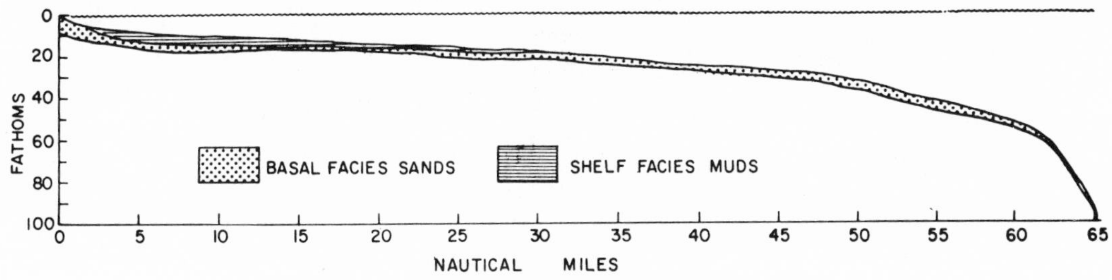

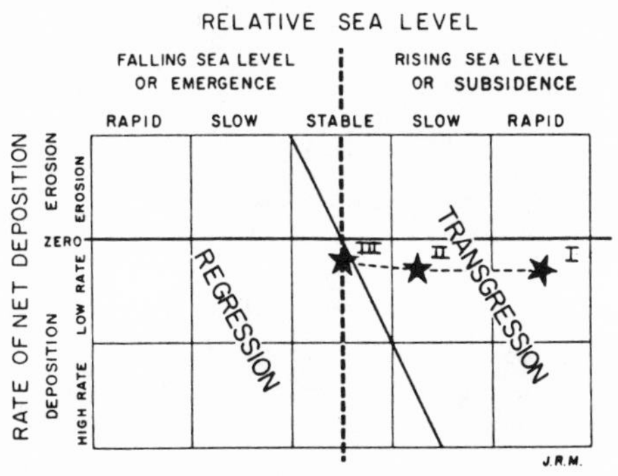

FIG. 10.4

Diagrammatic section of Holocene sediments in a line perpendicular to shore near Freeport, Texas. The basal transgressive sands were deposited during transgression phases I and II. The shelf facies muds were deposited during phases II and III, probably mainly during phase III. The coastline has prograded in depositional regression during the latter period, phase III, after filling of the Pleistocene channel of the rivers, the estuaries, and bays.

Continental Shelf off Freeport, Texas

The diagrammatic section in Figure 10-4 demonstrates the difference between the shelves off Rockport and Freeport, Texas. The rates of change of sea level were the same, but the rates of deposition have been different in these two areas. Rate of deposition along the shoreline was probably about the same during the transgression, but on the center shelf the rate of deposition was less off Freeport because of peculiarities of the current system distributing the shelf muds (Curray, 1960, pp. 231, 236, 264). As a result, the basal transgressive sands on the Freeport shelf are not as uniformly buried under shelf muds as on the Rockport shelf.

Rate of supply of sediment along the shoreline and on the inner shelf off Freeport is influenced by the fact that both the Colorado River and the Brazos River enter the gulf directly rather than deposit their sediments in bays as do the rivers near Rockport. Both river channels have been artificially altered during this century, though the Brazos had previously filled the bays near its mouth and prograded seaward as a deltaic bulge. The rate of supply of sediment is high enough that a slow depositional regression is in progress and shelf facies muds are being deposited on the inner 20 nautical miles of the shelf.

Mississippi Delta

Spectacular examples of the effects of high rate of supply on transgression and regression are demonstrated by the Mississippi Delta region of the Gulf of Mexico. Fisk and McFarlan (1955), Akers and Holck (1957), and McFarlan and Thomson (1957) have described the Quaternary sediments of the delta area by means of electric logs, cores, and cuttings, and have worked out the stratigraphy in terms of transgressive and regressive facies; and Shepard (1956b and 1960b), Shepard and Lankford (1959), and Scruton (1960) have discussed the stratigraphy of the Holocene sediments of portions of the delta. The rates of rise of sea level were the same as described previously, except for the effect of subsidence resulting from the compaction of the thick section of rapidly deposited underlying sediments. Rates of deposition have been extremely high in portions of the delta adjacent to active subdeltas, but with shifting of activity to other subdeltas the rate of deposition may become lower compared to the rate of subsidence resulting from compaction. This shifting results in the "destructive phase" sands of Scruton (1960) and Fisk (1955). As the old subdelta is lowered by compaction, wave action attacks the previously unsorted material and produces a sheet sand by winnowing. This can be called an erosional transgression, although it should be clearly understood that subsidence due to compaction aids the process of erosion.

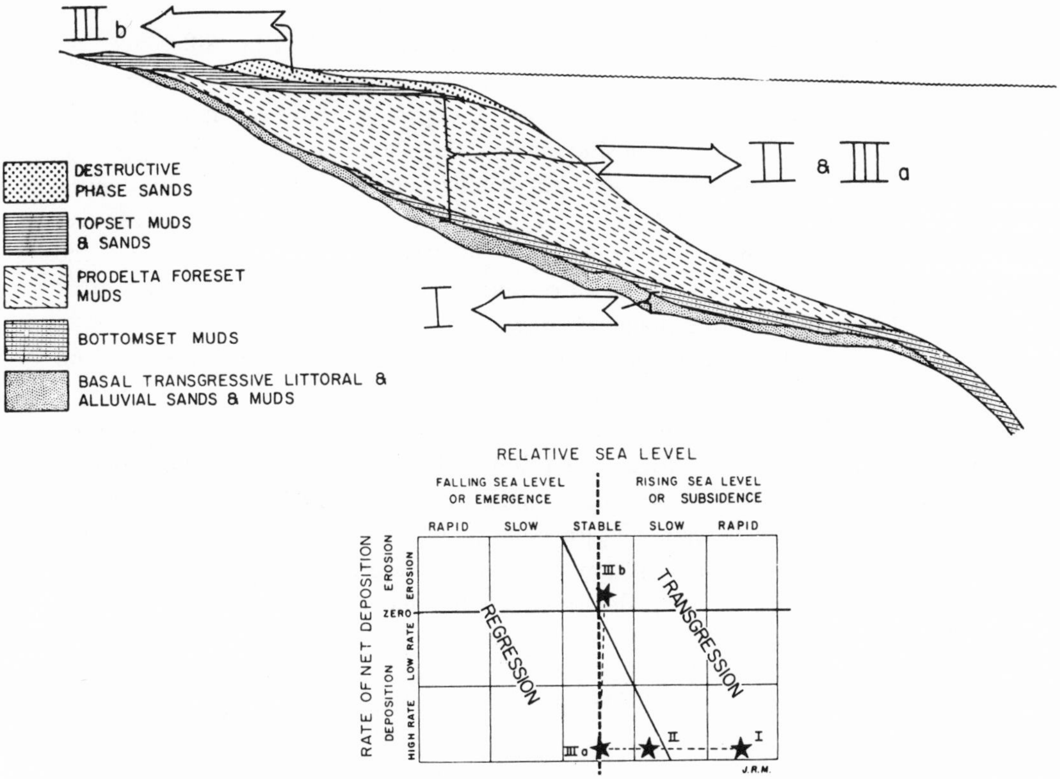

FIG. 10.5
Diagrammatic section through a hypothetical subdelta of the Mississippi delta. During phase I, basal transgressive deposits were emplaced, perhaps before a major distributory discharged toward this part of the shelf. During phases II and IIIa, a subdelta prograded across the shelf in depositional regression, by excess of sediment supply. During phase IIIb, this hypothetical subdelta was abandoned. Compaction and erosion produced a regression.

The Pleistocene alluvial deposits of the lower Mississippi Valley consist of a substratum of sandy deposits overlain by a topstratum of muddy deposits (Fisk, 1940, 1944, 1952, 1956). Each formation starts with a transgression across the weathered soils on top of the alluvium of the previous formation. Fisk and McFarlan (1955) have described the regressive deposits underlying the delta as primarily strandline sands deposited while the river channels were being deeply eroded across the shelf. These sands overlie offshore silts and clays in a regressive overlap. The strandline deposits, when preserved and not subsequently eroded away during lowered sea level, are covered by some alluvium and soil profiles. The marine transgression then moves across these weathered soils, forming a substratum of littoral sands and alluvial deposits, the base of

the new formation. These are in turn overlain by the top-set, fore-set, and possibly bottom-set beds of the delta (Figure 10-5).

Apparently, Akers and Holck (1957) have not separated the formations in exactly the same place. They correlated the regressions of the down-dip marine equivalents with the bulk of the substratum, and the transgressions with the upper substratum and topstratum. They included the basal sands of the transgression with the regressive sands described by Fisk and McFarlan, since their correlations were based primarily on electric logs in which it was impossible to recognize the transgressive unconformity.

A diagrammatic section through a composite, hypothetical Mississippi subdelta as described above is shown in Figure 10-5 (van Andel and Curray, 1960). Phase I was a period of eustatic transgression; phases II and IIIa were periods of depositional regression; and with abandonment of this hypothetical subdelta, phase IIIb was a period of erosional transgression. Local relative sea level was not constant during this last 3000 years because of local subsidence resulting from the compaction of the underlying sediments (Fisk and McFarlan, 1955; Curray and Shepard, 1959; and Shepard, 1960a).

Southwestern Louisiana Chenier Plain

The chenier plain of southwestern Louisiana presents an excellent example of the effect of the change of rate of deposition on lateral migration of the shoreline with stable sea level. This area, west of the Mississippi Delta, receives its sediment by westward transport from the river, or by later removal of deltaic deposits between the present delta and the chenier plain. Chenier (beach ridge) formation has been explained by alternation in conditions of supply of sediment from the Mississippi (Howe, et al., 1935; Russell and Howe, 1935; Fisk, 1948; Byrne, et al, 1959; and Gould and McFarlan, 1959). During periods of abundant sediment supply, the coast progrades by building of mudflats. During periods of reduced supply, wave action causes erosional transgression and formation of beach ridges of shell and sand.

Conditions of deposition during part of the period of Holocene geological history (Gould and McFarlan, 1959) are shown by the graph of Figure 10-6, and a diagrammatic sedimentary cross-section, also adapted from Gould and McFarlan, is shown in Figure 10-6. Conditions from about 5600 to 3000 B.P., during the period of slowly rising sea level, are shown by point II (Figure 10-6). The rate of deposition was probably intermediate, resulting in a slow transgression. Marsh and bay deposits (Figure 10-6) were laid down over the oxidized Pleistocene soils. Around 3000 B.P., the rise of sea level stopped and the shoreline started to prograde seaward in a depositional regression. Gould and McFarlan (1959, p. 3) suggest that progradation may even

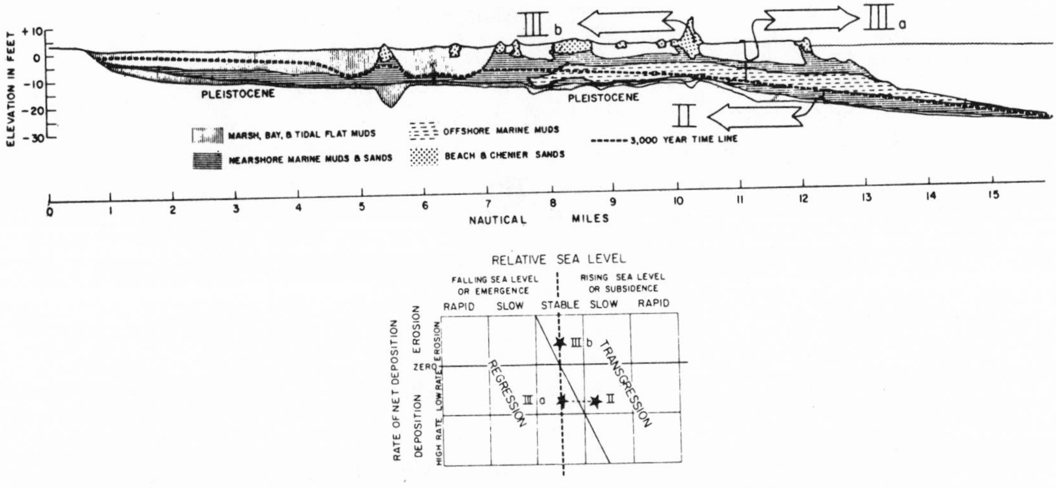

FIG. 10.6
Generalized section across the chenier plain of southwestern Louisiana (Byrne, et al, 1959; Gould and McFarlen, 1959). During phase II, transgression was slow. During phase III, the balance between erosion and deposition alternated with subdelta shifting of the Mississippi River. During periods of moderate rate of supply, depositional regression occurred as a result of the seaward prograding of tidal flats. During periods of reduced supply (IIIb), erosional regression produced beach ridges or cheniers of the sand and shell winnowed from the tidal flat sediments.

have started prior to this time while sea level was still rising, indicating a change in the balance between rate of rise and rate of supply of sediment.

Conditions from 3000 B.P. to present are indicated by two points, IIIa and IIIb. During times when supply of sediment was high, resulting from discharge of the Mississippi River in a more westerly direction, conditions were as represented by point IIIa: high rate of deposition and stable sea level. The result was sedimentary regression by the deposition of marsh-capped mudflats. Erosional transgression and chenier formation (point IIIb) resulted during alternate periods, when the Mississippi was discharging in a more easterly direction.

Pacific Coast of North America

The continental shelf off southern California presents a striking contrast to the shelf of the northwest Gulf of Mexico. The Gulf of Mexico is a broad open basin with a shelf width of from 54 to 130 nautical miles, bordering a wide coastal plain and a vast hinterland. The narrow southern California shelf, only one to thirteen miles wide, is located between a tectonically active mountainous coast and an offshore borderland of submerged basins (Emery, 1960). A given rate of rise or fall of sea level may result in a rapid migration

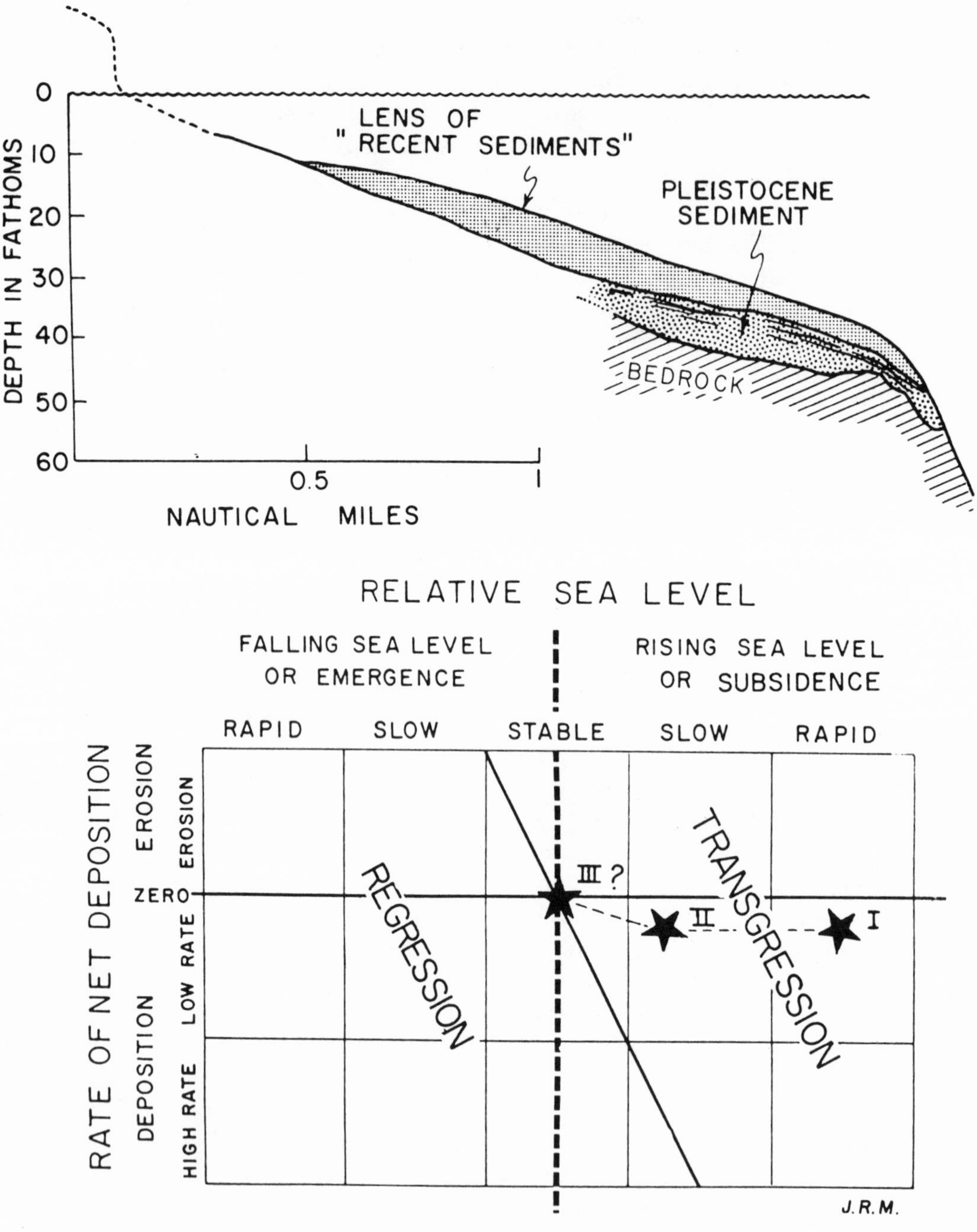

FIG. 10.7

Generalized section off Palos Verdes, California (Moore, 1960, pl. 3). Rate of deposition apparently decreased during phases II and III, or else subsequent erosion removed the deposits shoaler than 12 fathoms.

of the shoreline on a wide shelf, but only a slow lateral migration of the shoreline on a narrow shelf. The shoreline thus occupies a position on a narrow shelf for a longer period of time, resulting in more net deposition or erosion. A greater proportion or all of the width of a narrow shelf may, furthermore, be within the active zone of shelf deposition, while only perhaps the inner 20 nautical miles of a wide shelf may be an area of active deposition of significant rate.

Shelf width is a separate variable from those considered in the graph of Figure 10-1. The rate of rise of sea level refers to vertical rate rather than rate of lateral migration of the shoreline. Rate of deposition refers to instantaneous rate whereas total amount of deposition may be quite different for a rapid lateral migration than for a slow migration. The rather obvious effects of distance of shoreline migration should, therefore, be considered as a separate variable in attempting the interpretation of an ancient section.

Continental Shelf off Palos Verdes, California

The profile (Figure 10-7) off Palos Verdes, California (Moore, 1960, p. 3) shows the effect of a transgression across a very narrow shelf, approximately 1½ miles wide. The rate of supply of sediment was probably not very high at any time, but the lateral migration of the shoreline must have been slow even during the period of rapid rise of sea level. Sediment thicknesses are greater than in an area of the Gulf Coast with the same rate of deposition because: (1) the migration of the shoreline was slow, and (2) the shelf is narrow enough that all of it lies within the range of active deposition of sediment brought in from the coast. The thickest part of the wedge of sediment, identified by Moore as Recent, is in the middle of the shelf. The sediment thins at the edge of the shelf and pinches out at about 12 fathoms depth on the inner edge. Rate of deposition must have been higher during the early and middle periods of the transgression than during the last phase of 3000 years of stable sea level (Figures 10-2 and 10-7). The exact nature of the sediment units is unknown. The unit designated as Recent is probably shelf facies. It is acoustically homogeneous and shows no internal structure on Sonoprobe records. The unit designated as Pleistocene does show internal bedding, and may be of compound origin, consisting of transgressive basal littoral sands overlying lowered sea level alluvium (D. G. Moore, personal communication). Rate of supply and deposition must have been low during the last 12 fathoms of sea-level rise, or else sediment deposited there was subsequently removed and redeposited near the present shoreline.

Continental Shelf off La Jolla, California

In contrast to the Palos Verdes section, the La Jolla section (Figure 10-8)

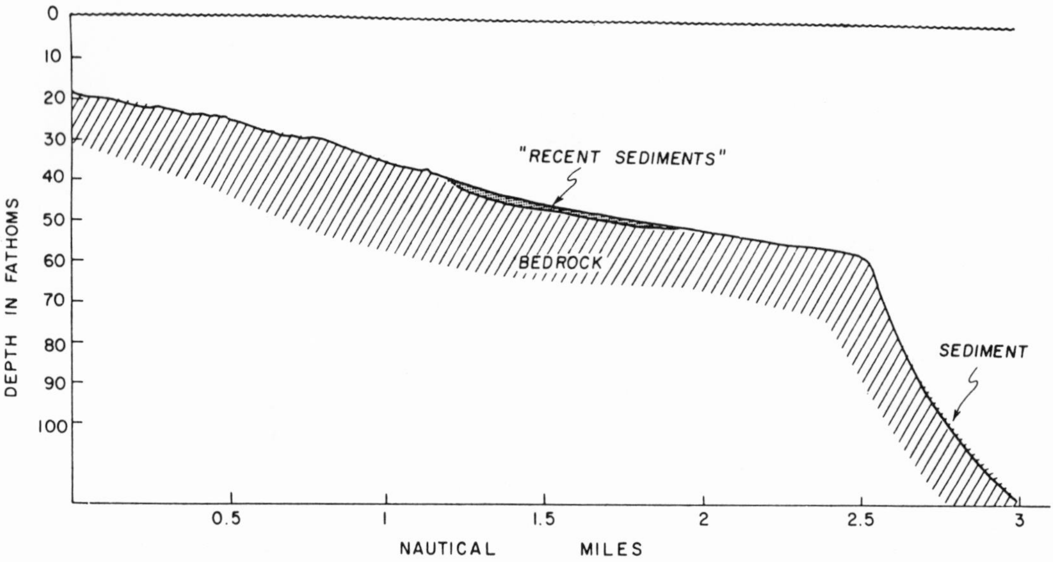

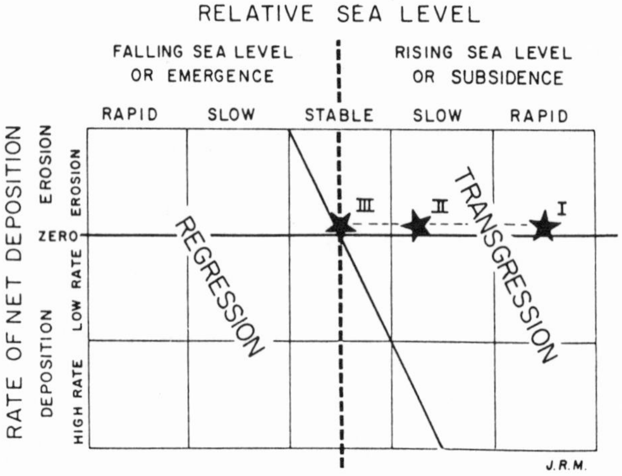

FIG. 10.8
Generalized section off La Jolla, California (Moore, 1960, pl. 1). Rate of supply of sediment has been low throughout the Holocene because of the submarine canyon lying to the north. Net erosion has probably occurred during the Holocene.

(Moore, 1960, pl. 1) shows the results of the rapid Holocene transgression with a low rate of supply of sediments. This profile was measured by Moore on a part of the La Jolla shelf where the sediment supply was derived primarily by local erosion from the country rock. Only a thin layer of sediment overlies a small part of the shelf; the remainder of the shelf consists of exposed bedrock. It is quite probable that erosion has exceeded deposition throughout the history of the Holocene transgression.

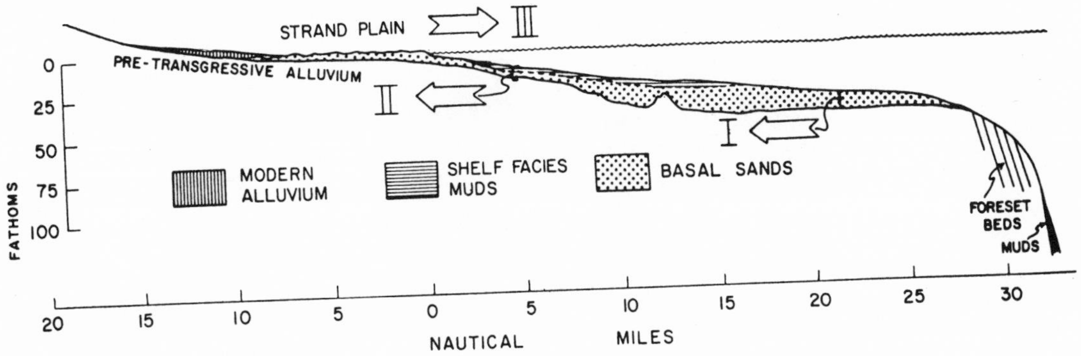

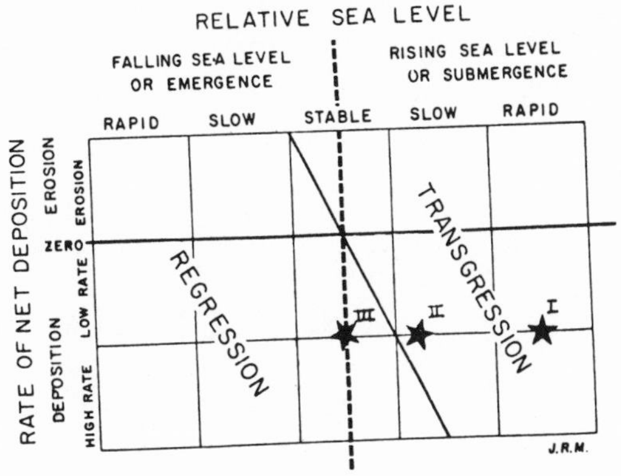

FIG. 10.9
Diagrammatic section off the Costa de Nayarit, Mexico. Rate of supply was fairly high during deposition of the thick wedge of probable basal sediments in phases I and II. During stable sea level (phase III), the coastline has prograded by deposition and successive accretion of off-shore bars to the shoreline. Younger alluvium has in turn started to cover some of these littoral sands.

Coast of Nayarit, Mexico

The coastal plain and continental shelf of the mainland coast of Mexico, between Mazatlan, Sinaloa, and San Blas, Nayarit, is currently under investigation by the writer and David G. Moore. This investigation has included bottom-penetrating sounding and sampling on the continental shelf, and auger and rotary drilling sampling on the coastal plain. Preliminary results and interpretations are shown diagrammatically in Figure 10-9.

The area consists of the late Pleistocene delta of the Rio Grande de Santiago. During the early Holocene rise of sea level (phase I), the surface of this delta was reworked by the transgression and partially buried by additional sediment contributed to the shelf from the river, producing a transgressive basal

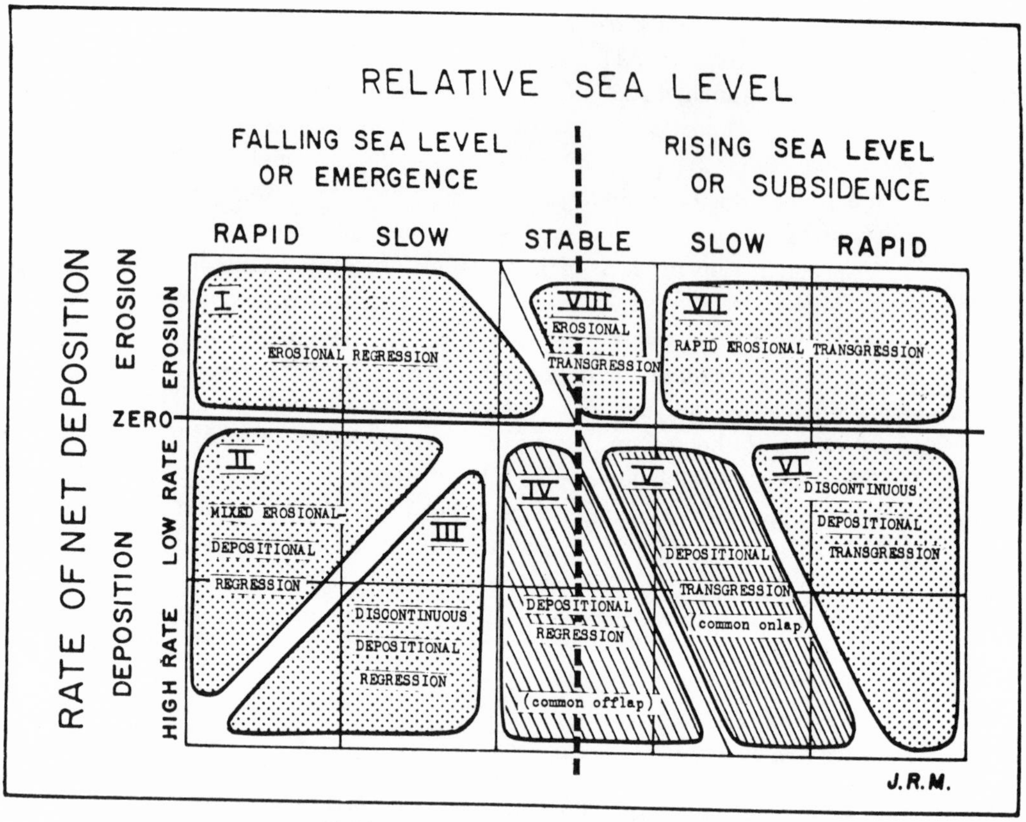

FIG. 10.10
Classification of transgressions and regressions.

sand overlying the old delta. During the slow rise of sea level (phase II), the transgression continued and basal sands were partially buried under shelf facies muds. At about the time when the rise of sea level ceased (Figure 10-9, point III), the rate of deposition balanced the rate of transgression caused by the rising sea level, the transgression ceased, and a progradation commenced in the form of seaward accretion of littoral sand as offshore bars. As each bar built upward, it captured the wave action, became a new beach, and isolated the former beach. Thus a depositional regression formed a strand plain, superficially resembling a chenier plain. During this depositional regression, seasonal flooding by rivers has started to spread a blanket of younger alluvium over the littoral sands by first filling the depressions between the beach ridges and, finally, by burying the beach ridges.

Discussion

Regressions

Most of the discussion thus far, although ostensibly devoted to both trans-

gressions and regressions, has in fact been devoted to transgressions. There are several reasons for this. The evidence in the surface and near-surface sidements of the continental shelves, beaches, bays, and lagoons of the world primarily concerns the Holocene transgression. Late-Wisconsin regressive sediments are buried under the cover of transgressive deposits and can be studied only by means of long cores in areas of thin cover, by drilling in areas of thick cover, or by chance sampling in the rare areas of exposure.

A second reason for lack of information on the late-Wisconsin regression is that regressive sequences, unless in areas of extensive alluviation or occurring on slow regressions, are at most thin and are furthermore subject to considerable erosion subsequent to deposition. Erosion may remove parts of a regressive sequence by gullying and cutting of river channels to adjust to the new low position of the sea and leave only a fragmentary record. The more rapid the regression and the lower the rate of supply of sediment, the higher the probability not only of nondeposition, but also of subsequent removal. This is probably true also of older regressive sequences, and it is suggested that the most common type preserved in the geological record is the "depositional regression" resulting from high rate of deposition with slow subsidence (Figure 10-10).

Relative Sea Level

The process of lateral migration of the shoreline is a result of a balance between two factors: direction and rate of change of relative sea level, and rate of deposition. These have been graphed diagrammatically in Figures 10-1 and 10-10. Change of relative sea level may be tectonic (local, regional, orogenic, or epeirogenic) or it may be caused by eustatic change of sea level through a net change in either the volume of the ocean basins or of the water filling the oceans. The Quaternary changes were eustatically controlled and modified locally by tectonics. Most changes throughout geologic time were probably tectonic, with the exception of the few periods of glaciation and the possible long-term changes in the volume of water and volume of the ocean basins. The latter were probably very slow with respect to tectonic uplift and subsidence. In general it is not possible to distinguish between the effects of these two causes of change in relative sea level, but tectonic effects are presumed to be appreciably more common.

In attempting to interpret ancient transgressions and regressions, we encounter the unknowns of rates of relative changes of sea level (subsidence and emergence) and rates of deposition. Without speculating further, suffice it to assume that most ancient transgressions and regressions for which a geological record is preserved were slow in terms of the classification used in this discussion. Some may even have been so slow as to compare with what is called

"stable" sea level in Figures 10-1 and 10-10. This assumption seems reasonable if it is indeed true that most ancient transgressions and regressions preserved in the geologic record were due to slow tectonic movements. Glacial eustatic changes probably represent an extreme in rate of change of sea level.

Modern Sedimentation

Deposition, as one of the two primary factors influencing transgression and regression, should be evaluated in terms of what is known of modern sedimentation. The continental shelves of today are not necessarily typical of the depositional environments of the past, but we can learn something of the mechanism of dispersion and deposition of sediments that might apply to ancient "equivalent" environments, whether they be epicontinental seas or geosynclines. An empirical model can be made of continental shelf sedimentation, and with suitable cautious re-evaluation and modification, this model can be applied to ancient "equivalent" environments.

Sediments of our modern continental shelves consist partly of modern sediments and partly of relict sediments (Emery, 1952), the latter not in equilibrium with their present environments. Most relict sediments are related to environments of the Holocene transgression, although some may be related to earlier Pleistocene conditions. Continental shelf sediments cannot, therefore, be indiscriminately considered as indicative of the present environment. Only the modern sediments can be so considered.

In an area with terrigenous sources consisting of both sands and muds (silts and clays), the sand and mud components are dispersed quite independently upon reaching the sea. The sands are generally deposited nearshore for dispersal by the longshore currents associated with wave action. Sands are transported in significant amounts to not over three miles from shore or ten fathoms depth off the Rhone delta (van Straaten, 1959), to about five fathoms and three miles from shore off the Texas coast (Shepard and Moore, 1955; Curray, 1960), and to two to ten fathoms, one to two miles offshore, depending upon exposure, around the southwestern part of Trinidad (van Andel and Postma, 1954). Exceptions surely exist, but these occur most commonly in minor amounts under unusual oceanographic conditions, or as results of slumping and/or turbidity currents in submarine canyons crossing continental shelves. Long-distance transport of sand by turbidity-current action on continental shelves is considered improbable except in steep-walled canyons terminating on the continental slopes and on other steep slopes rather atypical of depositional continental shelves.

The mud load of a river is dispersed independently of the sand, being kept in suspension either intermittently or continuously upon reaching the ocean. The mud is generally transported in the same direction as the sands but farther

from shore. The distance it is transported depends upon many factors, including the amount of load, type of clay, salinity and temperature change from river water to the sea, and turbulence. While the sands are deposited as a linear sand body parallel to shore, the muds are deposited as a blanket deposit seaward and down current of the sands.

Muds from the smaller rivers of the Gulf of Mexico are carried to a maximum of about 20 nautical miles from shore and are deposited in water less than about 15 fathoms deep. One significant exception is the mud load of the Mississippi River. The modern birdfoot delta has been built close to the edge of the continental shelf, where a vast amount of sediment is distributed to the edge of the shelf and at long distances parallel to the shelf edge by the semipermanent current system. Another exception in the northwest Gulf of Mexico is the area off Rockport, Texas (Figure 10-3). Due to a peculiarity of the current system, muds may be carried in small quantities to the middle or perhaps outer shelf for deposition. Despite these exceptions, however, the general rule of the area is probably that the rate of deposition is very low beyond about 20 nautical miles from shore.

Sedimentation during the Pleistocene was quite different, and the shoreline was near the present edge of the continental shelf. Deposition was directly on the continental slope rather than on the shelf, and slumps and turbidity currents distributed great volumes of sediment to the foot of the continental slope and to the deep ocean basins. Sediments remaining on these slopes today represent either old, compacted late Pleistocene or early Holocene sediments or else slowly deposited modern sediments carried across a narrow shelf. Slope sediments off the California coast have been shown by Moore (1960, 1961) to be stable. Slope sediments adjacent to a wide shelf such as the Gulf of Mexico are probably late-Wisconsin overlain by thin deposits of early transgressive muds laid down when the sources at the shoreline were still within less than 20 miles of the shelf edge (Greenman and Le Blance, 1956; Curray, 1960). Bypassing of sediment across a continental shelf to the slope is not considered by the writer to be of major significance, but may possibly apply locally on a small scale.

A striking contrast exists between the continental shelf of the northwest Gulf of Mexico and the southern California shelves. If the same model can apply, all of the California shelf width is within reach of muds from the coast. The model should perhaps be somewhat revised because the California intermittent rivers carry a much higher proportion of sand and lower proportion of mud than the Gulf coast rivers. Even with this modification, however, it is highly probable that the entire width of the southern California shelves is within reach of terrigenous muds, although the rate of mud deposition is slow. Depositions differ in many places because of the effects of the numerous submarine canyons which drain sediment off the shelves into the borderland basins (Shepard, 1951; Emery, 1960).

The shelf on the west coast of mainland Mexico is about 35 miles wide; therefore, the outer edge is not entirely within the reach of the terrigenous muds from shore. This is confirmed by preliminary results of sediment distribution studies which show sands on the outer shelves containing mixed faunas representing both shallow water and shelf-edge depths (Robert H. Parker, personal communication). The sands and shallow-water macroorganisms are relict from previous shoreline conditions and the shelf deeper-water faunas have been introduced under present conditions. A radiocarbon date (Curray, 1961) from predominately intertidal shells (Parker, personal communication) dredged at 62½ fathoms south of Mazatlan, Sinaloa, was 19,300 ±300 years B.P., showing that sea level was at approximately this level on the shelf during the late-Wisconsin glacial maximum.

If sea level would remain in its present position for a long period in the future, many of the coastlines of the world would start to prograde seaward in depositional regression. Marine deposition rate is low along many of our coasts because river sediment is deposited in estuaries and bays rather than directly into the ocean. These features, inherited from deeply eroded Pleistocene channels across the shelves, have not yet been filled and brought into a state of equilibrium with present conditions. After they are filled, the supply of sediment to the continental shelves would be increased, but deposition would still be confined to the inner parts of wide shelves. With sufficient time (short geologically, but very long by Pleistocene standards), many of these coasts would prograde and eventually cover the relict sediment with modern sediments in equilibrium with the environment. Possible deterrents to this eventuality are the probability of further glacial-controlled eustatic changes of sea level and the effect of the works of man. We are entrapping more and more of the water and sediments of rivers by our great demands for water supply. This decrease in sediment supply to the estuaries, bays, and oceans will greatly lengthen the time necessary to fill them and may eventually lead to starvation of some of our coastal beaches.

Application to Ancient Sedimentation

By the doctrine of uniformitarianism, it has long been assumed that the present is the key to the past. This concept must be applied with caution and with the restriction that though we believe that geological processes and principles are the same now as throughout geological time, we do not believe that they are necessarily occurring with the same rates and intensities. As a consequence of having inherited relatively large continental masses of great relief, deep ocean basins, and a glacial epoch, we now have high rates of erosion, transportation, supply, and deposition, and perhaps stronger winds and more intense oceanographic conditions than the average throughout geological time.

The processes are nevertheless the same, and, by suitable modification, the results of modern sedimentation studies can be applied to the interpretation of ancient sediments.

Many ancient marine sediments have been considered in two general classes: intracratonic or epeiric, and extracratonic or orthogeosynclinal. In consideration of ancient transgressions and regressions, we are concerned with the epeiric seas and the miogeosynclinal part of the orthogeosynclinal belt (Kay, 1951). Deposition presumably took place in a manner similar to modern sedimentation on continental shelves. Sand was deposited as linear sand bodies near shore, and muds (marine shales) were deposited in blanket-shaped deposits farther from shore. The rate of deposition of terrigenous sediments was probably negligible at a distance from shore exceeding about 20 miles, although this distance might vary by a factor of even two or three under some conditions.

Environmental conditions in epicontinental seas of the past probably differed from conditions on modern continental shelves. Dimensions of the basins were smaller than those of the present oceans, and fetch for generation of waves was accordingly less. Short fetch would have limited the waves to relatively short periods and therefore to shallow effective bottom surge. Tidal currents could have been high in inland seas with open connections to the ocean, but would probably have been low in enclosed basins. The zonation of sediments might, therefore, have been narrower and closer to the shore than on our present shelves. Widespread sheet sands, for example, had to be formed by lateral shifting of the shoreline back and forth across the area. Muds must either have been absent or were removed to elsewhere, or else the water must have been very shallow, much less than perhaps 20 to 30 feet, to maintain sufficient continuous turbulence to prevent their deposition and mixing with the sands.

Ancient orthogeosynclines perhaps more nearly resemble the continental terraces of today (Drake et al., 1959). Miogeosynclines are represented by the continental shelves and coastal plains, and the adjacent eugeosyncline by the deep sea at the base of the slope or perhaps in some cases by the adjacent trenches. The transgressions and regressions occurred back and forth across the shelf or miogeosyncline. Following a transgression, supply of sediment must have been low to the eugeosyncline; but conversely, following a regression, sediment was deposited directly on the slope, whence it slumped to flow into the eugeosyncline as turbidity currents. Sufficiently steep slopes to generate turbidity currents probably existed only on these "continental slopes," and sufficient instability in the sediment probably existed only during times when the shorelines lay close enough that active deposition took place on them.

The significant point to the analogy between modern and ancient sedimentation is simply confirmation of the hypothesis that the formation of widespread deposits of either sand or mud took place by migration of the shoreline rather than simultaneous deposition of the entire sediment body. This is par-

ticularly true in the case of the sheet sands of ancient sediments. Holocene counterparts may exist to some of these on the continental shelves in areas of nondeposition and in the destructive-phase sands of large compound deltas, but we know that these sands were formed during periods of shoreline migration. The continental-shelf sheet sands were formed as basal transgressive sands during the Holocene transgression by deposition of sands in a narrow band paralleling the shoreline. Destructive-phase sands of the Mississippi Delta were formed during erosional transgressions over the surface of abandoned subdeltas.

Whereas the direction and speed of migration of the shoreline is dependent upon the rate of deposition along the shoreline, the fate of the sediments may be dependent upon the rate of deposition at a distance from the shoreline. For example, with rising sea level and transgression, clean littoral sands may be better preserved as such only if there is a significant rate of shelf deposition of muds at a distance from the shoreline (Figure 10-3). In this case the sands will be rapidly buried, protected, and preserved. If the rate of deposition of shelf muds is negligible (Figure 10-4), the originally clean littoral sands will be reworked and the small amounts of mud brought to the area will be mixed with the sands (Curray 1960, p. 236, 240).

Classification of Transgressions and Regressions

By enlarging upon the diagrammatic representation of Figure 10-1 and by definition of arbitary groupings, transgressions and regressions can be classified (Figure 10-10). The groups will be outlined and discussed briefly in the following section, partly on the basis of Quaternary sediments, partly on the basis of the literature on ancient sediments, and partly by speculation and extrapolation from both.

Erosional Regression (Figure 10-10)

Previously called "emergence without offlap" (Dunbar and Rodgers, 1957), this type of emergence is caused by rapid uplift or relative lowering of sea level in an area of low rate of deposition. The newly exposed former sea floor or marine sediments are subjected to erosion and gullying. To be a true example of emergence with erosion, no littoral deposits should be present marking the former shorelines. Erosional regression is common in tectonically active areas, but is probably uncommon in the geologic record, partly because of difficulty of recognition and partly because it occurs in uplifting regions rather than in subsiding areas in which sediments will tend to accumulate and be preserved. Examples: Uplifted Pleistocene terraces without littoral deposits are common in many parts of the world.

Mixed Erosional and Depositional Regression

This is one of the transitional groups between *depositional regression* and *erosional regression.* This particular group occurs in the field of the graph suggesting predominance of deposition over erosion at the shoreline – i.e. deposition of at least regressive strand-line deposits. It differs from the next group, *discontinuous depositional regression,* in the balance between rate of deposition and rate of change of sea level. Regression occurs at a much higher rate than deposition. Thus the probability of subsequent removal by subaerial erosion is rather great and only a fragmentary record will be preserved if any. Examples: Some of the southern California Pleistocene raised terraces which show well-developed strand-line deposits (Emery, 1950). Possibly some of the Texas continental shelf with low rate of supply of sediment during the late-Wisconsin regression.

Discontinuous Depositional Regression

This is another transitional group, but somewhat farther along the line than the previous stage. This occurs primarily with high rate of deposition and slow regression. The deposits that result are the expected regressive sequency, but because of the rate of change of sea level involved, thickness is not great and the ultimate fate of the deposits will depend upon events subsequent to the emergence. With continued uplift the exposure, complete removal by erosion can result. With transgression following immediately, burial can protect the regressive sequence. Example: Late-Wisconsin regressive deposits under the present Mississippi Delta (Fisk and McFarlan, 1955; Akers and Holck, 1957).

Depositional Regression

Called "continental replacement" (Dunbar and Rodgers, 1957, p. 142), this is the classic regressive sequence or offlap of ancient sediments. It is most probable that this process occurs under conditions of relatively stable sea level with moderate-to-high rate of deposition. The probability of preservation is improved by slow subsidence and subsequent burial. Regression occurs as a result of seaward progradation of the shoreline by deposition, most commonly in a delta. Examples: The Mississippi Delta (Figure 10-5), the chenier plain of southwestern Louisiana (Figure 10-6), the coastal plain south of Mazatlan, Mexico (Figure 10-9), and many others. Most ancient examples of recognized regression belong to this group – e.g. the Catskill Delta (Chadwick, 1924, 1933; Cooper, 1930, 1933).

Depositional Transgression

This is the classic transgression (marine onlap) of ancient sediments caused

by dominance of subsidence over deposition. Most transgressive sequences recognized in the geological column probably belong in this group, with conditions of slow subsidence and moderate-to-high rate of deposition. The rate of subsidence may have been slow enough to belong in the arbitary "stable" category of the diagram, hence requiring only a moderate-to-low rate of deposition. Most transgressive-regressive alternations in the geological column occur in areas of slow subsidence. Periodic slowdown of subsidence or increase in rate of terrigenous sediment supply reverses the transgression to cause depositional regression. Examples: The Mississippi Delta (Figure 10-5), the Palos Verdes shelf (Figure 10-7), and the Rockport, Texas, shelf (Figure 3), which is a poor example. Ancient examples are numerous, such as the Mesaverde group in east-central Utah and west-central Colorado (Spieker, 1949), and the Cambrian formations in the Grand Canyon (McKee, 1945). These ancient examples show the alternations between depositional regression and marine onlap resulting from shifts in the balance between subsidence and deposition.

Discontinuous Depositional Transgression

This group grades from the previous group by either increase of rate of transgression or by decrease of rate of deposition. It is distinguished by the discontinuous nature of the sediment units, made so by the inability of deposition to keep pace with the rise of relative sea level. The usual result is a thin veneer of strand line or littoral sands spread over the underlying surface, with discontinuous deposits of shelf facies or marine muds (shales) overlying it. Lagoonal deposits may develop locally behind barrier-island sands. With sufficient supply, the barrier growth may for a while keep pace with the rise of sea level (or subsidence), but eventually the enlarged lagoon or estuary formed by the rise of the sea will trap too much of the river-derived sediment. The barrier is then overtopped and the lagoon becomes an open sound and eventually open ocean. The shoreline moves by steps in this way, adding to the discontinuous nature of the distribution of the transgressive sediment. The shelf facies muds (or marine shales) seaward of these littoral basal sands may be even more discontinuous because the source may move too rapidly to keep pace with deposition of a continuous blanket deposit and because the muds are deposited in lagoons and estuaries during part of the transgression. Examples: The shelf off Galveston, Texas, offers an excellent example (Figure 10-4). Details of the distribution of sediment on this part of the shelf (Curray, 1960) show the features discussed above: strand-line features exposed on the shelf are not yet buried under the shelf facies muds. Lagoonal shells are exposed in places at the surface because of the stepwise migration of the shoreline (Parker, 1960).

This type of transgression may not be very common in recognized examples

of ancient transgression, but it is very common on the continental shelves of today. The Atlantic Coast continental shelf of the United States is presumably of this type, with basal transgressive sands and relict shoreline ridges still exposed, not yet buried under shelf facies muds (Stetson, 1938). Many of the other continental shelves of the world are probably the result of this type of transgression.

Rapid Erosional Transgression

If marine deposition commences after subsidence or rapid rise of sea level, the resulting stratigraphic sequence can be called "overstep" (Swain, 1949; Dunbar and Rodgers, 1957). The requirement of the definition of this group is speed of submergence or rise of sea level so no littoral deposition occurs between the time of subaerial erosion and the time at which the area is in the marine environment. With subsequent deposition of marine facies, an unconformity will exist directly under the marine facies without facies change showing gradation in water depth.

The La Jolla profile (Figure 10-8) may be an incipient example, although no significant deposition of a marine facies has yet occurred. Dunbar and Rodgers (1957, p. 157) consider the Ordivician limestone overlying Pre-Cambrian rocks on the south margin of the Canadian shield to be an overstep.

Erosional Transgression

What has been previously discussed as *erosional transgression* is primarily due to the process of erosion under conditions of relative stability and negligible sediment supply. Wave-action intensity is great enough that net erosion and landward migration of the shoreline take place rather than shoreline stability or prograding. It is conceivable that this could occur even under conditions of falling sea level, if the erosion were rapid enough. As recognized in Quaternary deposits, however, the process of erosion is frequently aided by compaction of the underlying sediments to cause slow local subsidence. The term erosional transgression is nevertheless reserved for this phenomenon, because this occurs usually in regions of alternation between high and low rates of deposition. Compaction and subsidence aid the erosion, but the erosion and winnowing of previously unsorted, rapidly dumped sediments produces facies characteristic of the "destructional phase" of delta development. Such destructional facies are characteristic primarily of deltas building into an ocean, where the oceanographic conditions permit subdelta shifting, such as the Mississippi Delta (Scruton, 1960; Fisk, 1955). Examples: The Mississippi Delta (Figure 10-5) and the southwestern Louisiana chenier plain (Figure 10-6).

References

Akers, W. H. and Holck, A. J. J. (1957), "Pleistocene Beds Near the Edge of the Continental Shelf, Southeastern Louisiana": *Geol. Soc. America Bull.*, **68**, p. 983-992.

Byrne, J. V., LeRoy, D. O., and Riley, C. M. (1959), "The Chenier Plain and Its Stratigraphy, Southwestern Louisiana": *Trans. Gulf Coast Assoc. Geol. Societies*, **IX**, pp. 1-23.

Chadwick, G. H. (1924), "The Stratigraphy of the Chemung Group in Western New York": *New York State Museum Bull. 251*, pp. 149-157.

——— (1933), "Great Catskill Delta: and Revision of Late Devonic Succession (Upper Devonian Revision in New York and Pennsylvania)": *Pan-Am. Geologist*, **60**, pp. 91-107, 189-204, 275-286, and 348-360. (The editor severely mutilated this paper, changing the title and making more than 1000 other alterations; Chadwick was forced to deny responsibility for the paper and to distribute a list of the most important changes.)

Cooper, G. A. (1930), "Stratigraphy of the Hamilton Group in New York": *American Jour. Sci.*, 5th ser., **19**, pp. 116-134 and 214-236.

——— (1933), "Stratigraphy of the Hamilton Group of Eastern New York": *American Jour. Sci*, 5th ser., **26**, pp. 537-551; **27**, pp. 1-12.

Curray, J. R. (1960), "Sediments and History of the Holocene Transgression, Continental Shelf, Northern Gulf of Mexico," in Shepard et al., *Recent Sediments, Northwest Gulf of Mexico*, Am. Assoc. Petroleum Geologists, Tulsa, Okla., pp. 221-266.

——— (1961), "Late Quaternary Sea Level: a Discussion": *Geol. Soc. America Bull.*, **72**, pp. 1707-1712.

——— and Shepard, F. P. (1959), *Sea-level Rise Along the Texas Coast*, International Oceanographic Congress preprints: Washington, D. C., Am. Assoc. for the Advancement of Sci., pp. 609-610.

Drake, C. L., Ewing, M. and Sutton, G. H. (1959), "Continental Margins and Geosynclines: the East Coast of North America North of Cape Hatteras": *Physics and Chemistry of the Earth*, **3**, p. 110-198.

Dunbar, C. O. and Rodgers, J. (1957), *Principles of Stratigraphy:* New York, John Wiley & Sons, Inc., 356 pp.

Emery, K. O. (1950), "Ironstone Concretions and Beach Ridges of San Diego County, California": *Jour. Calif. Mines and Geology*, **46**, pp. 213-221.

——— (1952), "Continental Shelf Sediments of Southern California": *Geol. Soc. America Bull.*, **63**, pp. 1105-1108.

——— (1960), *The Sea off Southern California:* New York, John Wiley & Sons, Inc., 366 pp.

Fairbridge, R. W. (1958), *Dating the Latest Movements of the Quaternary Sea Level:* New York Acad. Sciences, ser. II, **20**, p. 471-482.

——— (1960), "The changing Level of the Sea": *Scientific American*, **202**, p. 70-79.

——— (1961), "Eustatic Changes in Sea Level": *Physics and Chemistry of the Earth*, **4**, pp. 99-185.

Fisk, H. N. (1940), "Geology of Avoyelles and Rapides Parishes, Louisiana": *Dept. Conserv. Geol. Bull.*, 18, pp. 32-40.

——— (1944), *Geological Investigations of the Alluvial Valley of the Lower Mississippi River:* Vicksburg, Mississippi River Commission, pp. 1-78.

——— (1948), *Geological Investigation of the Lower Mermentau River Basin and Adjacent Areas in Coastal Louisiana*, Appendix II, "Geology of Lower Mermentau River Basin": Dept. of Army Corps of Engineers, Definite project report, 41 pp.

——— (1952), *Geological Investigation of the Atchafalaya Basin and the Problems of Mississippi River Diversion:* Vicksburg, Mississippi River Commission, pp. 1-145.

——— (1955), "Sand Facies of Recent Mississippi Delta Deposits": *Proc. 4th World Petrol Congress,* sect. I/C, preprint 3, pp. 1-20.

——— (1956), "Nearsurface Sediments of the Continental Shelf Off Louisiana": *Proc.* 8th Texas Conference Soil Mech. and Found. Engin., Austin, 23 pp.

——— and McFarlan, E., Jr. (1955), "Late Quaternary Deltaic Deposits of the Mississippi River," in *Crust of the Earth,* Geol. Soc. America Special Paper 62, pp. 279-302.

Frye, J. C. and Willman, H. B. (1961), "Continental Glaciation in Relation to McFarlan's Sea-level Curves for Louisiana": *Geol. Soc. America Bull.,* **72**, pp. 991-992.

Gould, H. R. and McFarlan, E., Jr., (1959), "Geologic History of the Chenier Plain, Southwestern Louisiana": *Trans. Gulf Coast Assoc. Geol. Soc,* **IX**, pp. 1-10.

Grabau, A. W. (1906), "Types of Sedimentary Overlap": *Geol. Soc. America Bull.,* **17**, p. 567-636.

Greenman, N. N. and Le Blanc, R. J. (1956), "Recent Marine Sediments and Environments of Northwest Gulf of Mexico": *Am. Assoc. Petroleum Geologists Bull.,* **40**, pp. 813-847.

Howe, H. V., Russell, R. J., McGuirt, J. H., Craft, B. B., and Stephenson, M. B. (1935), *Reports on the Geology of Cameron and Vermillion Parishes:* Louisiana Geol. Survey, Geol. Bull. No. 6, 242 pp.

Jelgersma, S. and Pannekoek, A. J. (1960), "Post-Glacial Rise of Sea-Level in the Netherlands (a preliminary report)": *Geologie en Mijnbouw,* 39e Jaargand, NR 6, pp. 201-207.

Kay, M. (1951), *North American Geosynclines:* Geol. Soc. America Mem. 48, 143 pp.

Lahee, F. H. (1949), "Overlap and Non-Conformity": *Am. Assoc. Petroleum Geologists Bull.,* **33**, p. 1901.

Le Blanc, R. J. and Hodgson, W. D. (1959), "Origin and Development of the Texas Shoreline": *Trans. Gulf Coast Assoc. Societies,* **IX**, pp. 197-220.

Lovely, H. R. (1948), "Onlap and Strike-Overlap": *Am. Assoc. Petroleum Geologists Bull.,* **32**, pp. 2295-2297.

McFarlan, E., Jr. and Thomson, M. R. (1957), "Subsurface Quaternary Stratigraphy in Coastal Louisiana and Adjacent Continental Shelf": *Am. Assoc. Petroleum Geologists 31st. Annual Meeting, St. Louis, Mo.,* pp. 28-29.

McKee, E. D. (1945), *Cambrian History of the Grand Canyon Region, Part 1. Stratigraphy and Ecology of the Grand Canyon Cambrian:* Carnegie Institute, Washington, Publ. 563, pp. 3-168.

Melton, F. A. (1947), "Onlap and Strike-Overlap": *Am. Assoc. Petroleum Geologists Bull.,* **31**, pp. 1868-1878.

Moore, D. G. (1960), "Acoustic-Reflection Studies of the Continental Shelf and Slope Off Southern California": *Geol. Soc. America Bull.,* **71**, pp. 1121-1136.

———(1961), "Submarine Slumps": *Jour. Sed. Petrology,* **31**, pp. 343-357.

Parker, R. H. (1959), "Macro-Invertebrate Assemblages of Central Texas Coastal Bays and Laguna Madre": *Am. Assoc. Petroleum Geologists Bull.,* **43**, pp. 2100-2166.

——— (1960), "Ecology and Distributional Patterns of Marine Macro-Invertebrates, Northern Gulf of Mexico," in Shepard et al., *Recent Sediments, Northwestern Gulf of Mexico:* Am. Assoc. Petroleum Geologists, pp. 302-337.

Russell, R. J. and Howe, H. V. (1935), "Cheniers of Southwestern Louisiana": *Geog. Review,* **25**, pp. 229-261.

Scruton, P. C. (1960), "Delta Building and the Deltaic Sequence," in Shepard et al., *Recent Sediments, Northwestern Gulf of Mexico:* Am. Assoc. Petroleum Geologists, pp. 82-102.

Shepard, F. P. (1951), *Transportation of Sand into Deep Water: in Turbidity Currents and the Transportation of Coarse Sediments to Deep Water, A Symposium:* Tulsa, Okla., Soc. Economic Paleontologist and Mineralogists, spec. publ. No. 2, pp. 53-65.

——— (1956a), "Late Pleistocene and Recent History of the Central Texas Coast": *Jour. Geology,* **64**, pp. 56-69.

——— (1956b), "Marginal Sediments of Mississippi Delta": *Am. Assoc. Petroleum Geologists Bull.,* **40**, pp. 2537-2623.

——— (1960a), "Rise of Sea Level Along Northwest Gulf of Mexico," in Shepard et al., *Recent Sediments, Northwestern Gulf of Mexico:* Am. Assoc. Petroleum Geologists, pp. 338-344.

——— (1960b), "Mississippi Delta: Marginal Environments, Sediments, and Growth," in Shepard et al., *Recent Sediments, Northwestern Gulf of Mexico:* Am. Assoc. Petroleum Geologists, p. 56-81.

——— and Lankford, R. R. (1959), "Sedimentary Facies from Shallow Borings in Lower Mississippi Delta": *Am. Assoc. Petroleum Geologists Bull.,* **43**, pp. 2051-2067.

——— and Moore, D. G. (1955), "Central Texas Coast Sedimentation: Characteristics of Sedimentary Environment, Recent History and Diagenesis": *Am. Assoc. Petroleum Geologists Bull.,* **39**, pp. 1463-1593.

——— and ——— (1960), "Bays of Central Texas Coast," in Shepard et al, *Recent Sediment, Northwestern Gulf of Mexico:* Am. Assoc. Petroleum Geologists, Spec. Publication, pp. 117-152.

Spieker, E. M. (1949), *Sedimentary Facies and Associated Diastrophism in the Upper Cretaceous of Central and Eastern Utah:* Geol. Soc. America Mem. 39, pp. 55-82.

Stetson, H. C. (1938), "The Sediments of the Continental Shelf Off the Eastern Coast of the United States": *Papers in Physical Oceanography and Meteorology,* Massachusetts Inst. of Tech. and Woods Hole Oceanographic Inst., **5**, 4.

Swain, F. (1949), "Onlap, Offlap, Overstep, and Overlap": *Am. Assoc. Petroleum Geologists Bull.,* **33**, pp. 634-636.

Van Andel, TJ. H. and Curray, J. R. (1960), "Regional Aspects of Modern Sedimentation in Northern Gulf of Mexico and Similar Basins, and Paleogeographic Significance," in Shepard et al., *Recent Sediments, Northwestern Gulf of Mexico:* Am. Assoc. Petroleum Geologists, pp. 345-364.

Van Andel, TJ. H., Postma, H. et al (1954), "Recent Sediments of the Gulf of Paria," in *Reports of the Orinoco Shelf Expedition,* **I**, Kon. Nederl. Akad. Wetensch. Verh., **20**.

Van Straaten, L.M.J.U. (1959), "Littoral and Submarine Morphology of the Rhone Delta": *Second Coastal Geography Conf., Louisiana State Univ.,* April, pp. 233-264.

9

Reprinted from *Bull. Georgia Acad. Sci.* **22**(1):37–40 (1964)

THE RELICT-RECENT SEDIMENT BOUNDARY ON THE GEORGIA CONTINENTAL SHELF

ORRIN H. PILKEY and DIRK FRANKENBERG
The University of Georgia Marine Institute[1]

As a result of a recent bottom sampling program carried out on the Georgia continental shelf, it has been possible to delineate the boundary between relict or Pleistocene sediments and Recent sediments. The exact location of this boundary is of interest because its configuration gives clues to important sediment sources as well as to local current patterns.

Previous studies of the Atlantic continental shelf of the Southern United States have indicated the presence of a relatively narrow, near-shore band of fine sand and silty sediment flanked to the seaward by coarser sands (Stetson, 1938; Gorsline, 1963). Gorsline (1963) considers this band of fine-grained sediment to represent the extent of Recent sedimentation and the coarser sands to represent relict or Pleistocene sediments. If this hypothesis is correct, considerably less than one-fourth of the Georgia continental shelf is currently being subjected to sedimentation, and the sediments covering the remainder of the shelf were deposited during the ice ages at times of lowered sea level.

Methods

Bottom sediment samples were obtained with a Pierce box dredge operated from the University of Georgia research vessel. Eight east-west transects were made consisting of samples collected every 1 to 4 miles (Figure 1A). The median grain sizes of both the total sample and acid insoluble residue of each sample were determined by standard techniques.

Results

Recent sediments along the Georgia coast are dominantly fine grained sands with some silts and very fine sands. These include the modern beaches and beach ridges. It is probable that much of the fine material, particularly silt and clay, presently being contributed to the area is either being trapped in salt marshes behind the barrier island system or is being carried off the continental shelf altogether. The Pleistocene or relict sediment is medium to coarse sand.

[1]Contribution No. 63 from the Sapelo Island Research Laboratory. This research was supported by the National Science Foundation under Grant No. GB871.

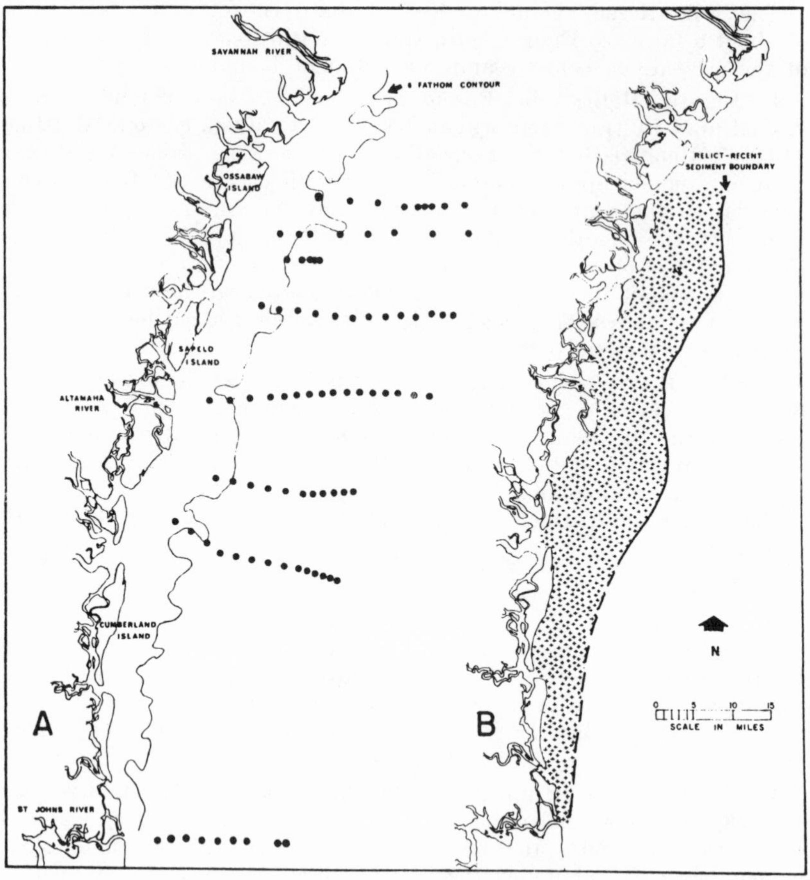

FIGURE 1. *Maps showing the Georgia coast study area, sampling locations, and location of the relict-Recent sediment boundary.*

The Pleistocene sediment is characteristically yellow in color due to iron-stained quartz grains. Phosporite is fairly common, and coarse shell material is abundant. The recent sediment is usually more gray in color and contains less calcareous material. Pilkey (1963) noted that the heavy mineralogy of the relict and Recent sands was similar.

The boundary line between the two types of sediment is quite sharp in most of the area studied. It is believed that this boundary is marked by a zone of interfingering sediment types, 2 to 4 miles wide. There seems to be little mixing of the two sediment types as would be indicated by intermediate grain sizes. The relict-Recent boundary on the northern-most transect could not be satisfactorily located on the basis of available samples, possibly because a wider zone of interfingering sediment types occurs at this location.

The relict-Recent boundary occurs quite consistently at a depth of about 6 fathoms. Figure 1 illustrates the gross similiarity of location of the six fathom contour and the sediment boundary.

In Figure 1B the relict-Recent boundary has been extended (note dashed line) by tracing approximately the six fathom contour. Gorsline (1963) estimated that the clear distinction between relict and Recent sediments ended approximately at the mouth of the St. Johns River, Florida. In the course of the present study a sampling transect was made off the mouth of the St. Johns River and no relict-Recent boundary was observed. The median grain sizes of these samples followed the classic distribution with coarse sands nearshore and increasingly fine sands seaward. Such a size distribution has not been observed elsewhere in the study area.

North of the St. Johns River the nearshore bottom slopes very gently and a wide expanse of bottom shallower than 6 fathoms is present. South of the St. Johns River the bottom slope is greater and only a very narrow strip of bottom of less than six fathoms depth is present. Hence, both sedimentological and bathymetric data support the hypothesis that the distinct relict-Recent boundary ends near the mouth of the St. Johns River.

Discussion

There are three possible sources of Recent detrital sediments on the Georgia Coast. These are (1) present day rivers, (2) nearshore transportation from areas or sources to the north, and (3) winnowing and transportation of fines from outer shelf to inner shelf. Of course, more than one major source may be involved.

Offshore winnowing appears to be the least likely possibility, particularly in view of probable energy gradients. That is, material winnowed out under the low energy conditions of the outer shelf cannot be deposited under the higher energy conditions of the inner shelf.

Within the study area is the Altamaha River which is one of the largest rivers in terms of annual mean discharge south of Cape Hatteras on the U. S. Atlantic coast. The presence of active sand bars in the lower reaches of the river and in Altamaha Sound indicates that sand is being transported. However, preliminary sampling in Altamaha Sound and in the river reveals that sand in these locations is usually coarser than the Recent nearshore sediment and bears a closer resemblance in grain size to the Pleistocene outer shelf sediments. However, nothing is known of the suspended load of the river and it is possible that fine sand is being winnowed from the bed load.

If the Altamaha is an important source of sediment, one might expect some deflection in the shape of the relict-Recent boundary. However, no "bulge" occurs in the boundary off the mouth of the Altamaha, and the boundary is found at a reasonably constant depth. This indicates that in this area the distribution of the fine sand is probably a function

of local energy conditions. The Recent sediment band may well correspond to the "active surf lens" described by Dietz (1963) in his concept of the development of the continental terrace.

Grain size considerations as well as the location and shape of the relict-Recent boundary seem to favor the hypothesis that the Recent sand of the Central Georgia coast is derived from the North, where possibly the Savannah river is an important sediment contributor. However, additional information concerning the nature of the suspended load of the Altamaha river will be necessary in order to establish sediment sources with a greater degree of certainty.

Literature Cited

Dietz, R. S. 1963. Wave-base, marine profile of equilibrium wave built terraces: A critical appraisal. Geol. Soc. Amer. Bull. 74:971-990.

Gorsline, D. S. 1963. Bottom sediments of the Atlantic shelf and slope off the Southern United States. Jour. Geol. 71:422-440.

Pilkey, O. H. 1963. Heavy minerals of the U. S. South Atlantic continental shelf and slope. Geol. Soc. Amer. Bull. 74:641-648.

Stetson, H. C. 1938. The sediments of the continental shelf off the Eastern coast of the United States. M. I. T. and W. H. O. I. Papers in Phys. Ocean. and Meteor. 5:1-48.

10

Reprinted from pp. 63–100 of *Lagunas Costeras, Un Simposio*, A. A. Costonares and F. B. Phelger, eds., Universidad National Autónoma de Mexico, 1967, 686 pps.

HOLOCENE HISTORY OF A STRAND PLAIN, LAGOONAL COAST, NAYARIT, MEXICO *

J. R. Curray,
F. J. Emmel and
P. J. S. Crampton **

ABSTRACT

The Nayarit coastal plain consists of lagoons interspersed between the flood plains of rivers and bordered on the seaward side by a complex strand plain barrier up to 15 Km wide. The strand plain contains about 280 subparallel ridges formed by successive accretion to the shoreline of low narrow beach ridges overlying longshore bars. The sands of the ridges coalesce to form a continuous sheet sand overlying pre-transgressive alluvium and lagoonal deposits. The sands are in turn locally overlain by marsh, lagoon, and younger alluvial deposits of the regressive sequence.

The rise of eustatic sea level prior to 7,000 B.P. (years before present) caused rapid transgression of the shoreline across the continental shelf. When the eustatic rise slowed, the transgression was locally balanced by sand deposition along the coastline at various times between 3,600 and 4,750 B. P., and progradation of the coastline (depositional regression) commenced shortly thereafter while eustatic sea level was still slowly rising. Strong prevailing longshore drift during this period of coastal progradation has given these lagoons long and sinuous tidal inlets which migrated longshore in response to changes in transport direction.

Climatic changes during the regression are manifested as realignments of the coast, or unconformities, with local truncation and erosion of older beach ridges. At about 3,600 B.P., the direction of longshore transport of sand along the northern part of the area reversed from north-flowing to south-flowing. This is interpreted as possible regional climatic cooling which caused a change of wind and wave regime and longshore drift. At about 1,500 B.P., the longshore transport direction changed to northward again, probably due to a change back to a warmer climate. This latter change was accompanied by a large immigration of Indians to the coast. Shortly before the Spanish conquest of the area, about 500 B.P., the major river of the area, the Río Grande de Santiago, shifted its mouth from its old flood plain and confluence with the Río San Pedro to a new distributary to the south. This resulted in local erosion of the strand plain, and rapid progradation of approximately 120 Km^2 of new strand plain adjacent to the new river mouth.

Introduction

The coastal plain and continental shelf of slowly subsiding continental terraces are the sites of significant accumulations of sediments. The instantaneous position of the shoreline results from the interplay between rate of subsidence and rate of influx of sediment, producing alternate periods of transgression and regression. The present day surface of the continental shelf and coastal plain differs from that of most of past geological time in that it has resulted from rapid eustatic sea level fluctuations during

* Contribution from the Scripps Institution of Oceanography, University of California at San Diego, La Jolla, California.

** Scripps Institution of Oceanography University of California at San Diego, La Jolla, California, U.S.A.

the Quaternary, while most pre-Quaternary changes of relative sea level were caused by subsidence or emergence of the continental margin. A rapid worldwide transgression from the edge of the continental shelves at approximately 125 m depth occurred within the past 20,000 years. The rate of rise of sea level slowed approximately 7,000 years ago and has approached its present day level within the past few thousand years, possibly, according to some workers, with fluctuations above and below present level. Many shore lines of the world ceased transgressing and stabilized their position within this period of 7,000 years. Where the rate of influx of sediment was high, the transgression reversed to become a depositional regression with progradation of the shoreline by littoral or deltaic sediments. The coastal plain of the region of investigation represents an excellent example of reversal from transgression, due to eustatically rising sea level, to depositional regression by littoral sedimentation.

Results of previous studies of sediments, morphology, structure and history of this continental shelf and coastal plain have been published by Curray (1959), Curray and Moore (1963, 1964a and b), and Moore and Curray (1964). The present paper will deal with the detailed morphology of the coastal plain and sedimentological history of approximately the last 7,000 years. The adjacent continental shelf will be discussed in further detail in a future publication (Curray *et al*, in preparation).

This study has resulted from work both at sea on the continental shelf and from a series of sampling trips on land starting in 1959. These investigations have been variously supported by the American Petroleum Institute, the Office of Naval Research, the National Science Foundation, Chevron Research Corporation, Jersey Production Research Company, and the Instituto de Geología, Universidad Nacional Autónoma de México. Many associates have contributed to this investigation. We acknowledge particularly the assitance of G. P. Salas, A. Ayala-Castañares, A. Arellano, F. B. Phleger, D. G. Moore, R. H. Parker, Tj. H. van Andel, J. R. Moriarty, B. Swope, and D. Tuthill.

Regional Setting

The region of investigation lies on the coastal plain of most of the State of Nayarit and the southern part of the State of Sinaloa, on the Pacific coast of Mexico, south of the Gulf of California (Figure 1). Most of the rivers entering the area are of limited discharge and drainage area, rising only west of the central plateau of Mexico (table 1; figures 1 and 2). The one large river of the area is the Río Grande de San-

TABLE 1

River	*Drainage Basin Area Km²*	*Runoff $10^6 m^3/yr$*
Presidio	4,825	844
Baluarte	5,383	807
De las Cañas	1,635	261
Acaponeta	6,101	1,068
San Pedro	29,300	3,457
Grande de Santiago (plus Río Lerma)	122,850	10,900
Santa Cruz	200	—
Chico	730	—
Chila	535	—
Ameca	13,995	2,520

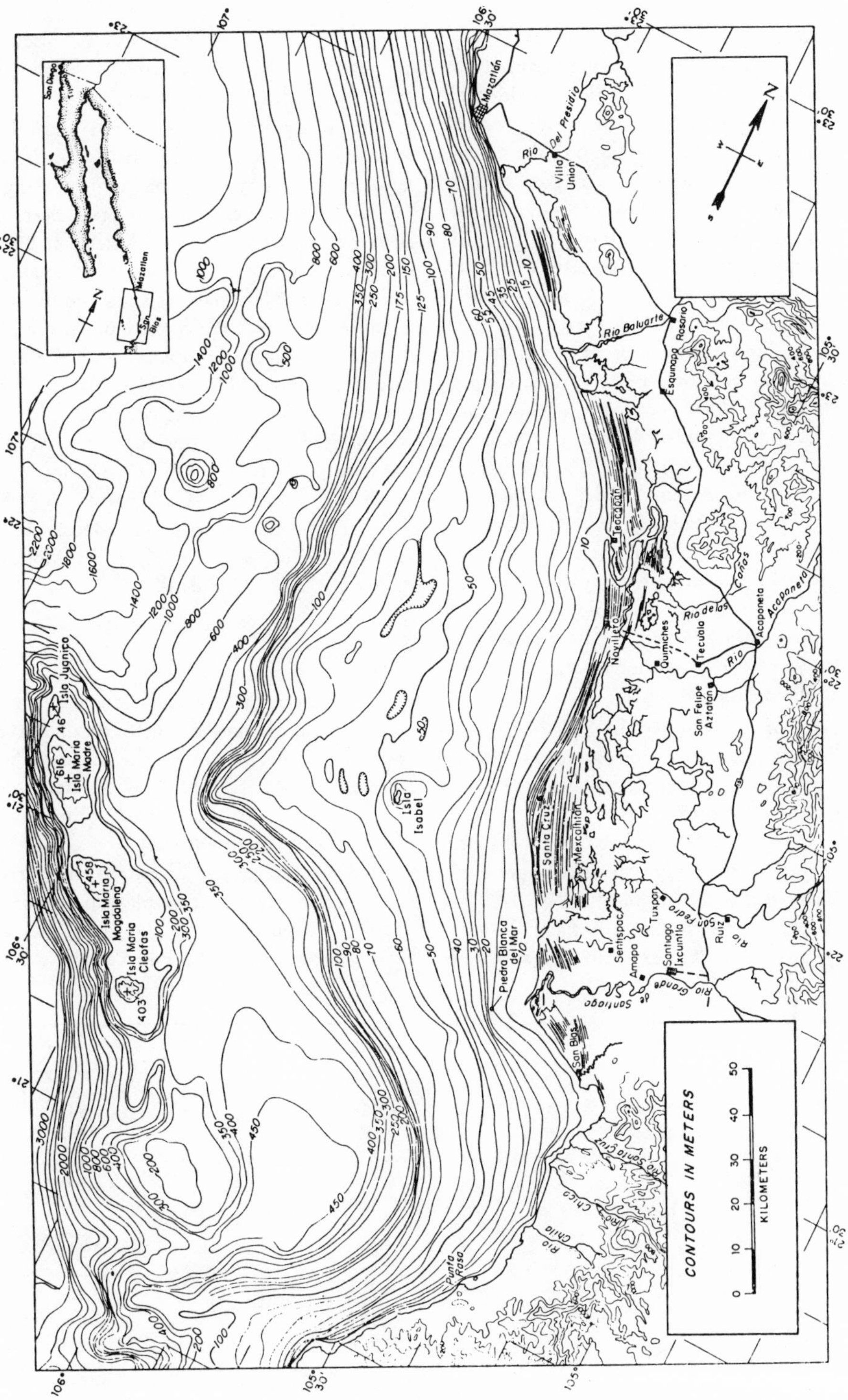

Figure 1. Chart showing topography of the region of study, adjacent foothills, and continental terrace,

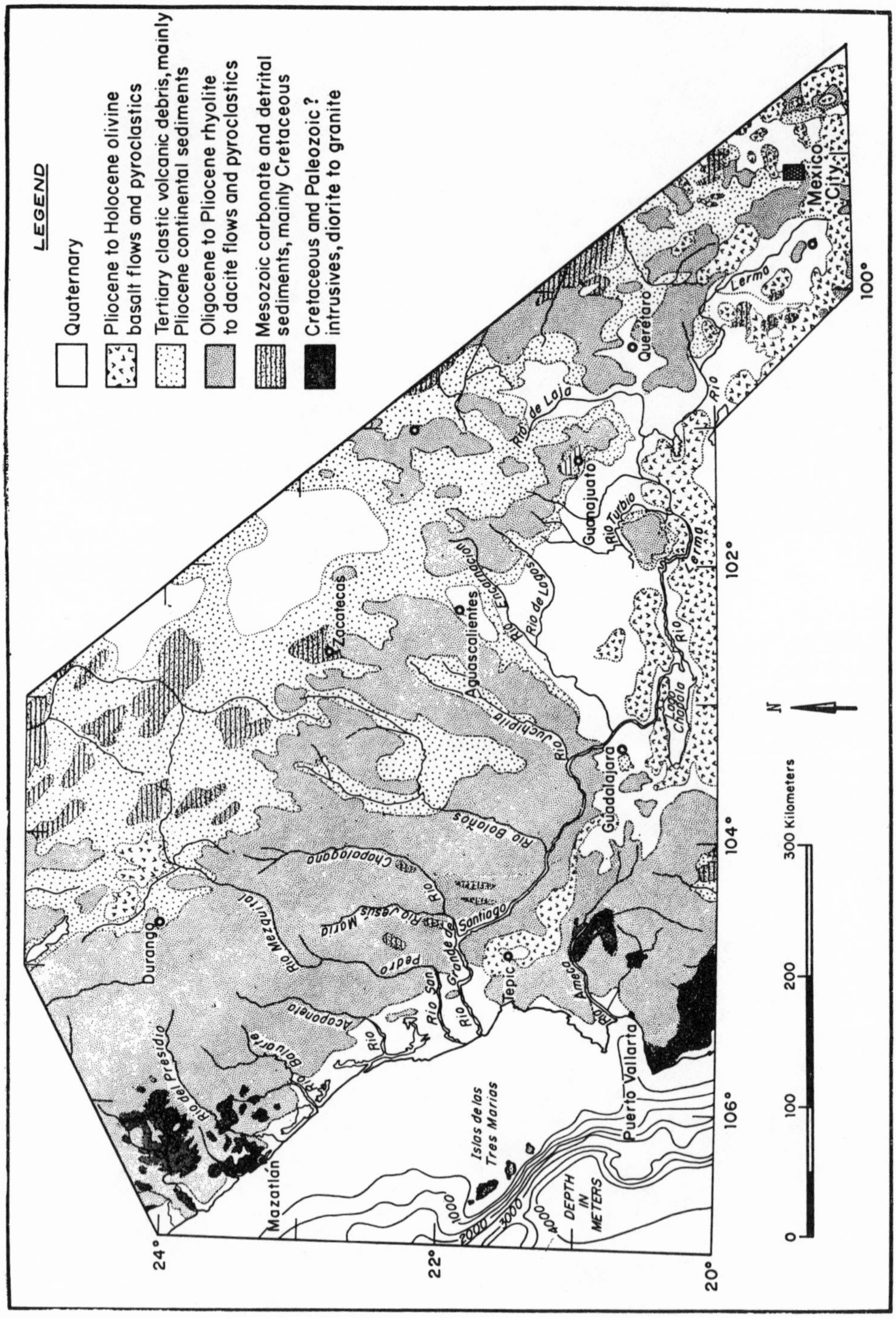

Figure 2. Geology of drainage area, simplified from De Cserna (1961).

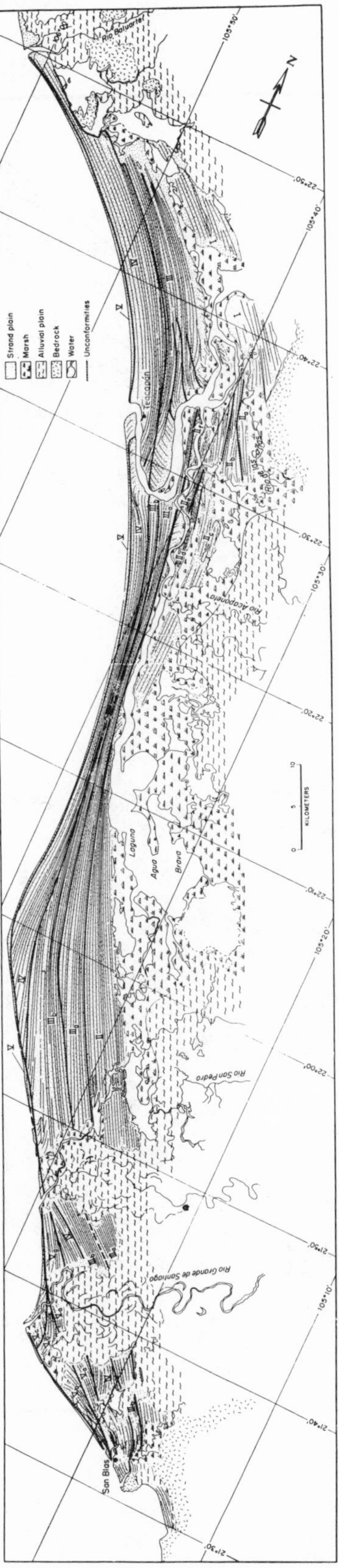

Figure 3

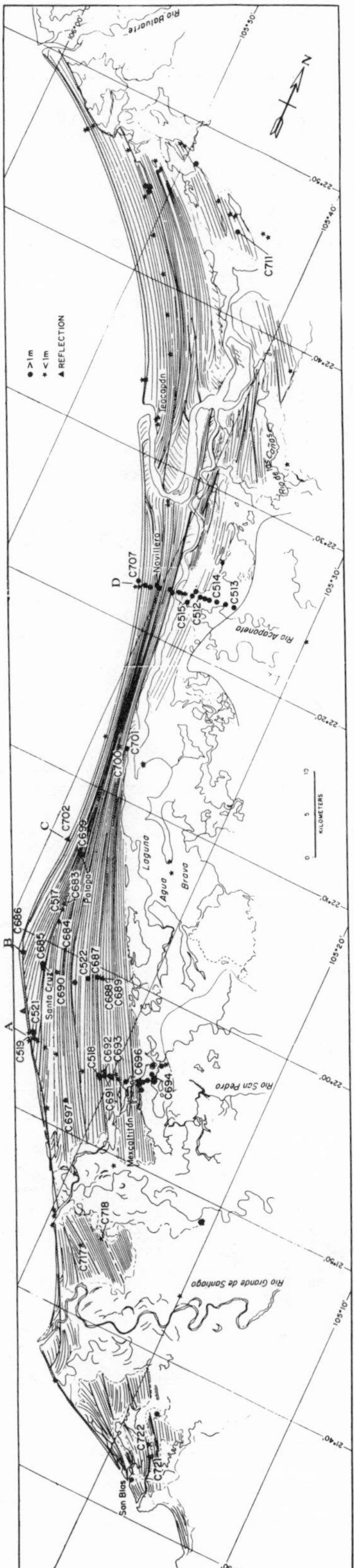

Figure 4

tiago, one of the largest rivers of the west coast of Mexico. The Río Grande rises in Lake Chapala near Guadalajara, figure 2. which is in turn fed by the Río Lerma coming from near Mexico City. This river has had a profound effect upon the nature of the region. It has supplied most of the sediments comprising the coastal plain, and during Pleistocene lowered sea level, it prograded the edge of the continental shelf with thick wedges of deltaic sediments (Curray and Moore, 1964b; Moore and Curray, 1964).

The geology of the drainage basins of the rivers of the area is shown in figure 2. Mineralogy of the sands of some of these rivers and also of some of the samples collected in this study was described by van Andel (1964). Heavy mineral assemblages of the principal rivers of the area are sufficiently different to enable identification of river source and longshore transport direction.

The climate of the region is subtropical to tropical, with a mean annual temperature over 25°C. Annual rainfall ranges from about 850 mm at the northern edge in Mazatlán (figures 1 and 2), to approximately 1200 mm at Tepic (figure 2), and as high as 1660 mm in the southern coastal plain near San Blas (figure 1). Prevailing winds are from the northwest in the winter months and from the west to southwest in the summer. This is a regime of sea breezes in the afternoon, decreasing after sunset. Most of the rains fall in late summer and early autumn, frequently accompanying tropical storms or *chubascos* coming from the south. Mean spring tidal range is 1.25 m in Mazatlan and about 0.98 m in San Blas; mean tidal range is respectively 0.85 and 0.70 m. Tides are mixed semi-diurnal.

Coastal Plain Physiography

The detailed physiography of the region of investigation is represented diagrammatically in figure 3, prepared from study of a large number of vertical and oblique air photographs, supplemented by fieldwork. The region consists of isolated hills of bedrock, largely late Tertiary volcanics, protruding through Quaternary sediment of the coastal plain. This Quaternary sediment consists of Pleistocene and Holocene alluvial floodplain deposits, lagoons, marsh, and a wide strand plain of subparallel elongate beach ridges. This strand plain extends along the coast with variable width from Mazatlán to south of San Blas (figure 1), a distance of over 225 Km, and reaches a maximum width of 15 Km. Locations of all coastal plain samples are shown in figure 4.

Air photos of typical portions of this strand plain are shown in figure 5, and photos of the region on the ground are shown in figure 6. The surface of the strand plain is furrowed by abandoned beach ridges, ranging in width from 15 to 200 m crest to crest, and averaging 50 m in width. Some individual ridges can be traced continuously for a distance of more than 50 Km. Relief of the surface ranges typically from less than 1 m above mean sea level on the ridges to less than 50 cm below mean sea level in the elongate depressions between adjacent ridges. The highest, best developed ridges within the strand plain itself reach a maximum of from 1 to 2 m elevation. Dunes immediately landward of the present beach produce the highest portions of the entire strand plain, locally attaining an elevation of 4 m.

Parallelism of the beach ridges is very striking. At the widest portion of the strand plain near Santa Cruz (figure 4), approximately 280 individual ridges can be counted on the air photographs. Along most of the strand plain the present coastline is parallel to the ridges landward of it and represents the youngest in the sequence of beach ridges. The postulated mechanism of formation of these ridges (Curray, 1959; Curray and Moore, 1964a) is the successive accretion of new beach ridges to the coast by up-

a.

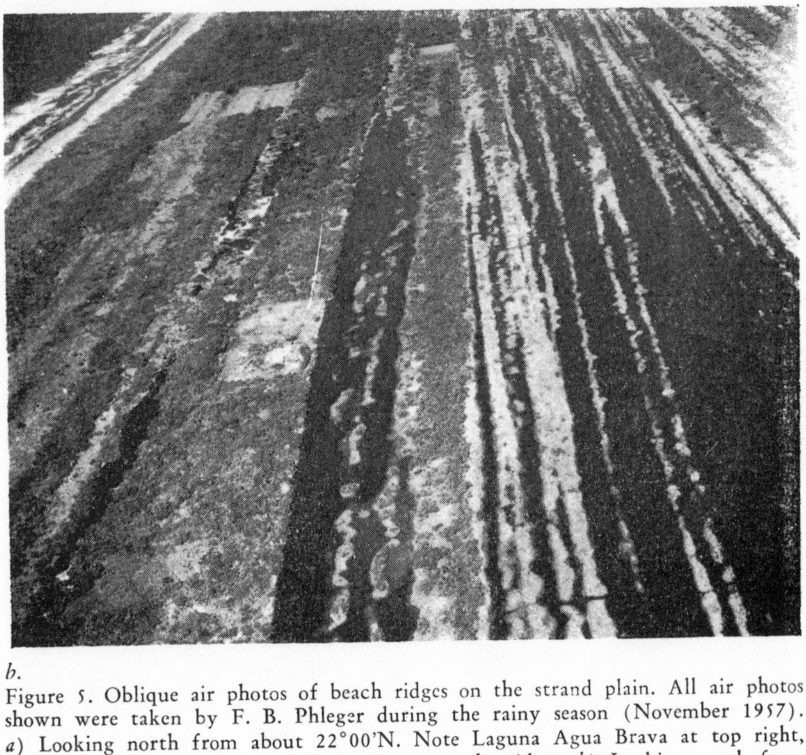

b.

Figure 5. Oblique air photos of beach ridges on the strand plain. All air photos shown were taken by F. B. Phleger during the rainy season (November 1957). *a*) Looking north from about 22°00'N. Note Laguna Agua Brava at top right, the Pacific Ocean, top left, and mangroves on the ridges. *b*) Looking north from about Palapa (figure 4, 22°05'N) near the present beach.

a.

b.

Figure 6. Views of depressions between adjacent beach ridges, looking south from Section B. Santa Cruz Line (figure 4) *a*) Near boring C-685. *b*) East of boring C-686.

building and emergence of longshore bars (figure 7). The nearshore zone in this region has a high rate of influx of sand. A longshore bar or low tide terrace exists beneath the surf zone throughout most of the region. It is postulated that after sufficient influx of sand into the near-shore zone, the submerged longshore bar is built upward until it emerges above sea level during conditions of low wave action. With continued low wave action, this bar is enlarged, until it captures the wave action and becomes a new beach ridge, thereby isolating the former beach and leaving an elongate depression between the new and the old ridges. If conditions of high wave action follow, the newly created beach ridge is destroyed, and the sand either pushed up on the face of the old beach ridge or else removed to deeper water. With continued conditions of low waves, however, more sand is piled on the seaward face and on top of the newly emerged beach ridge, and it eventually becomes a permanent beach ridge with dune capping. During this process, washover and wind transportation of sand partially fill the depression between new and old ridges. The average spacing of beach ridges in the strand plain is similar to the distance between the submerged bar and the present beach ridge.

Although the present beach is parallel to former beach lines immediately landward along most of this coast, there are sectors of the coastline, where former beach ridges are being truncated at an angle and eroded back (figure 8). This represents a realignment of the coastline in response to change in oceanographic conditions or rate of sediment supply. Such realignments of the coastline have occurred many times in the past during the formation of the strand plain as shown in figures 8 and 9. New periods of deposition with normal formation of successive beach ridges follow these brief periods of erosion.

Each erosional period of realignment of the coastline produces an unconformity, and surface traces of these unconformities can be correlated and mapped on air photos of the strand plain. On the basis of these correlations, the strand plain has been divided into five major units as shown in figure 3. In addition to these five major units, some subdivision within these major units can be carried through portions of the area, further subdividing the sequences. Each of these unconformities is assumed to approximate a time line. Differences in the conditions of deposition during these sequences and postulated causes for the realignments will be discussed in a later section describing the historical development of the region.

Subsurface Investigations

Methods

Subsurface investigations of the coastal plain have been conducted by means of deep auger samples, by use of a small rotary drilling rig, and by use of a bailer type drilling rig. The most successful sampling was performed with this latter equipment shown in figure 10a. Cores and uncontaminated cutting samples taken at frequent intervals in the bailer holes were used for analyses of minor structures, lithology, and faunal content.

Locations of all samples taken on the coastal plain are shown in figure 4, which differentiates near surface samples from those samples which penetrated to more than one meter below the surface. Surface samples were generally collected by digging a pit through the soil zone with a shovel, but also included collection of representative shells from some of the many kitchen middens of the area (figure 11).

In addition, information on depth of interfaces within the sediment column was provided by acoustic reflection. A portable, self-contained, single pulse, seismic reflection system (figure 10b) was used on one

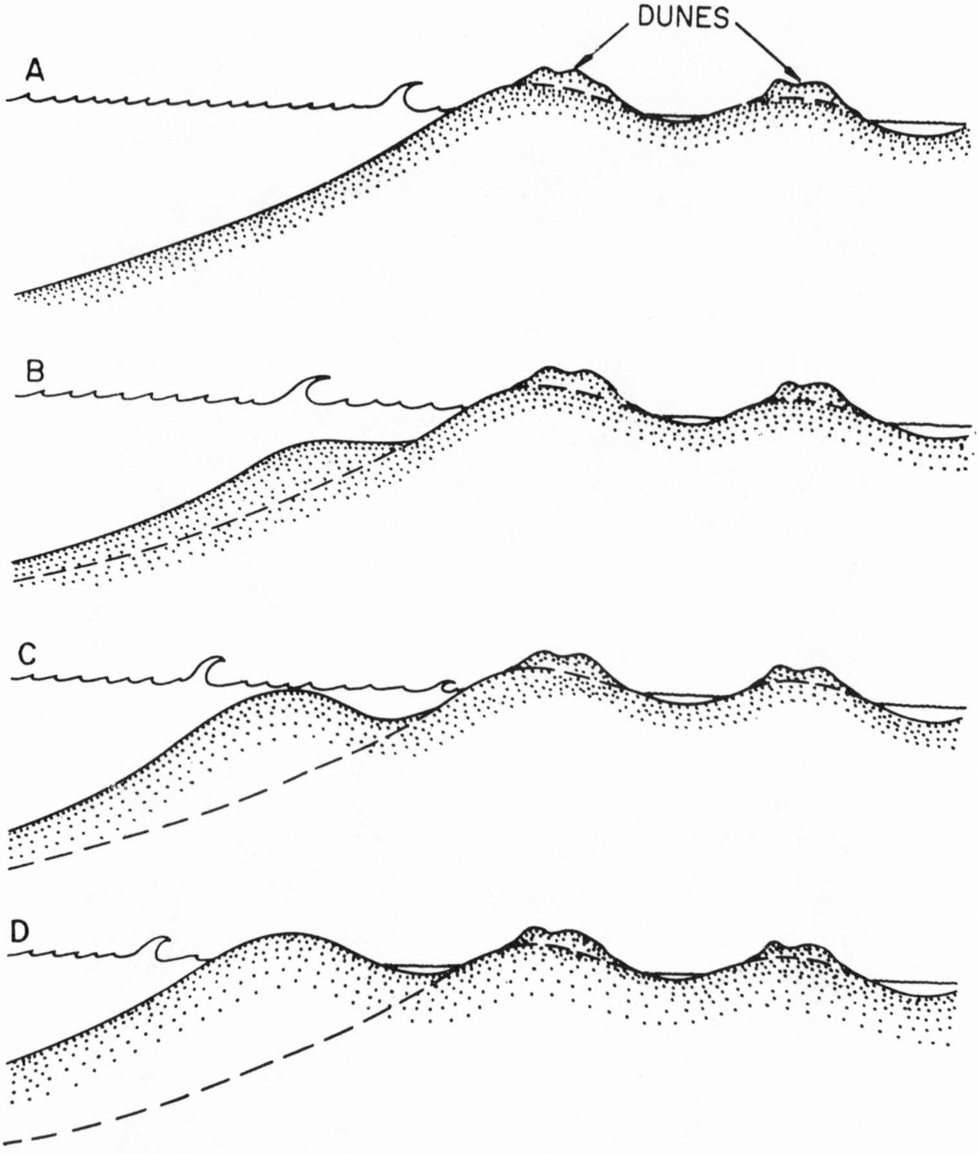

Figure 7. Diagrammatic sequence of events in formation of beach ridges. A) Initial conditions, with no longshore bar or low tide terrace. Ridges are shown with some capping of dune sand. B) Low tide terrace is formed of sand derived from either longshore drift or, in early stages of regression, from landward transport of basal transgressive sand of inner continental shelf. C) Longshore bar is built initially at plunge point of breakers. D) During continued influx of sand and conditions of low surf action, bar is built to sea level. With fall in tide and continued low surf, bar emerges, and builds higher. Unless it is destroyed by more intense surf, emergent bar becomes a new beach ridge, thereby isolating former beach and creating a narrow lagoon which is subsequently partially filled.

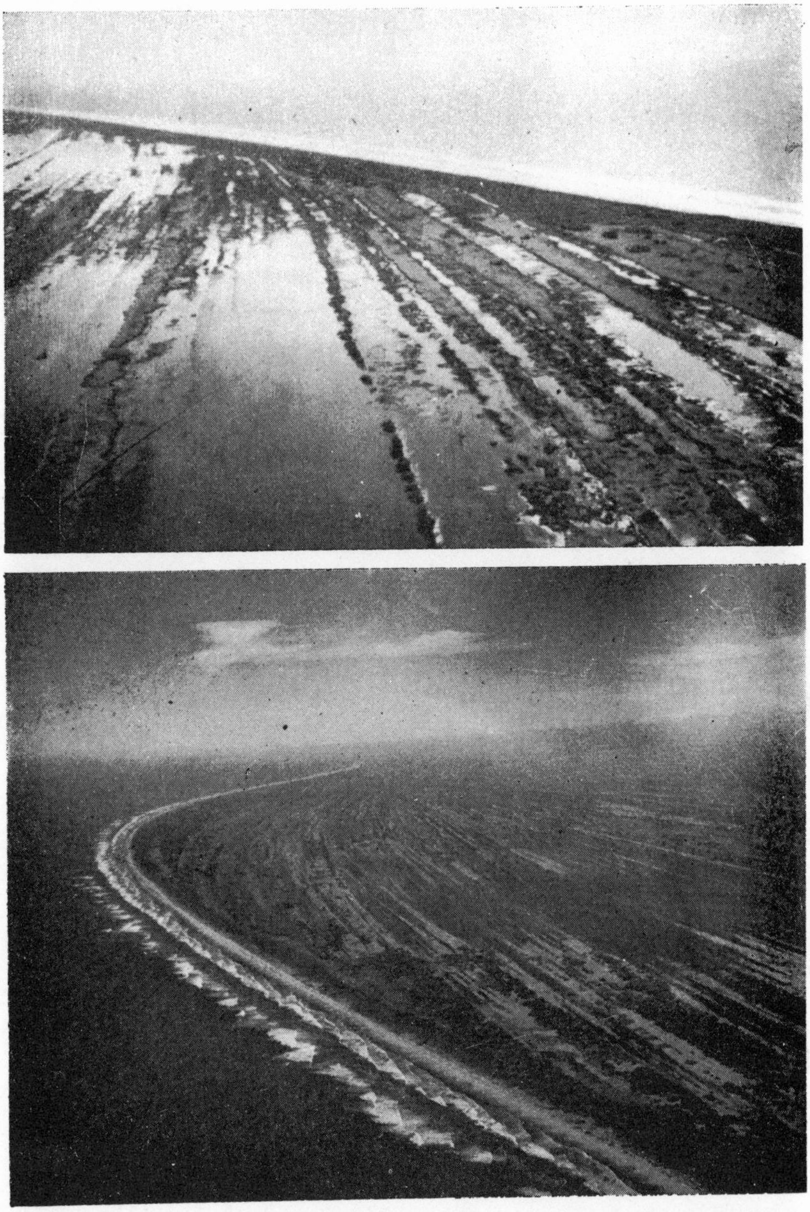

a.

b.

Figure 8. Air photos showing reorientation of the coast and truncation of ridges following period IV. *a*) Looking south at the present coast from about 22°N. *b*) Looking north from about 21°55'N. Note the truncation of ridges (group IV) in the foreground and the parallel relationship of the present beach (group V) to the group IV ridges in the distance. Note also the low area in the center right side of the photo where the beach ridges are more flooded by interconnected lagoons. This will be discussed in the section on geological history, the transgression following 7,000 B.P.

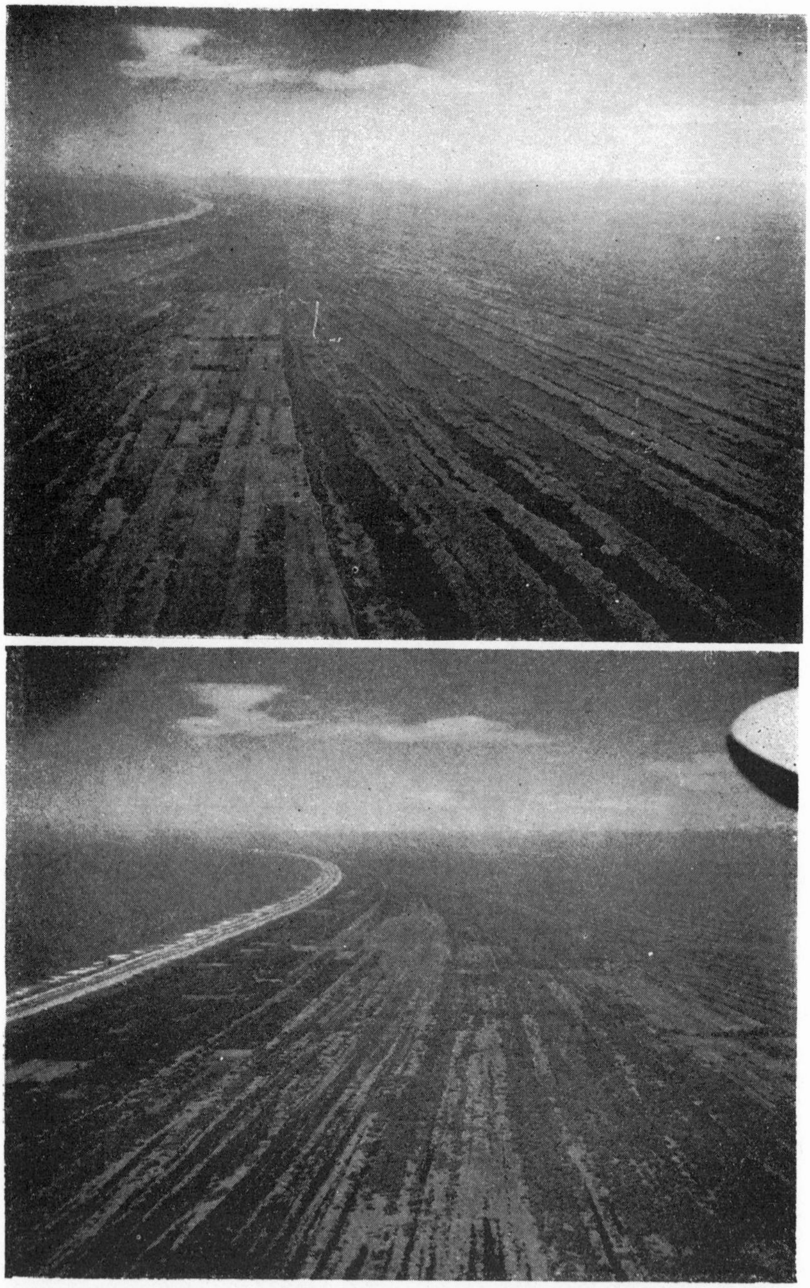

a.

b.

Figure 9. Air photos showing truncation of group III b ridges by group IV ridges. *a*) Looking north from about 22°05'N. Village of Palapa is on seaward side of the unconformity halfway in distance. *b*) The same area as 9a, taken from slightly farther toward the ocean. Village of Palapa is in right center. Series IV ridges in foreground.

a.

b.

Figure 10. *a*) Bailer equipment used for making borings in the coastal plain. Hole is cased continously. Cutting samples are reasonably uncontaminated. Cores were taken below bottom of the casing. *b*) Self-contained, portable acoustic reflection equipment used through courtesy of D. G. Moore, USNUC, San Diego.

a.

b.

Figure 11. *a*) Midden mound of shells, station C-690, composed primarily of *Tivela* sp. This is larger than most kitchen middens in the region. *b*) Section of a small shell midden, C-517, composed largely of *Tivela* sp.

trip in the area through the courtesy of D. G. Moore, U. S. N. U. C. San Diego. A 10 KHz pulse of acoustical energy produced by a transducer penetrated the sediment column and reflected off acoustical interfaces. These correlated with recognizable lithologic changes in the adjacent borings.

Four sections are shown in figures 12-15 and located in figure 4. Datum for each section is approximate mean sea level, as determined by tide staff observation on the beach. Level lines were run from mean sea level on the beach, most of the way across the strand plain on two of the lines, Line B, Santa Cruz (figure 13), and Line D, Novillero (figure 15). These two surveys demonstrate the flatness of the strand plain and the relative relief of the ridges and adjacent depressions. Levels of the adjacent lines have been estimated by assuming horizontality of interconnected waterways. The seaward ends of the sections have been correlated with Sonoprobe surveying and coring on the adjacent continental shelf.

Sediment Facies

Facies determinations have been made from the samples in these borings and from the intermediate surface samples by the examination of minor structures, lithology and faunas. The following facies have been distinguished and associated with their respective environments of deposition, as indicated in the physiographic chart (figure 3):

1. *Littoral Sand*

The bulk of the strand plain deposits consists of a fine to medium grain, well sorted, sand. Locally it contains varying amounts of shells, shell fragments, pumice, peat, and wood, depending upon the specific environment of deposition. Mineralogically the sand consists typically of about 40% quartz, 40% volcanic rock fragments, 10% plagioclase, and 10% potash feldspars.

Stratification ranges from fair to poor. Beach sands in the area are not well laminated and cores from borings beneath most of the coastal plain show, at best, crude stratification. Locally, particularly near the base of the section, some alternation of beds of mud, sand, and peat layers occurs. Grain size of the sands varies with the specific environment of deposition. The finest sands are deposited on the shore face and longshore bar, and beach and dune sands are progressively coarser. These grain size variations have aided in specific environmental interpretations. Typical molluscs of the littoral sands are *Tivela* sp, *Donax carinatus*, *D. contusus*, *D. punctatostriatus*, and *Mulinia pallida*.

Indian kitchen middens are common on the beach ridges of the strand plain (figure 11). These consist primarily of *Tivela* sp, which lives in the intertidal sands of the present beach. Dates from these middens have proven to be very useful in interpreting the history of the area.

2. *Alluvium*

Sediments grouped in the sections as alluvium include both pre-transgressive, older alluvium (deposits laid down before rise of sea level brought the shoreline into this area), and alluvium deposited after sea level reached approximately its present position. These sediments consist predominantly of mud or clayey silt with lesser amounts of silty sand. Locally cobbles and coarse sand were encountered in the borings. These are interpreted as channel fill, although control is not sufficient to establish the dimensions and trends of these channels.

3. *Lagoon and Marsh*

For purposes of the sections, lagoon and marsh deposits have been grouped together.

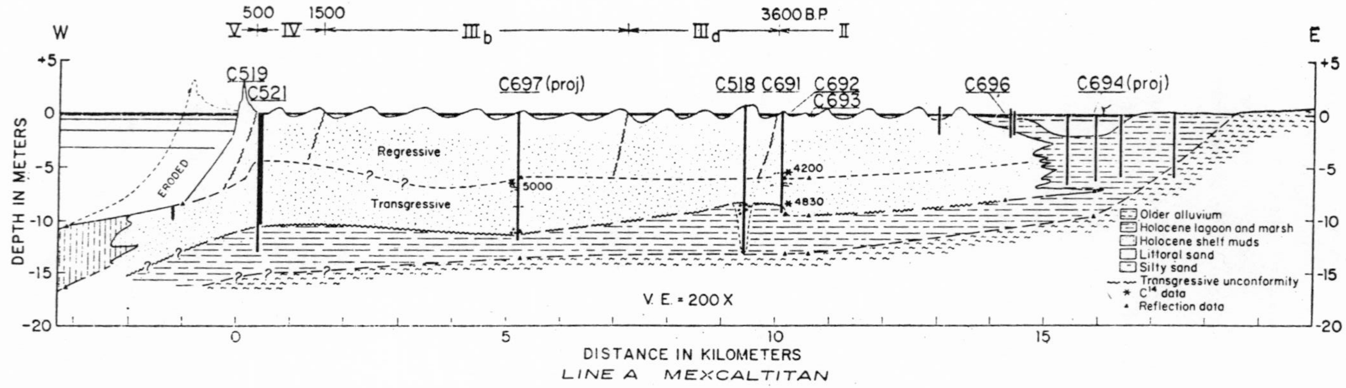

Figure 12. Section A, Mexcaltitán Line (see figure 4). See text for description and table 2 for details of radiocarbon dates. Boring C-697 projected to line of section approximately 3 Km.

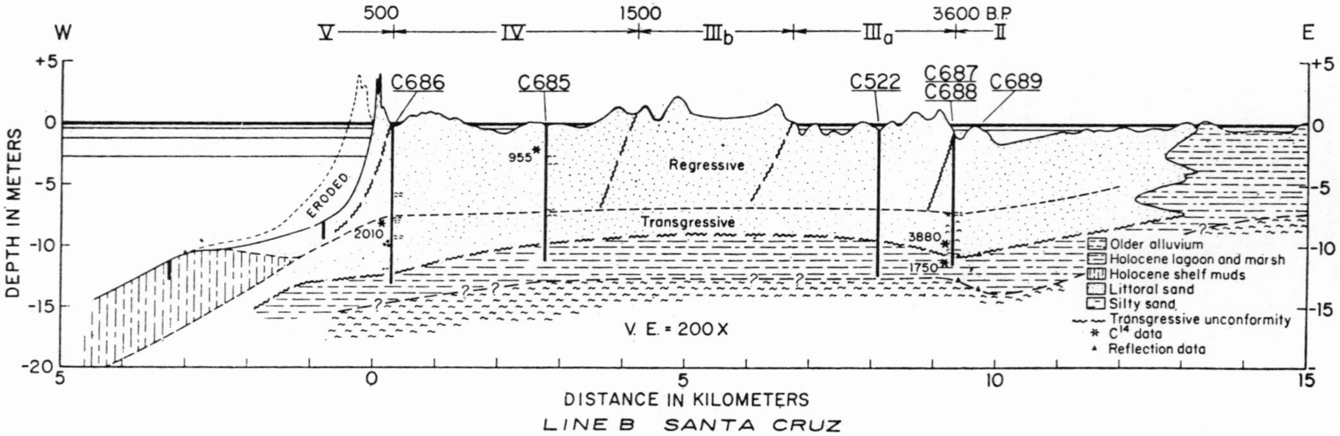

Figure 13. Section B, Santa Cruz Line (see figure 4). See text for description and table 2 for details of radiocarbon dates.

They consist largely of alternating strata of sandy and muddy sediments containing abundant peat layers and concentrations of molluscs of lagoon habitat. Present lagoons of the area (figure 3) are bordered by mangrove, and generally consist almost entirely of silty clays with a paucity of sandy layers. Lagoons and marshes formed ahead of the advancing shoreline during the transgression were perhaps more restrictive, and their deposits contain interbeds of sands from washover or blowover from the advancing shoreline. A typical mollusc of these lagoonal deposits is *Crassostrea columbiensis*.

4. *Shelf Muds*

The shelf facies muds underlying the seaward edge of the strand plain in one section (figure 15, Line D) are essentially the same as those sampled on the adjacent inner continental shelf. These are clayey silts and silty clays which were deposited seaward of about 10 m water depth. Typical fauna of the shelf facies includes *Chione gnidia, Anadara obesa, Laevicardium elenense*, and *Nassarius* cf. *N. versicolor*.

5. *Silty sand*

The silty sand encountered in section D, Novillero, figure 15, represents a transition zone between the littoral sands and shelf facies muds. Such a transition zone is found frequently both at the inner and outer margins of the lens of the shelf deposits on continental shelves. In this case, it probably represents a zone of former shelf muds reworked during the regression of littoral sands over the top of the shelf muds.

Radiocarbon dates

All boring, surface, and midden samples cointaining peat or shell material were studied in detail in order to isolate sufficient material for radiocarbon dating. Results of these analyses are listed in table 2. Location of samples from which dates have been run are shown in figure 4, and the positions of dates in the borings are indicated in the four sections. In most cases an attempt has been made to date shells of only a single species with a precisely known living depth range or habitat. Samples of peat were picked carefully to eliminate contaminating material. No corrections have been made on any dates and all are reported with the identification of the analyzing laboratory exactly as reported by that laboratory.

Description of Sections

Section A, Mexcaltitán (figure 12)

This line extends from landward of Station C-694 on the coastal plain to C-519 near the shoreline and ties in with the inner end of one of the shipboard Sonoprobe and coring traverses on the continental shelf. The town of Mexcaltitán lies at approximately kilometer 13½ in the section. No leveling was done along this line because of the network of interconnected waterways, assuring horizontality and close control of the back part of the section from about C-691 through C-694. Low tidal fluctuation was observed in these waterways during the course of drilling operations and mean sea levels were determined from these tidal observations.

Inhabitants of Mexcaltitán state that a maximum tidal range of approximately 1 m is observed in the lagoon around the town during spring tides of the dry season when lagoons are quite brackish. During the rainy season when the rivers and lagoons are flooded, the lagoon waters become fresh and no tidal fluctuation is observed. The coastal plain between Station C-691 and the coastline does not contain waterways which are permanently connected with the open ocean. During the dry season many of the depressions between beach ridges are

TABLE 2

Sample Number	Lab. 1/ Number	Material Dated	Depths Below M.S.L. meters	Age years B.P.	Comments
		Borings-Line A-Mexcaltitán			
C-697	LJ-1365	Mixed shells: *Donax punctatostriatus; D. contusus; Crassostrea corteziensis; Mulinia pallida; Tivela* sp.	—7.10	5000±500	Intertidal sand, mixed with reworked lagoon. Base of regression IIIb. Appears anomalously old.
C-691	LJ-1362	*Mulinia pallida*	—5.90	4200±300	Intertidal sand; base of regression II.
C-691	LJ-GAP51	Peat (mangrove?)	—8.95	4830±200	Transgressive lagoon and marsh.
		Borings-Line B-Santa Cruz			
C-686	I-3073	*Mulinia pallida*	—8.03	2010±170	Intertidal sand; base of regression IV.
C-685	I-3072	*Donax punctatostriatus*	—2.53	955±170	Intertidal sand. Regressive shoreface IV.
C-687	I-2228	*Mulinia pallida*	—9.84	3880±105	Intertidal sand; transgressive. Appears anomalously young.
C-687	LJ-GAP50	Peat	—11.58	1750±200	Transgressive lagoon or marsh. Anomalously young.
		Borings-Line C-Palapa			
C-699	I-3074	*Tivela* sp	—6.00	1670±105	Intertidal regressive, late IIIb.
C-699	I-2230	*Mulinia pallida*	—8.00	2180±115	Intertidal; base of regression IIIb.
C-701	I-3076	*Tivela* sp and *Donax punctatostriatus*	—3.20	3340±260	Intertidal regressive, late II.
C-701	LJ-GAP52	Peat	—7.00	7200±500	Lagoon or marsh interbed in stationary barrier early II.
C-701	I-3077	*Mulinia pallida*	—9.60	4720±200	Intertidal sand; early II stabilized barrier.

TABLE 2 (continuation)

		Borings-Line D-Novillero			
C-707	I-2232	*Mulinia pallida, Tivela* sp.	—4.36	205±90	Intertidal regressive V.
C-512	LJ-568A	Peat, carbonized	—5.27	5430±300	Lagoon and marsh.
C-512	LJ-568B	*Polymesoda* sp; *Donax* sp; *Natica* sp; *Mitrella* sp.		5000±300	Transgressive beach and marsh.
		Midden and Surface Shell Samples			
C-721	LJ-1363	*Protothaca metadon* (Pilsbry and Lowe, 1932). 100 cm deep in shell pit above M.S.L.		1700±200	Wave built shell accumulation on unconformity between IV/IIIb. Faunal assemblage intertidal rocky to nearshore shelf.
C-722	I-3079	*Tivela* sp 150 cm deep in midden above M.S.L.		1170±100	High midden mound containing potsherds. Assemblage mainly *Tivela*, plus other intertidal and some lagoon mouth forms; along unconformity IV/IIIb.
C-717a	I-2233	*Agaronia testacea*		1260±95	Surface veneer midden aprox. at unconformity IV/IIIb. Assemblage mainly *Tivela.*
C-717b	LJ-1358	*Tivela* sp from surface midden		1700±300	Surface veneer midden approx. at unconformity IV/IIIb. Assemblage mainly *Tivela.*
C-718	I-3078	*Tivela* sp from surface midden		1300±100	Surface veneer midden in IIIb. Assemblage mainly *Tivela.*
C-690	I-2229	*Tivela* sp from 67 cm deep in midden above M.S.L.		1310±95	Large midden almost entirely shell, mainly *Tivela*; in IIIb near unconformity IV/IIIb. See Fig.
C-684	LJ-1361	*Tivela* sp from surface midden		1500±200	Surface veneer midden along unconformity IV/IIIb, mainly *Tivela.*
C-517a	LJ-518	*Agaronia testacea*, from 12 cm deep in midden above M.S.L.		3175±220	Thin shell midden along IV/IIIb unconformity with potsherds. C-517a date rejected as anomalous. Mainly *Tivela.*
C-517b	LJ-1360	*Tivela* sp from 12 cm deep in midden above M.S.L.		1670±200	
C-683	I-3071	*Tivela* sp from surface midden		1320±100	Surface veneer midden along IV/IIIb, mainly *Tivela.*
C-700a	LJ-1384	*Donax punctatostriatus*, from 69 cm deep in midden above M.S.L.		600±100	Shell midden mound in IIIa, but near to coast and lagoon both. Mainly *Tivela* and *Donax* intertidal sand forms. Evidently represents late occupation site and does not date ridge.
C-700b	I-3075	*Tivela* sp from 69 cm deep in midden above M.S.L.		620±100	

1/ *L.J.* Samples were dated by the La Jolla Radiocarbon Laboratory, Scripps Institution of Oceanography, La Jolla, California,
I. Samples were dated by Isotopes, Inc., Westwood, N. J.

dry or locally contain hypersaline water. Elevations of the borings in this portion of the section are assumed, and the topography of the upper surface of the section is diagrammatic. Note also that Well C-697 is projected approximately 3 Km to the line of the section.

Pre-transgressive alluvium slopes seaward at a mean gradient of approximately 1 m/Km. Overlying this are several meters of lagoon and marsh deposits, consisting of alternating sands and muds containing brackish water lagoon faunas. These were the coastal lagoons which formed ahead of the transgressing shoreline. Overlying the lagoon and marsh facies are basal transgressive sands of the Holocene transgression, which closely resemble the sands of the present coastline.

The transgression moved the shoreline inland to approximately the Km 14 mark, Station C-696, with sea level at approximately —5 m. The shoreline presumably stabilized at this location and built upward during the later slower rise of sea level. Regression was initiated at the beginning of Period II and has moved the shoreline seaward by almost 15 Km, during Periods II, III, IV and V. Some erosion and cutting back on the coastline occurred between Periods II to III a and IV to V, especially along this portion of the coast. Erosion of the coastline of more than 1 Km is postulated here during Period V.

Chronology of the transgression and regression will be discussed following descriptions of the sections.

Section B, Santa Cruz (figure 13)

This section was controlled by a leveling survey carried from mean sea level, established at the beach, back to the wide lagoon at C-687. Within the limits of this survey the morphology of the coastal plain is shown simplified and diagrammatic. East of C-687, the line is extrapolated across the marsh and partially filled lagoon area, going toward the outcropping older alluvium many kilometers further landward. The back edge of the strand plain is placed by correlation with the aerial photograph physiography, although thicknesses and depths are entirely assumed by correlation with Line A and by assumption of approximate uniformity of slope toward the outcropping older alluvium.

The section here is essentially the same, with pre-transgressive alluvium, Holocene lagoon and marsh which were formed ahead of the advancing shoreline during the marine transgression, and basal transgressive sands overlain in turn by the regressive littoral sands. The initial stabilization of the shoreline is assumed to have occurred here at the most landward beach ridge recognized in the aerial photographs. Rate of sea level rise was decreasing by the time the shoreline reached this point, and the shoreline apparently built upward in position before initiation of the depositional regression at approximately 4,500 B.P.

The seaward edge of this section appears to have been eroded back by several hundred meters during realignment of the coast during Period V.

Section C, Palapa (figure 14)

This section runs across one of the narrowest parts of the coastal plain and includes information projected a long distance from Well C-701. The section, therefore, represents a composite of information in the general vicinity where a multiple ridge barrier lies seaward of the large Laguna Agua Brava.

No leveling was done to construct this line, but Laguna Agua Brava shows a low tidal fluctuation because of its connection with the open ocean. By assuming an approximate uniformity of slope between the outcropping older pre-transgressive alluvium at the landward edge of Laguna Agua Brava and from the Sonoprobe information on

the continental shelf, it again appears that the slope of this alluvial surface is approximately 1 m/Km. This is overlain by lagoon and marsh sediments, basal transgressive sand, and regressive littoral sand of the present strand plain. The shoreline of this region was probably first stabilized in the vicinity of C-701. Apparent reversal of ages of dates occurs between the two oldest dates in Well C-701. This could have occurred either because one of the dates is anomalous or because the date of 7,200 years was run on older peat which was reworked into the section of beach sand building up in the stabilized shoreline.

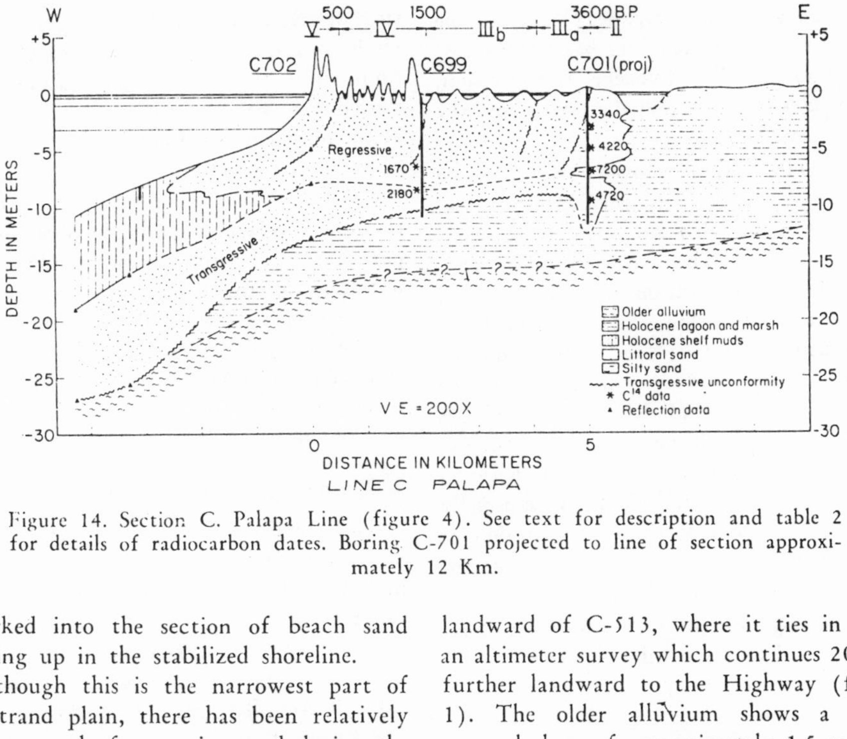

Figure 14. Section C. Palapa Line (figure 4). See text for description and table 2 for details of radiocarbon dates. Boring C-701 projected to line of section approximately 12 Km.

Although this is the narrowest part of the strand plain, there has been relatively little removal of regressive sand during the erosional and nondepositional periods. Deposition and progradation have been continuous, but at a slower rate than in the vicinity of Sections A and B. Seaward of C-702 the regressive sand has been locally deposited on top of the Holocene shelf facies muds which occur on the inner continental shelf adjacent to this section.

Section D, Novillero (figure 15)

This is one of the more thoroughly studied sections of the region, although sample recovery was sparse in the borings. The Series 500 wells were all drilled with a rotary rig, and only one sample from a core could be dated. Cuttings from these wells did not recover shell material and were badly contaminated by caving and intermixing in the hole. A level survey was run from mean sea level at the present beach near C-707 back across the coastal plain to landward of C-513, where it ties in with an altimeter survey which continues 20 Km further landward to the Highway (figure 1). The older alluvium shows a mean seaward slope of approximately 1.5 m/Km.

This line differs from the other sections in that it runs onto the foot of the alluvial plain of one of the major rivers of the area, the Río Acaponeta, while the other sections all run into the depression between adjacent alluvial plains, where the major lagoons of the area have formed. Interpretation of the section shows, however, thick lagoon

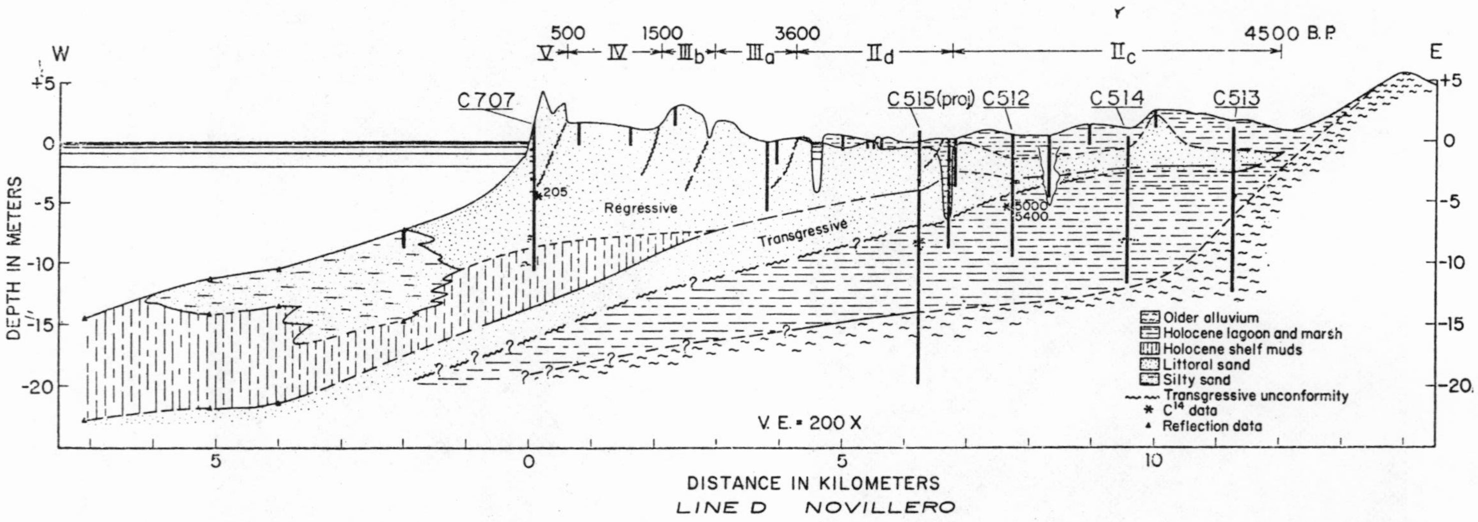

Figure 15. Section D, Novillero Line (figure 4). See text for description and table 2 for details of radiocarbon dates.

a.

b.

Figure 16. Ridges of Series I of probable late Pleistocene age. *a*) Air view looking NW at about 22°45'N. Note wide indistinct character typical of ridges. Note also the hills of volcanic rock, lagoons, and Pacific Ocean in the background. *b*) Typical edge of a Series I ridge, near C-711. Ridge is composed of Pleistocene alluvium containing limonite nodules. Depression is filled with modern alluvium. Boring C-711 penetrated alluvium to 9 m, where beach rock was encountered.

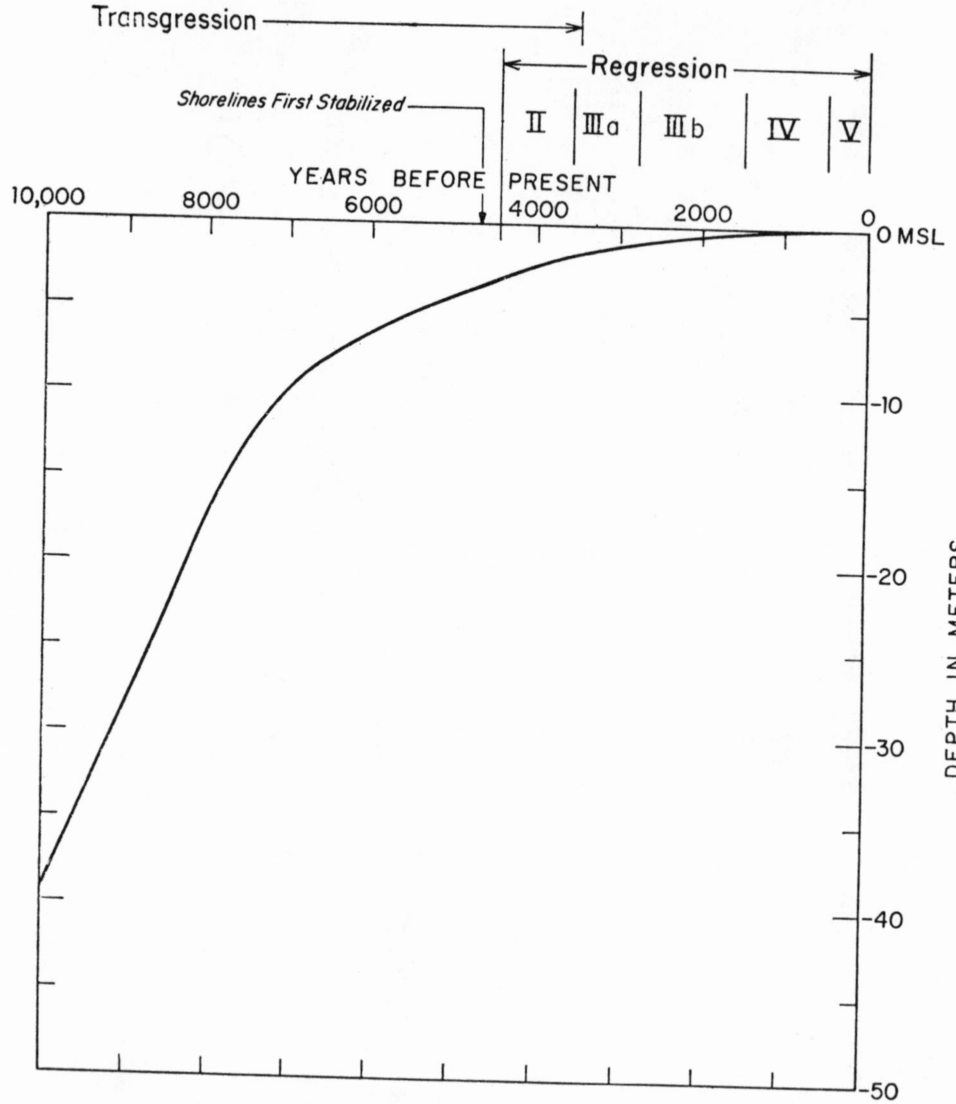

Figure 17. Curve of eustatic sea level changes of past 10,000 years slightly modified from Curray (1965) and Shepard and Curray (1967). Other authors have proposed fluctuations of sea level above and below present level within past 6,000 years. Evidence from this investigation favors the curve as shown.

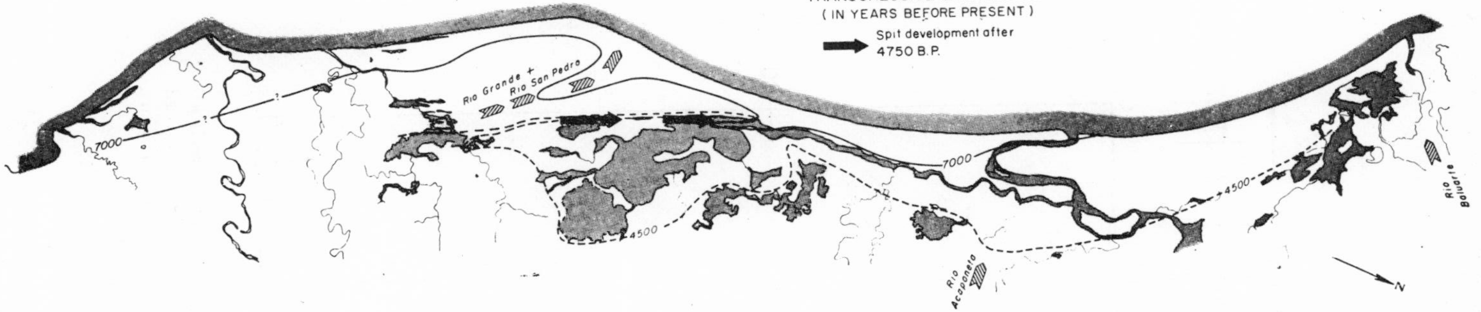

Figure 18. Shoreline changes during period of transgression between about 7,000 B.P. and 4,500 B.P. Shoreline is believed to have stabilized first at about 4,750 B.P. at a point near C-696 (Section A, figure 12) and a spit built northward to enclose the ancestral Laguna Agua Brava by 4,500 B.P.

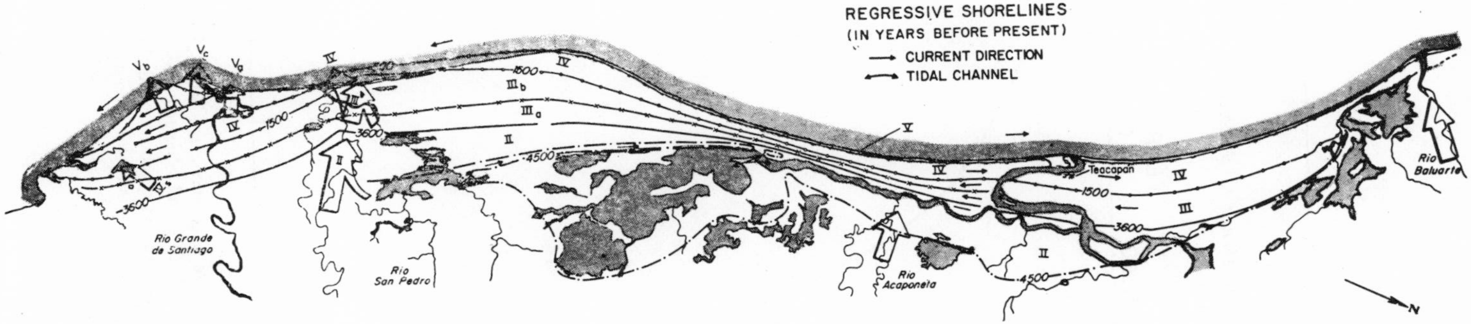

Figure 19. Shoreline changes during period of regression from about 4,500 B.P. to present. The shorelines indicated are the unconformities at start of Periods II through V. Note that Laguna Agua Brava must have formed originally by 4,500 B.P. by northward longshore drift. Drift in northern part of area was northward during Period II, southward during Period III, and northward during periods IV and V. Mouth of Río Grande de Santiago formed a new distributary (IV s) in the middle of Period IV, abandoned its confluence with Río San Pedro at end of Period IV, and formed the present delta during Periods Va, Vb, and Vc.

and marsh deposits overlying the older pre-transgressive alluvium. The basal transgressive sand appears to continue all the way across the region at least to between wells C-514 and C-513. The sands at the pinch-out lying landward of this point are probably dune sands formed by blow-over from the beach line. Note that the upper surface of all of these sands lies beneath mean sea level in this vicinity and that the upper part of the overlying section consists of younger lagoon, marsh and alluvial fill.

This section shows convincing evidence of the regression having proceeded well over the top of the inner shelf facies muds. Faunas obtained at the deepest point of penetration in well C-707 are definitely inner shelf forms. Control of the seaward end of this line on the inner continental shelf is through shipboard coring and Sonoprobe survey.

Note the three channels in this section, respectively at Km 4½, 6½ and 8½. The present tidal inlet to Laguna Agua Brava and the mouth of the Río Acaponeta is the channel at Km 4½. Strong tidal currents are observed here, keeping the channel open. The older inlets to this lagoon system appear to be the old filled channels. On aerial photographs the 6½ Km channel can be seen to be the tidal inlet to Laguna Agua Brava between Periods II c and II d (figure 3).

Geological History

Geological history has been interpreted from all aspects of this study, including physiography of the coastal plain, correlation of unconformities, radiocarbon dates on middens from the strand plain surface, radiocarbon dates and facies interpretations from the borings, and from study of the adjacent continental shelf. Pleistocene and early Holocene transgressive history of the adjacent continental shelf has already been reported in part by Curray, and Moore (1964b), and further detail and revision will be reported by Curray *et al* (in preparation). All events are summarized in Table IV.

Pleistocene

The Series I ridges shown in figure 3 have not previously been discussed. It was noticed early in the field investigations and studies of aerial photographs of the region that Series I ridges are characteristically different from other ridges in the area. They are broader, less continuous along the crest (figure 16 a), and consist of alluvium (figure 16 b), rather than the littoral sand of the remainder of the coastal plain. One bailer boring was put down in the Series I ridge area, C-711, at approximately 22°45'N, 105°45'W. This well was located in the depression between two adjacent ridges, but the section penetrated for the upper 9 m was essentially the same as that observed in auger holes, pit samples, and in the eroded edges of the ridges elsewhere in the Series I region. The sediments are a reddish brown, sandy-silt to silty-sand, containing abundant nodules of limonite up to 1 cm in diameter. At a depth of 9.5 m, calcareous beach rock and cocina were encountered, consisting of well-sorted, fine sand, with abundant recrystallized shells (mainly *Chione gnidia*). No radiocarbon dates have been attempted on this beach rock because of the recrystallized nature of the shell material.

The upper 9 m of this portion of the section are quite typical of the oldest pre-transgressive alluvium found elsewhere in the area. This is interpreted as Pleistocene alluvial fill which is draped over the top of a series of Pleistocene beach ridges. The present topography represents the results of this draping, differential compaction, and subsequent erosion of the alluvial surface during the flood season when the depressions between these adjacent ill-defined ridges are

filled with water. The depressions and ridges are accentuated by formation of low escarpments. It is believed that the older underlying ridges now cemented to beach rock and cocina are of late Pleistocene age, representing an interglacial or interstadial. By comparison with interpreted geological history of the adjacent continental shelf, it is suggested that this may be either of Sangamon age or of mid-Wisconsin age, the latter correlating with a stand of sea level postulated for approximately 25,000 to 35,000 years Before Present (Curray, 1961) and a warm period during the same time (e.g. Flint and Brandtner, 1961; Frericks, 1968).

A very clear discontinuity or unconformity can be observed between the Series I beach ridges and the Holocene beach ridges of Periods II and III lying seaward of them. These Series I ridges are observed only in this northern part of the area, but it is possible that a similar sequence lies buried under younger material elsewhere in the region. This portion of the coastal plain receives relatively less water discharge and sediment. The main river of this portion of the coastline, the Río Baluarte, flows directly into the ocean, and the lagoons are apparently filled with water only during flood periods. During the dry season the lagoons of this region are either dessicated or contain hypersaline water which is locally harvested for salt. Rainfall is less in this area than it is in the southern part, near San Blas and the Río Grande de Santiago.

Holocene Transgression, 7,000 to 3,600 B.P.

Studies of sea level changes in other parts of the world suggest that the rate of rise of sea level (figure 17) was relatively rapid from approximately 18,000 years B.P., until approximately 7,000 B.P. During this rapid rise of sea level the shoreline transgressed the continental shelves of the world to approximately the present shoreline. Many shorelines stabilized or regressed during the slower rise of the past 6,000 to 7,000 years (Curray, 1964, 1965).

Comparison of the oldest dates of the central part of this area suggests that transgression first stopped sometime between 4,830 B. P. (C-691, Line A) and 4,720 B. P. (C-701, Line C). The former is from the top of the lagoon-marsh deposits, while the latter is in intertidal sands overlying the marsh facies. Approximately 4,750 B. P. would then be the time at which the shoreline first stabilized. Older dates, C-512, Line D, were obtained on both peat and shells from lagoon facies underlying the basal transgressive sand. The 7,200 year date (C-701) is on peat which may have been reworked and physically emplaced into the section above the underlying sands. The depths found for the oldest dates are not entirely in agreement with the sea level curve of figure 17, perhaps as a result of some differential compaction.

It appears from studies of the adjacent continental shelf and from morphology of the alluvial and lagoonal surfaces under the strand plain, that during lowered sea level the Río Grande de Santiago and the Río San Pedro entered the Pacific Ocean together, far to the north of the mouth of either river today (figure 18). This crosses a low area of the strand plain which is more flooded and covered with lagoon and mangroves than adjacent portions (figure 8b). The shoreline apparently first stabilized around 4,750 B.P. at a point near C-696, Line A, near Mexcaltitán, at the mouth of the river confluence. Longshore drift built a spit northward from this point, fed by the sands from the rivers and by derivation of basal transgressive sand from the inner continental shelf. By approximately 4,500 years B.P. this spit extended to C-701 and had enclosed the ancestral Laguna Agua Brava. In the meantime, the continued slow rise of sea level had enlarged the lagoon to that shown in figure 18.

The shoreline of the northern part of the area, near and north of Line D Novillero, lay at the back edge of the Series II beach ridges. The transgression or landward migration of the shoreline continued here until at least 4,500 B.P., and at the north and south edges of the area of study until about 3,600 B.P. Where the transgression stopped earlier and the shoreline stabilized, a barrier built upward in place and a spit built laterrally to enclose a lagoon. Where the transgression continued until later, no lagoon was formed.

Longshore transport along the northern portion of this coastline, near the tidal inlet of Teacapán, is toward the north today, as shown by the direction of spit growth and the site of inlet erosion. This represents the net direction of transport by waves approaching the coastline, and is probably a subtle balance between northward approaching and southward approaching swell. The wind regine today represents a balance between the winter and summer winds. During the winter wind regime, more winds approach from the northwest and produce generally southward longshore transport; in contrast, the south and southwesterly winds of the summer period produce a northward longshore transport. Wind regime of the period 4,500 to 7,000 B.P., therefore, must have been similar to that of today and this must have represented a relatively warm period in history. No other indications of temperature differences have been obtained from thorough examination of the evidence, such as faunal changes, and this relative temperature is postulated solely on the basis of wind patterns.

Holocene Regression

Transgression stopped at the mouth of the confluent Ríos Grande and San Pedro at 4,750 B.P., elsewhere in the central to north central part of the area by 4,500 B.P., but continued until 3,600 B.P. at the north and south edges of the area. Following these times, regression has occurred until the present day with intermittent interruptions represented by unconformities. Postulated positions of the shoreline are shown in the paleogeographic reconstruction of figure 19.

Period II, 4,500 to 3,600 B.P.

At the beginning of Period II, the rivers Baluarte and Acaponeta entered the ocean approximately along their present flood plains. The Ríos Grande and San Pedro entered the ocean together south of the present course of the Río San Pedro at about 21° 48'N (figures 3 and 19). The oldest Holocene beach ridges detectable in the field and from photographs are located north of the confluence of the Ríos Grande and San Pedro, and north of the Río Acaponeta in the embayment, today lying landward of the tidal inlet of Teacapán. Where the shoreline was either still transgressing or had only just stabilized in position, no Period II beach ridges can be located (north of the tidal inlet of Teacapán or south of the present course of the Río San Pedro).

During Period II, the dominant longshore transport was again toward the north in the northern part of the area (figure 19). Continuation of northward growth of the spit joined the entrance of Laguna Agua Brava and the mouth of the Río Acaponeta and moved them to about 22°35'N, inland from the present inlet at Teacapán (figures 3 and 19). The Río Acaponeta has flowed through this tidal inlet until the present day. Sands forming the beach ridges were derived from both by longshore transport from the river mouths and by derivation from basal transgressive sands of the inner continental shelf. Deposition must have been nearly continous north of the mouth of the Ríos Grande and San Pedro. No unconformities can be correlated throughout the Series II beach ridges in this area. North of the Río Aca-

poneta, however, the sequence is very complex, and on the basis of unconformities has been broken into Periods II a, through II f (figure 3).

The relative position of sea level throughout this period was probably slightly below that of the present, although resolving power of these studies is not sufficient to give any absolute amount of difference.

Measurements of strand plain width, counts of ridges, estimates of sand volumes, and rates of progradation or regression are listed in table 3. Note that the 74 ridges in the widest development of Series II are 57 m in average width and formed on the average in 12.2 years apiece. The mean rate of progradation, both in terms of rate of accretion of the coastine and in volume of sand added, was higher during this period than later, even though sea level was still apparently rising slowly. The high rate of sand accretion must reflect the ready supply of sand available by reworking of basal transgressive sand from the inner continental shelf, as well as by longshore drift from the river mouths.

TABLE 3

Sequence [1]		II		III		IV		V	
Time, years B. P.	4500	to	3600	to	1500	to	500	to	0
Time interval, years		900		2100		1000		500	
Maximum width of sequence, km		4.25		6.3		3.25		0.5	
Number of ridges at widest part of sequence		74		127		80		—	
Mean width of ridges at widest part of sequence, m		57		50		41		—	
Mean years per ridge		12.2		16.5		12.5		—	
Rate of coastal progradation, m/year		4.7		3.0		3.2		1.0	
Volume of regressive sand, 10^6 m^3		1600		1820		730		—	
Volume of sand per year 10^3 m^3/yr.		1770		870		730		—	
Volume of sand per ridge, 10^6 m^3		21.6		14.3		9.1		—	
Relative climate		warm		cool and stormy?		warm		warm	
Direction of longshore transport at Teacapán		N		S		N		N	

[1] All measurements and estimates are made between Río San Pedro (21°50'N) and Teacapan (22°30'N). Sequence V is not included in the estimates of volume and number or ridges because of the amount of coastal erosion which is postulated to have occurred here during this period.

Period III, 3,600 to 1,500 B.P.

A major change occurred in the area between Periods II and III. This was apparently a change in climate which resulted in a change of wind regime. Notice the inlet at Teacapán 22°30'N (figure 3). During Period II, longshore transport was northward, as shown by the trend of the tidal inlet to Laguna Agua Brava and by the orientation of the hooked spits. During Period III transport was southward. The change which occurred in this area at about 3,600 years B.P. must have been a subtle shift in wind balance which produced the striking difference in longshore transport shown here. The southward transport of sediment along the coast lasted through Period III until approximately 1,500 B.P.

The Río San Pedro and the Río Grande de Santiago still entered the ocean together south of the present flood plain of the Río San Pedro. The portion of the coastline near this river mouth and immediately to the north was prograded rapidly during this period, suggesting that longshore transport of sediment may have been northward immediately north of the mouth of these rivers. In this vicinity a local unconformity subdivides Period III into Periods III a and III b. This discontinuity can be traced southward to the edge of the area and northward as far as the tidal inlet at Teacapán, but cannot be distinguished north of the tidal inlet. The cause of this realignment of the coastline may then have been some fluctuation of the discharge pattern of the river mouth.

The relative position of sea level cannot be determined in this area with any confidence, but information from leveling surveys on Lines B, Santa Cruz, and D, Novillero, suggests that sea level was lower than present, at least at the beginning of Period III, by approximately the amount indicated in the sea level curve (figure 17). The possibility still exists, however, that sea level could have been near present level, or could have taken a jump upwards to present level at the intermediate point in Period III, between III a and IIIb, or at the end of Period III.

Comparison of the data in table 3 suggests that the rate of progradation to the coastline was lower in Period III than in Period II, resulting in a significantly longer period of time for formation of each ridge, 16.5 years, as compared with 12.2 years for Period II. The rate of progradation in cubic meters of sand per year between the latitudes of 21°50'N and 22°30'N is reduced from what it was in Period II, suggesting reduction in the availability of sediment supply from the inner continental shelf. Climate is thought to have been relatively cool throughout all of this period. Perhaps the climate was also stormier, and the process of formation of new beach ridges was more difficult. With a stormier climate, longshore bars could be destroyed by intense wave action before being built up into new beach ridges.

Period IV, 1,500 B.P. to 500 B.P.

The change from Period III to Period IV was one of the most important events in the development of this area. Longshore drift in the Teacapán tidal inlet area shifted from southward to northward, and has so remained since that time. This could have been due to a general warming of the climate along the coast.

Accompanying the change in longshore transport direction and climate was a realigment of the coastline which can be traced from one end of the area to the other. This particularly striking unconformity, illustrated in figure 9, produced some of the highest portions of the entire strand plain, and many of the roads and villages of the area are now located along this discontinuity.

Also of significance is the concentration of Indian shell middens located along this discontinuity. These range from thin veneers of shell material, especially *Tivela*, to higher, well developed mounds as illustrated in figure 11a, located immediately landward of the discontinuity. These shell middens have been of great value in determining the chronology of the region.

It is admittedly risky to attempt to date the age of a beach ridge by the age of the shell middens located on that beach ridge. The possibility always exists that the beach ridge is significantly older than the midden, and that the Indians carried the intertidal *Tivela* clams from the existing beach at the time, landward to this high well developed ridge before utilizing them as food. One might then expect some range in the ages of middens, from the oldest, possibly coinciding with the age of the ridge, to those

which are younger, representing shells transported some distance inland. The 10 dates run on midden shell from on or near this ridge have, however, shown surprisingly little scatter, suggesting that the live clams were not transported far from the beach, and that this particular beach ridge was the site of occupation of a large population of Indians. Of these 10 dates, one (C-517a, 3175 B.P.) is rejected as anomalous because of potsherds contained in the same level of the midden. The other 9 dates average 1436 years B.P.

The middens lying on older ridges landward of this discontinuity are very rare, suggesting that this period of reorientation of the coastline, accompanied by a climatic change, marked the time of influx of a large population of Indians to the coastal region. This Indian migration could have been a result of the climatic change affecting living condition in the inland portions of Nayarit. Radiocarbon dates on burial and occupation sites in the foothill regions of Nayarit predate this period by a few centuries, averaging about 1,700 years B.P. (Furst, 1965).

The sea during Period IV was, within limits of the resolution of this study, at present level.

Table 3 shows a steady trend in the dimensions and rates of progradation with time, from Period II to Period IV. The mean width of the ridges decreased with time; the rate of progradation to the shoreline decreased; the volume of sand comprising each ridge decreased with time. These trends probably reflect two factors. First, the availability of sand supply decreased because the shoreline prograded over the top of shelf facies muds. Early in the regression virtually the entire shelf was surfaced with basal transgressive sand, and the sand in the nearshore zone was available for reworking toward shore for incorporation into the shoreline sand bodies. With time, stability of sea level, and regression of the shoreline, this sand supply was buried beneath shelf facies muds, and no longer furnished a supply for reworking. From the figures of table 3, it is apparent that half, or slightly more than half, of the sand comprising Series II ridges was derived from the inner continental shelf, if it is assumed that all of the sand of Series IV ridges was derived by longshore transport.

Second, as the regression proceeded, the shoreline built into slightly deeper water. Rather than requiring more sand to form each ridge, however, the volume per ridge decreased. This probably in part reflects variability in lateral continuity in the estimates, but primarily demonstrates the decrease in mean width of the ridges. With time and building into slightly deeper water the ridges became narrower and closer together, requiring less sand per ridge.

By the end of Period IV the coastline had built somewhat further seaward than it is today near the present mouth of the Río San Pedro (figure 19).

The confluence of the Río San Pedro and the Río Grande de Santiago had produced a delta lying on what is today the inner continental shelf. This was probably a cuspate delta very much like the present delta of the Río Grande, resulting from balance between rate of deposition of sediments at the river mouth and the intensity of wave action on this rather exposed portion of the Pacific coast.

During about the middle of Period IV a new minor distributary of the Río Grande broke through toward the south and entered the ocean at about 21°35'N, 105°20'W (figures 3 and 19, river mouth IV s.). The present flood plain of the Río Grande had not yet formed, so this distributary flowed across pre-Transgressive alluvium and the strand plain of Period III.

Period V, 500 B.P. to present

The change from Period IV to Period V was probably caused solely by shift in the

flood plain of the Río Grande de Santiago. Prior to approximately 500 B.P. the Río San Pedro and Río Grande de Santiago entered the ocean together along what is today an abandoned flood plain at about 21°48'N (figures 3 and 19). This confluence has been demonstrated independently, by study of affinities of freshwater fishes of the central mesa of Mexico. C. C. Barbour and S. C. Balderas (personal communication) have both shown a relationship between the endemic freshwater fishes of the Río Mezquital, a tributary to the Río San Pedro, and those of the Río Grande de Santiago. All freshwater connections between these two rivers were severed when the Río Grande shifted its discharge to its present course south of the mouth of the Río San Pedro.

This shift in discharge pattern upset the depositional balance along the entire southern portion of the coastline. The former cuspate delta of the river confluence was eroded back because of the reduction in the rate of influx of sand to the coastline. Realignment of the coastline is illustrated in figures 8 and 19, and suggested amounts of erosion are shown in Sections A and B (figures 12 and 13).

The center of deposition was shifted southward to where the new delta of the Río Grande de Santiago was formed. Location of mouths of rivers can be readily determined at various periods of time by the alignment of the individual beach ridges within each group. After the Río Grande de Santiago abandoned its former course, it flowed into the ocean across the ridges of Periods III and IV, to enter the ocean at probably two places. The southward distributary (IV s, figure 19) was still active early in Period Va, and a new major distributary (Va, figure 19) formed at 21°40'N. Subsequent mouths of the river are shown by the trends of the beach ridges in Periods Vb and Vc.

The time of the Period IV to V shift cannot be determined by radiocarbon dating. The only date available for Period V ridges is located at the present beach at Novillero, Line D. This date indicates the very young age of the shell material deep within the present beach ridge, approximately 200 years B.P. The other approach to determining age of this shift, then, is to examine old charts of the area as drawn by early Spanish explorers.

The earliest recorded entry of Spanish Conquistadores into this area was the conquest by Nuno de Guzmán in 1530 (Brand, 1958). He traversed this region on land, crossed the Río Grande de Santiago at Santiago Ixcuintla (figura 1, 21°50'N, 105°12' W), crossed the Río San Pedro at Tuxpan (figure 1, 21°56'N, 105°18'W), and spent the rainy season at San Felipe Aztatán (figure 1, 22°22'N, 105°25'W). His troops suffered severe losses from floods and disease during this rainy season before proceeding on to the north. In 1538, this region became part of the province of Nueva Galicia, with Coronado named as the first governor. The oldest available chart drawn by the Spanish (Brand, 1958) is dated 1550. The positions of towns and villages can be located, but little can be determined about the courses of the rivers. De Guzman's report of crossing the Río Grande and the Río San Pedro in separate places, however, probably indicates that the shift in river discharge had occurred before 1530 A.D. or more than 400 years B.P.

The earliest available informative chart of the area (figure 20), was drawn in 1763. This chart clearly indicates separate courses of the Río Grande de Santiago and the Río San Pedro. The city of Tepic (figure 2, 21°30'N, 104°50'W) can be located on one of the tributaries of the Río Grande. The city of Ixcuintla, where de Guzman crossed the Río Grande, is downstream of this tributary and many other of the towns shown in figures 1 and 3 can be located. The Río San Pedro is shown as entering the

lagoon system near Mexcaltitán (called Temescaltitlán) as it does today. Of interest, however, is the tributary shown to the Río San Pedro near its present mouth. This represents the abandoned flood plain of the Río Grande de Santiago which today is very nearly blocked off by deposition. The Río Grande de Santiago is shown as entering the Ixcuintla. The main coastal port for the Spanish was at that time not San Blas, but was the embayment lying south of San Blas at 21°30'N, known as Ens. de Matanchel (figure 20). The port of San Blas was not founded by the Spanish until 1768, when the temporary facilities at Matanchel were abandoned. The waterways now existing at

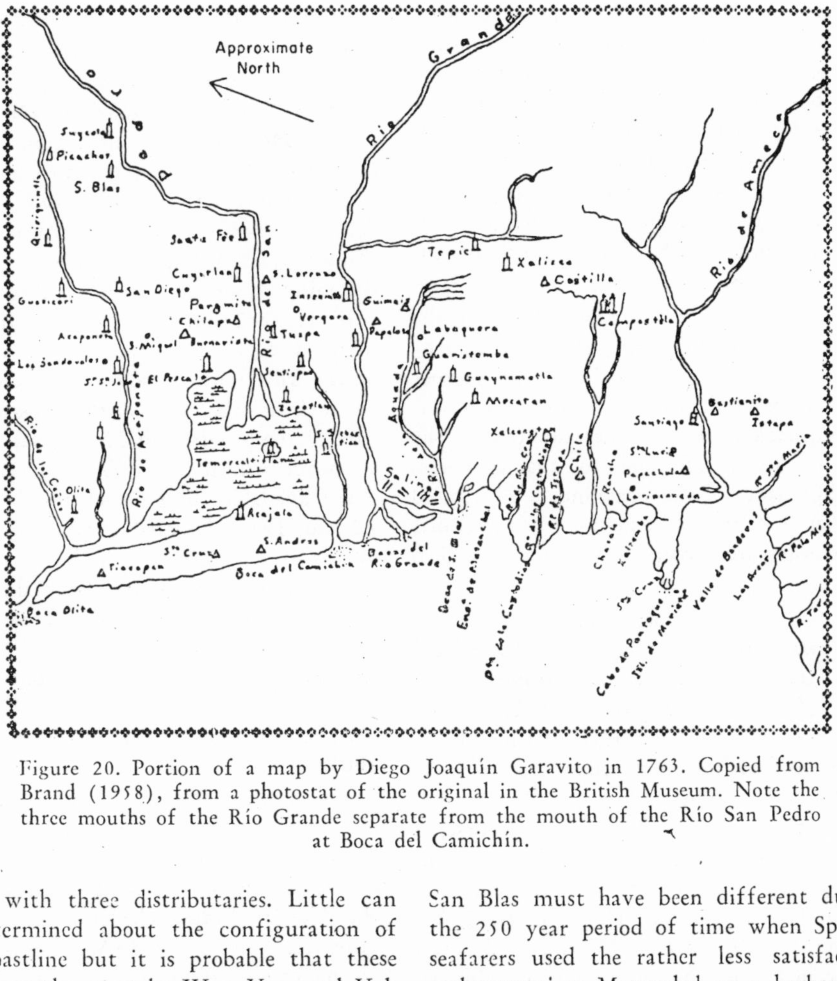

Figure 20. Portion of a map by Diego Joaquín Garavito in 1763. Copied from Brand (1958), from a photostat of the original in the British Museum. Note the three mouths of the Río Grande separate from the mouth of the Río San Pedro at Boca del Camichín.

ocean with three distributaries. Little can be determined about the configuration of the coastline but it is probable that these three mouths must be IV s, V a, and V b (figure 19).

Some of the distributaries of the Río Grande de Santiago are known to have been navigable by ships of the period, because a shipyard was erected in 1644 at Santiago San Blas must have been different during the 250 year period of time when Spanish seafarers used the rather less satisfactory embayment at Matanchel as a harbor and boatyard. The earliest available reliable survey of the port of San Blas was made by the British in 1822. This survey, updated by use of aerial photography, still forms the basis for present charts published by navies

TABLE 4

SUMMARY OF GEOLOGICAL HISTORY

Epoch or Age	*Dates*	*Events*	*Relative Climate*	*Longshore Transport Teacapán*	*Rio Grande Mouths*
Late Pleistocene Interglacial or *Interstadial*	125,000 BP, 80,000 BP or 25,000-35,000 BP	Relatively high sea level; deposition of beach ridges under Sequence I. Could be Sangamon or Mid-Wisconsin interstadial.	warm	—	—
Late Wisconsin glacial and regressive maximum	ca. 18,000 BP	Sea level ca. —125 m; shoreline near edge of present continental shelf; deposition of delta and shoreline deposits dated by Curray and Moore (1964b).	cold	—	northward
Holocene Transgression Early	18,000-7000 BP	Rapid sea level rise to ca. —10 m (Fig. 17); rapid transgression across soils and alluvium on continental shelf and deposition of basal sands.	warming	north?	northward (Fig. 18)
Late	7000-3600 BP	Slower rise of sea level to ca. —2 m (Fig. 17); slower transgression and start of blanketing of inner shelf by marine muds.	warm	north?	northward
Stabilization of shorelines	4750-3600 BP	First stabilization of shoreline in this area at 21° 50' N, 105° 25' W at ca. 4750 BP. Start of spit growth northward from this point. At north and south ends of area shoreline not stabilized until later.	warm	north	II
Holocene Regression Period II	4500-3600 BP	Earliest regression by beach ridge formation in central part of area of study. Transgression still continuing at north and south of here. Laguna Agua Brava formed by 4500 BP with inlet at 22° 15' N.	warm	north	II
Climatic change	3600 BP	Postulated local cooling of climate, with more northerly winds and storms, producing change in longshore transport direction to southward and slowing of process of beach ridge formation.			

TABLE 4 (continuation)

Period III	3600-1500 BP	Beach ridge formation along entire area of study.	cool and stormy	south	III
Climatic change	1500 BP	Postulated amelioration of climate; change to present wind patterns and northward longshore transport; resumption of high rate of beach ridge formation. Important immigration of large Indian population to coastal region.			
Period IV	1500-500 BP	Beach ridge formation along entire region of study. About middle of period a new distributary to Río Grande broke through toward south near San Blas (IVs). Main channel not abandoned until end of period.	warm	north	IV and later IVs
Shift of Rio Grande	ca. 500 BP	Abandonment of main channel (IV) of Río Grande de Santiago. Initiation of erosional transgression of coastline adjacent to former mouth and delta.			
Period V	500 BP to present (before 1530 AD)	Formation of new delta of Río Grande and erosion of coastline of old delta. Rate of coastal progradation appears to be slow.	warm	north	
Subperiod Va	ca. 500-200 BP ca. 1450-1750 AD	Firste and entry of Spanish into area 1530 AD with indications of previous separation of Río San Pedro and Río Grande. Earliest informative chart (1763) shows 3 mouths to Río Grande (IVs, Va, and Vb), at about end of subperiod Va and start of Vb.			IVs, Va
Subperiod Vb	ca. 200-100 BP ca. 1750-1850 AD	Founding of port of San Blas and abandoning of Matanchel (1768) British chart of San Blas of 1822 is during subperiod.			Vb
Subperiod Vc	ca. 100-0 BP after ca. 1850 AD	Present conditions. No basis for estimation of time of formation of river mouth Vc, but there was probably a period of use of both Vb and Vc.	warm	north	Vc

of the world. As nearly as can be determined by comparison of the 1822 chart with recent aerial photographs, the coastline at that time lay approximately in the middle of Sequence V b. Beach trends in Sequence V b suggest that the river mouth at IV s no longer existed. Thus the 1763 chart was drawn at about the end of Period V a, where the Río Grande de Santiago had a connection with the northern inlet of San Blas. The 1822 British chart was drawn in the middle of Period V b. The rivers had separated from their confluence before 1530, but we surmise that this may not have occurred very much before that time, and estimate the shift at approximately 500 years B.P. In that 500 year period, a new subaerial delta of approximately 120 km^2 has formed with loss by erosion of a lesser amount of old deltaic and strand plain material from Periods III and IV ridges farther to the north.

The coastline has prograded very little during Period V except in the delta region. The mean Period V width between the Río San Pedro and Teacapán (table 3) is of the order of 500 m. This suggests a net progradation of only about 1 m/year as contrasted with the former rates of progradation from 4.7 to 3.0 m/yr. No estimate of sand volume can be made because much of this coastline has shown net erosion rather than net deposition. Rate of formation of new beach ridges along the coastline cannot be estimated because of insufficient numbers of aerial photographs for a long enough period. With the exception of the delta region, however, it appears improbable that any new beach ridges have been added to the coast since the aerial photography of World War II.

Events in the geological history are summarized in table 4.

Discussion

The Nayarit coastal plain is a many faceted problem. It can be considered a type of delta, or a type of barrier, lagoon, strand plain coast. It can be studied as an example of the turning point from transgression to regression. It can be used as a modern model for deposition of regressive sheet sands so important in the geological record. And finally, it can be used as a model for evaluating coastal climatic changes of the past few thousand years. These various facets will be considered individually.

Deltaic Coasts

The Nayarit coastal plain is an example of a type of deltaic coast which is especially important in the geologic record. Many studies of modern deltas have been made on deltas of the very large rivers of the world, such as the Mississippi, Orinoco, Niger, Rhone, Rhine, Colorado, and Copper. Such very large river deltas may not necessarily be most typical of deltaic deposits of past geological time. While large rivers and deltas must have existed throughout geological time, perhaps the smaller, coalesced deltaic coastal plains, such as Nayarit, were also of very great importance.

The region of present investigation gains its distinctive character from concentration of deposition of sediments entering the ocean from the various rivers, Río Grande de Santiago, San Pedro, Acaponeta, and Baluarte. The coalescence of the deltas of these rivers has been vastly complicated by changing condition of sea level, climate, direction of longshore transport, and abundance of sand supply. These effects are superimposed upon the factors which are commonly considered in formation of deltas. The shape of a delta (Bernard, 1965) is a fuction of the ratio between rate of sediment influx and intensity of oceanographic conditions. With a high ratio, a birdfoot type of delta results, such as the Mississippi. With progressively lower ratios, the form becomes lobate, cuspate, or arcuate. Finally with low enough rate of sediment influx or intense conditions of wave and tidal action, no subaerial deltaic

form is evolved, and deposition is distributed uniformly along the coast and out into the continental shelf. In this region, the only external deltaic form is the cuspate mouth of the Río Grande de Santiago superimposed on a broad arc or bulge in the shoreline. Should conditions remain constant for the next several tens of thousands of years, without change in relative sea level, this process of deposition will probably continue. The entire coastline will continue to prograde at a rate somewhat lower than that of the last 4,500 years, and ultimately a large rounded bulge in the coastline, that is an arcuate shaped, coalesced delta, will result with a single cusp marking the instantaneous position of the mouth of the Río Grande de Santiago.

Barrier-Lagoon-Strand Plain Coasts

A voluminous literature exists on morphology and origin of barrier islands, lagoons and strand plains (see Shepard, 1960; and Hoyt, 1967, for recent reviews). Most such investigations have dealt with the type of barriers found on the east and the Gulf coasts of the United States, or with the chenier plain of southwestern Louisiana. Multiple beach ridges of the type which exist in Nayarit, have been noted in the literature, but surprisingly little attention has been given to complex, beach-ridge coasts of this type, even though they are probably not rare on a worldwide basis (see, for example, Lewis, 1932; Gierloff-Emden, 1959; Allen, 1965; Moore, 1966; Hay, 1967; Zenkovich, 1967). There is real need for comparison of the characteristics and study of genesis of simple, large barrier islands, chenier plains and regressive beach-ridge strand plains in the light of sea level and climatic changes of the past few thousand years. Such a synthesis is beyond the scope of the present paper, but some simple relationships should be pointed out at this time.

Depositional shoreline forms can be subdivided in many ways. Let us first consider the case of the chenier plain as opposed to common beaches and barriers.

1. Chenier plain. The chenier plain, named from the excellent example of southwestern Louisiana (Howe *et al*, 1935; Gould and McFarland, 1959), is formed by alternation of transgression and regression due to changes in rate of influx of sediment. When rate of sediment influx to that portion of the coastline is high, mud flats are prograded seaward into the Gulf of Mexico. When the Mississippi river turns and discharges its sediment farther away, wave action winnows formerly unsorted sediment, piling up beach ridges (cheniers) during a period of transgression. A chenier plain typically contains relatively few individually traceable beach ridges (e.g. 10) lying in a coastal marsh.

2. Beaches and barriers. Beaches are differentiated from barriers here only by the absence or presence of a coastal lagoon.

a. Transgressive. When the effects of rise in relative sea level exceed the effects of deposition along the coastline, net transgression or landward migration of the shoreline occurs. This can take place either with or without a coastal lagoon, but a common case is a beach migrating landward across a seaward sloping alluvial plain with, at most, only a narrow coastal lagoon, such as occurred in advance of the transgressive sands of Nayarit. These beaches and barriers are generally simple, in the sense that they do not consist of multiple ridges as in Nayarit or Galveston Island, Texas (Bernard *et al*, 1959).

b. Stable. If the effects of influx of sediment exactly balance any effects of rising relative sea level, neither transgression or regression result. In the general case, this occurs on a subsiding continental margin with a barrier island building upward in place. Slight change in balance of conditions might result in brief periods of transgression or regression, but generally the resulting barrier

will be simple, that is, without multiple beach ridges.

c. Regressive. Where rate of influx of sediment exceeds the effects of rising relative sea level, the shoreline progrades. Regression or progradation can occur by several different mechanisms. Uniform accretion to the shoreface and foreshore of the beach or barrier can build the shoreline seaward without showing apparent periodicity. A complete spectrum exists from such uniform progradation, to the regressive strand plain of the Nayarit type and to even more complex forms as shown on other coastlines. Galveston Island, Texas, e.g. is a single barrier island but with multiple, close-spaced ridges showing the trend of progradation. It could be considered as an intermediate stage between the uniform accretion and the Nayarit types.

The type of progradation shown in Nayarit evidently requires submerged longshore bars to build up above sea level and become new beach ridges. A recent review of barrier island genesis by Hoyt (1967) rejects the old hypothesis (De Beaumont, 1945; Johnson, 1919) that barrier islands have formed by emergence of formerly submerged longshore bars. The Nayarit type coast apparently represents an example of such a process, although the features which are formed are not individually barrier islands, but represent only beach ridges with average spacing of 50 m.

Along many of the other coastlines of western Mexico, multiple barrier islands exist, some with multiple beach ridges within each barrier island. Formation of multiple barrier islands is generally due to fluctuation of conditions in relative sea level, climate, or in rate of influx of sediment. Thus all combinations apparently exist, from 1) single simple barrier islands, to 2) single complex (multiple ridge) barrier islands, to 3) multiple simple barrier islands (without multiple ridges), to 4) multiple complex barrier islands.

Regressive Sheet Sands

Regressive sheet sands are known to be important in sections of ancient sediments. Many of the very extensive blankets or sheets of sand on the Colorado Plateau are possibly of regressive origin, alternating with sections of marine and continental facies and transgressive sands. To understand these sands we need to know something of the possible modes of origin of such very extensive blanket shaped deposits. Conditions of episodic and long continued subsidence must be necessary for formation of thick sections of sediment. Balance between subsidence and rate of influx of sediments produces alternate transgression and regression. The mode of formation of the littoral regressive sands of Nayarit is one possible mechanism for formation of extensive regressive sands. Detailed comparison of characteristics between well exposed ancient regressive sands and the more difficulty studied modern regressive sands of the Nayarit type will aid in understanding both ancient and modern records. Possible analogy of one of the Cretaceous sands with a Nayarit type sand was suggested by Scruton (1961).

Climatic changes

Coastal plains of the Nayarit type represent a unique opportunity for study of changes in wind and wave conditions in the past few thousand years. They are unique in that unconformities, which are usually difficult to recognize in cores through modern sediments, are here well exposed, cutting the multiple ridges of the coastal plain. Climatic changes can be judged by direction of longshore transport, character of the ridges, fauna and flora. The present study has proposed significant local climatic changes at about 3,600 and 1,500 years B.P. We need now to extend our studies to adjacent regions to evaluate the regional

significance of these climatic changes. These studies can also be profitably extended to nearby areas of multiple barrier islands, where these same climatic changes may have manifested themselves in different ways.

LITERATURE CITED

ALLEN, J. R. L. 1965. "Coastal geomorphology of Eastern Nigeria: Beach-ridge barrier islands and vegetated tidal flats." *Geol. Mijnbouw* 44 Jgr.(1):1-21.

BERNARD, H. A. 1965. "A resume of river delta types." (Abstract) *Bull. American Assoc. Petrol. Geol.* 49:334-335.

BERNARD, H. A., MAJOR, Jr., C. F. and PARROT, B. S. 1959. "The Galveston barrier island and environs: a model for predicting reservoir occurrence and trend." *Trans. Gulf Coast Assoc. Geol. Soc.* 9:221-224.

BRAND, D. D. 1958. *Coastal Study of Southwest Mexico.* In two parts, Univ. Texas, Austin, 279 p. and 140 p.

CURRAY, J. R. 1959. Coastal plain-continental shelf studies, in Study of Recent sediments and their environments in the Gulf of California. Univ. Calif., Inst. Mar. Res., IMR Ref. 59-7, Quart, Rept. API Proj. 51.

———. 1961. "Late Quaternary sea level: a discussion." *Bull. Geol. Soc. America* 72:1707-1712.

———. 1964. "Transgressions and regressions." *In: Papers in Marine Geology.* (Shepard Commem. Vol.) Macmillan, N. Y.: 175-203.

———. 1965. "Late Quaternary history, continental shelves of the United States." *In:* Wright, H. E., Jr. and Frey, D. G. (Eds), *The Quaternary of the United States.* Princeton Univ. Press: 723-735.

CURRAY, J. R. and MOORE, D. G. 1964 a. "Holocene regressive littoral sand, Costa de Nayarit, México." *In*: Van Straaten, L. M. J. U. (Ed), *Deltaic and Shallow Marine Deposits.* Elsevier, Amsterdam: 76-82.

CURRAY, J. R. and MOORE, D. G. 1964 b. "Pleistocene deltaic progradation of continental terrace, Costa de Nayarit, Mexico." *In*: Van Andel, Tj. H. and Shor, G. G. Jr. (Eds), *Marine Geology of the Gulf of California.* American Assoc. Petrol. Geol., Tulsa, Oklahoma: 193-215.

DE BEAUMONT, E. 1845. *Leçons de géologie pratique.* Paris: 223-252.

DE CSERNA, Z. 1961. "Tectonic Map of Mexico." *Geol. Soc. America.*

FLINT, R. F. and FRIEDRICH, B. 1961. "Climatic changes since the last interglacial." *American J. Sci.* 259:321-328.

FRERICHS, W. E. 1968. "Pleistocene-Recent boundary and Wisconsin glacial biostratigraphy in the Northern Indian Ocean." *Science* 159:1456-1458.

FURST, P. T. 1965. "Radiocarbon dates from a tomb in Mexico." *Science* 147:612-613.

GIERLOFF-EMDEN, H. G. 1959. Die kuste von El Salvador. Eine morphologisch-ozeanographische Monographie. (Acta Humboldtiana, Ser. Geogr. et Ethonographica Nr. 2, mit 38 Textabb., 38 ABB. a. Taf. u. 14 ktn., Wiesbaden.

GOULD, H. R. and MCFARLAN, E., Jr. 1959. "Geologic history of the chenier plain, southwestern Louisiana." *Trans. Gulf Coast Assoc. Geol. Soc.* 9:1-10.

HEY, R. W. 1967. "Sections in the beach-plain deposits of Dungeness, Kent." *Geol. Mag.* 104:361-370.

HOWE, H. V., RUSSELL, R. J., MCGUIRT, J. H., CRAFT, B. C. and STEPHENSON, M. B. 1935. "Reports on the geology of Cameron and Vermillion Parishes." *Louisiana LGS, Geol. Bull.* (6):242.

HOYT, J. H. 1967. "Barrier island formation." *Bull. Geol. Soc. America* 78:1125-1135.

JOHNSON, D. W. 1919. *Shore processes and shoreline development.* John Wiley and Sons, Inc., New York, 584 p.

LEWIS, W. V. 1932. "The formation of the Dungeness foreland." *Geogr. J.* 80:309-324.

MOORE, D. G. and CURRAY, J. R. 1964. "Sedimentary framework of the drowned Pleistocene delta of Rio Grande de Santiago, Nayarit, Mexico." *In*: van Straaten, L. M. J. U. (Ed), *Deltaic and Shallow Marine Deposits,* Elsevier, Amsterdam: 275-281.

SCRUTON, P. C. 1961. "Rocky Mountain Cretaceous stratigraphy and regressive sandstones." *Wyoming Geol. Assoc. Symp. 16th Ann. Field Conf.*, Green River: 242-249.

SHEPARD, F. P. 1960. "Gulf Coast barriers. *In:* Shepard, F. P., Phleger, F. B. and van Andel, Tj. H. (Eds), *Recent Sediments, Northwest Gulf of Mexico.*" American Assoc. Petrol. Geol., Tulsa, Oklahoma: 197-220.

ZENKOVITCH, V. P. 1967. *Processes of coastal development.* Interscience Publ., New York, 738 p.

11

Reprinted from *Sedimentology* **16**:221–250 (1971)

TEXTURAL DIFFERENTIATION ON THE SHORE FACE DURING EROSIONAL RETREAT OF AN UNCONSOLIDATED COAST, CAPE HENRY TO CAPE HATTERAS, WESTERN NORTH ATLANTIC SHELF

DONALD J. P. SWIFT, ROBERT B. SANFORD, CHARLES E. DILL JR.[1] AND NICHOLAS F. AVIGNONE

Institute of Oceanography, Old Dominion University, Norfolk, Va. (U.S.A.)

(Received September 14, 1970)

(Resubmitted October 30, 1970)

ABSTRACT

SWIFT, D. J. P., SANFORD, R. B., DILL JR., C. E. and AVIGNONE, N. F., 1971. Textural differentiation on the shore face during erosional retreat of an unconsolidated coast, Cape Henry to Cape Hatteras, western North Atlantic Shelf. *Sedimentology*, 16:221–250.

In order to evaluate a model of Holocene shelf sediment distribution requiring a nearshore modern sand facies and an offshore relict sand facies, we have undertaken a textural reconnaissance of the Virginia–North Carolina Coast between Capes Henry and Hatteras. Grab samples were subjected to grain-size analysis by means of a modified Woods Hole Rapid Sediment Analyser. Textural provinces were erected with the aid of factor vector analysis. These include medium-grained sands of the beach and surf zones; seaward fining, fine-grained sands of the shore face, and heterogenous sands of the sea floor. In this latter province, grain size is controlled by a ridge and swale topography, with coarser sand on the crests. Coast-wise grain-size trends on the beach and shore face can be explained by assuming that wave heights increase toward the south, and that the Pleistocene sediment source is exposed higher on the shore face in the north than it is in the south. The shore face is retrograding, except in the vicinity of Diamond Shoals. There is textural evidence for a former Albemarle River channel, which bisects the study area. A model for sediment fractionation on a retreating barrier coast with low sediment input is proposed, based on studies which indicate that on such coasts: (*1*) barrier superstructures retreat more or less continuously by upper shore-face erosion and storm washover; and (*2*) lower shore-face erosion results in an equal-volume aggradation of the adjacent sea floor, and forms the leading edge of the Holocene transgressive sand sheet. The nearshore "modern" sands and offshore "relict" sands are both present in the study area, but the terms are unnecessarily restrictive. Both are "relict" in the sense of being derived from a Pleistocene substrate, and both are "modern" in the sense of having undergone adjustment to a modern hydraulic regime. While modern and relict are useful general terms, it is convenient in this area to refer to a Holocene barrier sand prism, versus a Holocene transgressive sand sheet.

---

[1] Present address: Geology Department, University of Delaware, Newark (Del.).

INTRODUCTION

A standard conceptual model for the sedimentary provinces of a modern shelf includes three major facies (Emery, 1952, 1968; Curray, 1965; Swift, 1970; Swift et al., 1971). A *nearshore modern sand prism* consists of mainland beach, spit, or barrier island, and a seaward thinning and fining apron of fine sand; a *shelf relict sand blanket* deposited during Pleistocene low stands of the sea, and

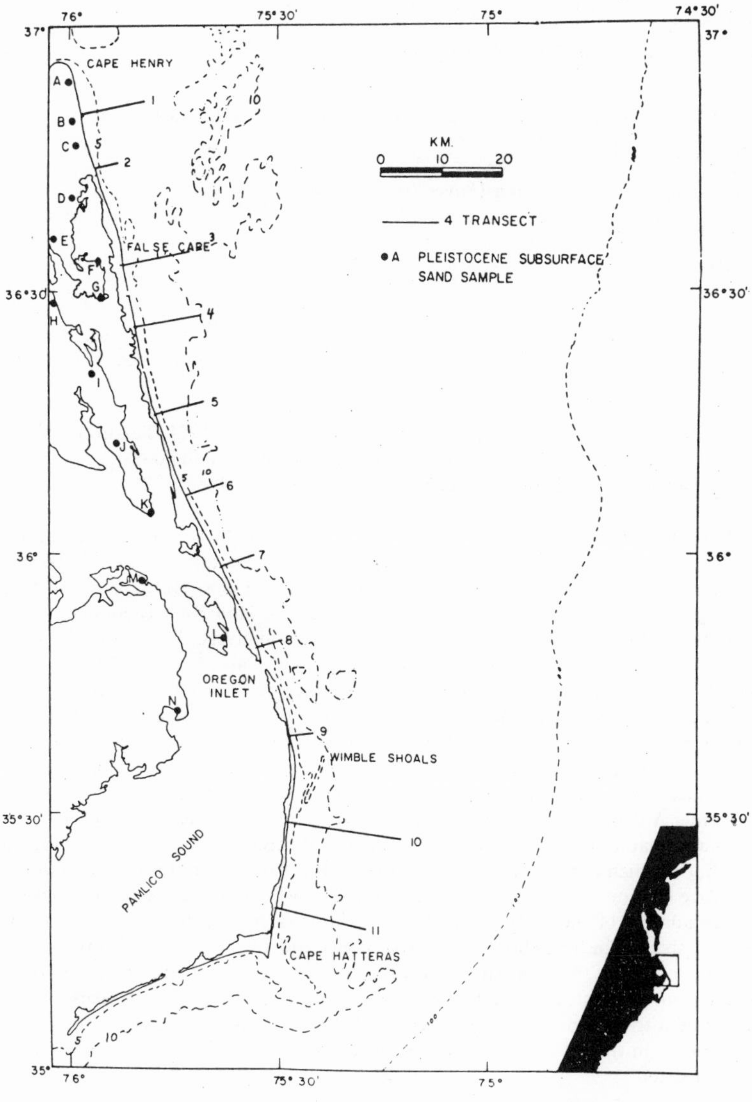

Fig.1. The study area showing profiles along which grab samples have been collected, and stations (lettered) at which the Pleistocene has been sampled.

where fine sediment input is high, a *shelf modern mud blanket* overlying the relict sand. The depositional locus of the two sand facies is the shoreline, moving across the shelf through time. In order to resolve the mechanisms by which these sand facies are generated, we have examined in detail the inner North Atlantic Shelf between Cape Henry, Virginia, and Cape Hatteras, North Carolina.

METHODS

Sample plan

We sampled along 11 transects, each 20 km apart, oriented normal to shore (Fig.1). Samples were collected on the berm of the beach, and at 2 m depth increments until the horizontal spacing between stations became 1 km. Samples were then collected at 1-km intervals until at least two successive samples consisted of the coarse, iron-stained sand considered relict (EMERY, 1968), or until a distance of 14 km from shore was reached. In addition the beach was sampled three times between transects, so that the sample spacing on the beach was 5 km. A total of 60 beach samples and 146 offshore samples were collected. A sonic bottom profile was obtained for each transect. Finally, 17 samples were collected from Pleistocene outcrops and subcrops of the Virginia Beach area, and from the inner shore of the modern lagoon system to the south (Fig.1). Pleistocene samples were collected with a post hole digger. Beach samples were collected with a stovepipe sampler (MCMASTER, 1954). Samples at 2 m depth (surf zone) were collected by hand, by a Scuba diver. Offshore samples were collected with a Shipek grab sampler, from the r.v. "Albatross", a 21-m T-boat. Navigation was conducted by means of Loran C, radar, horizontal sextant angles, range poles erected on the beach, bottom topography, and dead reckoning.

Laboratory methods

Samples were split to 200 g, sieved to obtained the —1 to 4 phi fraction, weighed, decalcified with HCl, weighed, split to 3 g, weighed, wet sieved at 62 μ, dried, and weighed again.

Size-frequency distributions were determined by means of a Woods Hole Rapid Sediment Analyser (R.S.A.), manufactured by Benthos Inc. It was necessary to repair the pressure transducer supplied by Benthos before the analyser would work. The analyser tube's height is 1 m and its diameter is 15 cm. Fall velocities were converted to phi sizes by means of a Schlee overlay (SCHLEE, 1966). Comparison of our results with sieving of the same samples suggested that settling values bear a linear or near-linear relationship to sieve values. The settling method systematically overestimated the modal diameter of our finer sand samples by as much as 0.28 phi, and underestimated the diameter of our coarser sand samples by as much as 0.41 phi, while samples of intermediate diameter underwent little

or no deviation (SANFORD and SWIFT, 1971). When replicate sample splits were run on the R.S.A., the critical factor limiting reproductactability appeared to be not the R.S.A. but the Otto microsplitter. The results of size analysis are presented in the appendix.

LATE QUATERNARY HISTORY OF THE VIRGINIA–NORTH CAROLINA COAST

The youngest Pleistocene unit of the Virginia–North Carolina Coast is Sandbridge Formation (OAKS and COCH, 1963; OAKS, 1964). The portion of this formation within our study area has been referred to by Oaks and Coch as the

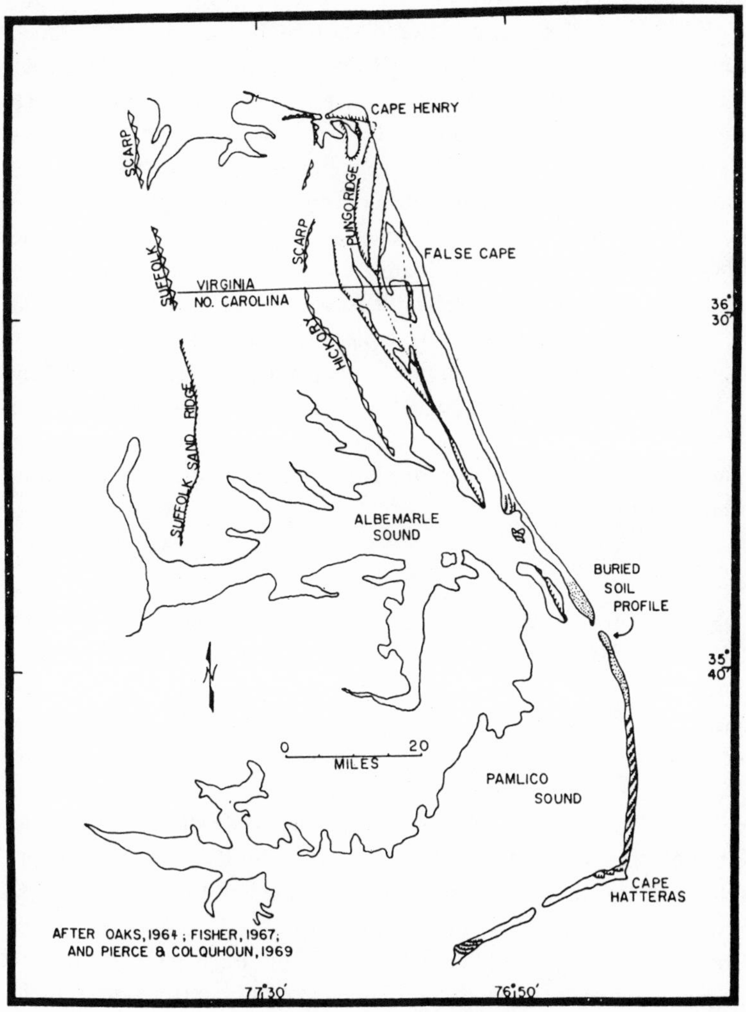

Fig.2. Quaternary framework of the study area.

"sand ridge and mud flat facies" (Fig.2), a chenier-like facies deposited by a regressive sea, possibly during Late Sangamon–Early Wisconsin time. The Sandbridge Formation is overlain by Holocene marine-marginal deposits, including a mainland beach prism in the north portion of the study area (Fig.2), Currituck Spit in the central portion; and Hatteras Island in the south, a barrier island separated from Currituck Spit by Oregon Inlet.

FISHER (1967) has suggested that the Atlantic shelf barriers were initiated by the Late Holocene reduction in the rate of sea-level rise of 4,000–7,000 years ago (see also CURRAY, 1969, p.JCII–16). NEWMAN and MUNSART (1968) have suggested a date of 5,500 years for the initiation of the Virginia Coast further to the north. There may have been an early period of barrier progradation, as indicated by beach ridge sequences (FISHER, 1967; NEWMAN and MUNSART, 1968; DOLAN, 1970). During the last century, however, the Virginia sector of the study area receded at an average rate of 22 cm/year (Felton, unpublished manuscript, Norfolk district Corps of Engineers), a rate comparable to the retreat of some other sectors of the Middle Atlantic Bight. A study by LANGFELDER et al. (1968) suggests that the North Carolina sector is similarly retreating. The rate of retreat for the Virginia Coast, when extrapolated back to 5,500 years ago yields a total retreat of 1.2 km. Stratigraphic evidence suggests a higher value. Back-barrier marsh peats, and stumps of back-barrier forests are intermittently exposed in the intertidal zone along of the barrier-front beaches of the northern part of the study area. A single stump has been dated at 725 + 70 years ago (OAKS, 1964). The nearest modern back-barrier trees are 0.32 km to the west, suggesting a retreat rate on the order of 45 cm a year and total retreat on the order of 2.5 km. However, recent vibracores collected in the False Cape sector indicate that watery muds tentatively identified as "Currituck Sound muds" underlie the shore face and sea-floor sand ridges as far offshore as 9.25 km. If this identification is correct, then either the retreat rate was yet higher, or else the retrograding barrier system is here much older than 5,500 years.

FISHER (1967; 1968a) examined the geomorphology of our study area, and reports that the southern portion of Hatteras Island is crossed by transverse beach ridges, suggesting that at least this portion of the barrier system originated as a coast-parallel, distally-prograding spit. Fisher has noted the proximity of the mainland to the barrier system in the vicinity of Oregon Inlet, and suggests (personal communication) that the ancestral Hatteras spit grew south from a former "Bodie Island headland". This suggestion has gained support from the drilling study of PIERCE and COLQUHOUN (1970). They note that on both sides of Oregon Inlet, and on the northern half of Hatteras Island, the Holocene sequence of barrier sand over lagoonal mud is separated from the underlying Pleistocene by a fossil soil profile (Fig.2). North and south of this sector, Holocene rests on Pleistocene with no intervening soil. They conclude that the barrier over the soil profile is a "primary barrier" (primary shoreline; COLQUHOUN, 1969). HOYT (1967)

has suggested that such barriers start as mainland beaches which are detached from land by a rising sea level; the drift-nourished beach grows upward, while the swale behind it floods.

The origin of Currituck Spit, north of Oregon Inlet, is problematic. Insufficient evidence is available to indicate whether it, too, is a detached mainland beach, or whether it is a genetic, as well as geomorphic spit.

AREAL VARIATION OF TOPOGRAPHY AND GRAIN SIZE

The general relationships between grain size and topography may be seen in Fig.3, 4, 5 and 6. Fig.3 and 5 indicate that at most transects, two well defined geomorphic provinces are present. The nearly planar shore face has a gradient of 8 m/km down to 14 m depth. Seaward of the shore face an undulating shelf floor slopes seaward at 1 m/km. Perusal of ZENKOVITCH's (1967, p.205) survey of coastal profiles suggests that such sharply angulated profiles are characteristic

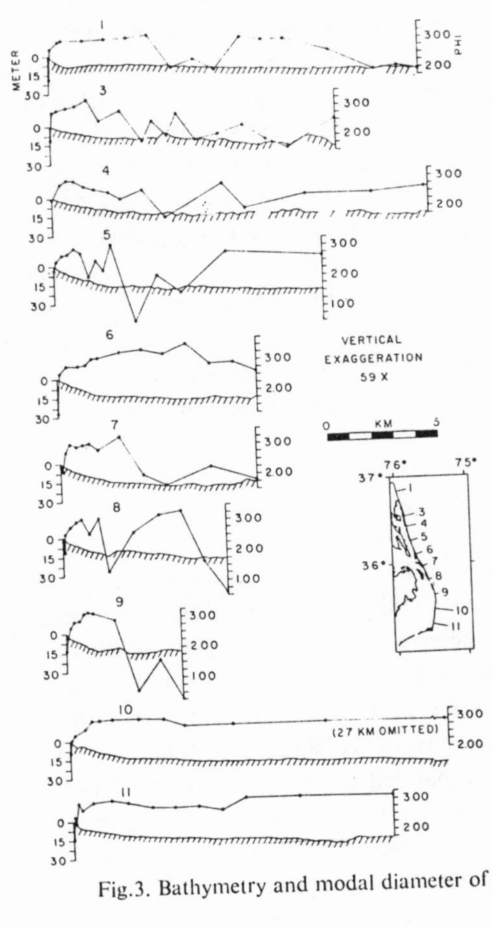

Fig.3. Bathymetry and modal diameter of transects.

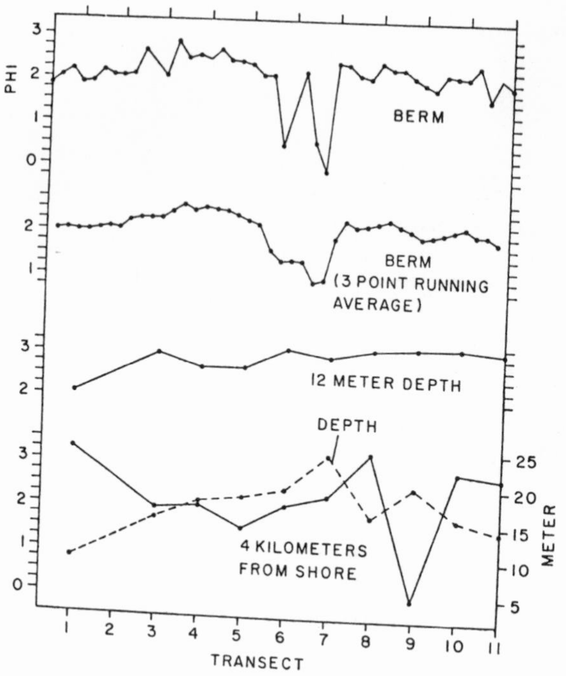

Fig.4. Longitudinal profiles of modal diameter at berm, 12 m depth, and 4 km from shore.

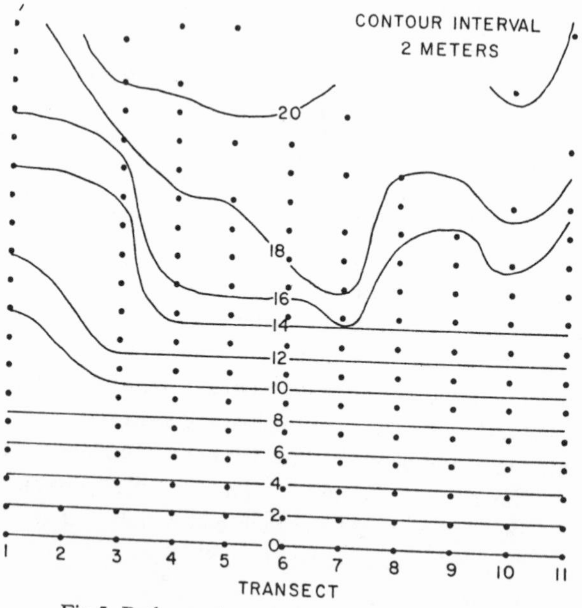

Fig.5. Bathymetric map of study area. The sample net in this and following figures has been systematically distorted in order to more clearly reveal regional variation, by designating an arbitrary, fixed interval for the east–west axis, and a second arbitrary interval for the north–south axis.

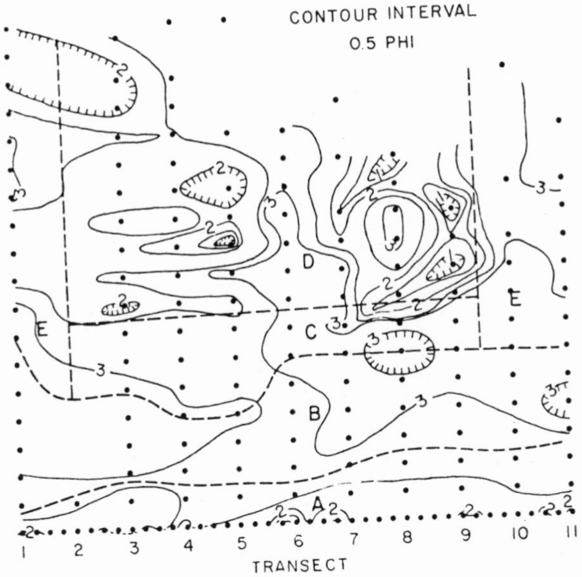

Fig.6. Phi modal diameter map and geomorphic textural provinces. For explanation, see text.

of retreating, unconsolidated coasts. Reconnaissance vibracoring in the False Cape area suggests that the inflection point occurs at the intersection of the profile with the contact between soft Holocene lagoonal muds and the Pleistocene "basement". Fig.5 and 6 suggest that the area may be divided into five provinces on the basis of morphology and modal diameter. These are (Fig.6): *A*. the coarse beach and surf sand; *B*. the fine, seaward-fining sand of the upper shore face; *C*. the heterogenous sand of the lower shore face; *D*. the heterogenous sand of the sea floor; and *E*. the fine sand of the terminal shoals. The Pleistocene substrate, presumed source of sediment, comprises a sixth province.

In order to assess the distribution of grain sizes within the study area by an independent, more quantitative technique, the modal diameter data were subjected to factor vector analysis (KLOVAN, 1966). This analysis permitted the resolution of each sample in terms of four, hypothetical, end-member size distributions. Since the program did not compute normalized factor-loading scores, it was not possible to directly assess these end numbers. Instead, cumulative curves of real samples were grouped for comparison by dominant end number (Fig.7). In addition, a map of the study area was prepared for each end member, and contoured for the absolute value of the 0.5 factor loading (Fig.8). The sand types are described below in order of decreasing abundance.

Factor-1 sands are fine to very fine sands of the shore face and Diamond Shoals. Although it is not always apparent on the cumulative curve envelopes, the curves of this (and other factors) commonly consist of three straight-line

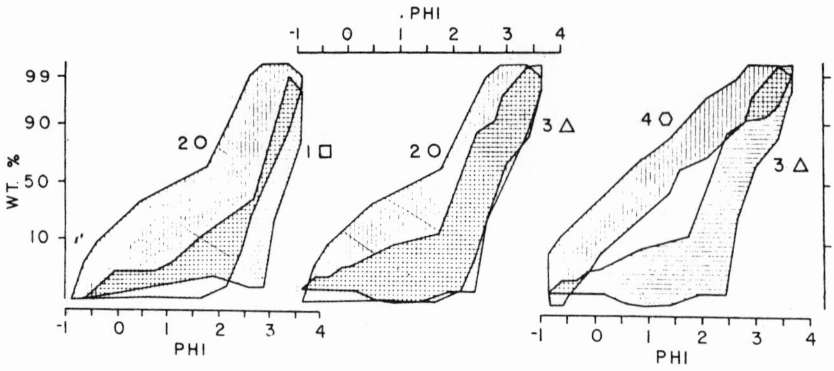

Fig.7. Cumulative curve envelopes for the 4-factor sediment classes.

segments, indicating the presence of three log-normal subpopulations. In Table I, we have adopted the premise of VISHER (1969) that these populations were deposited by rolling or dragging, saltation, or from suspension, and have indicated the weight percent assigned to each. Such an analysis indicates that factor 1 sand comprises 1% traction load, 11% saltation load, and 80% suspension load. Table I indicates that this is the finest, best sorted, most strongly fine-skewed, and muddiest of the sand types.

Factor-2 sands are medium- to fine-grained sands found in the beach and

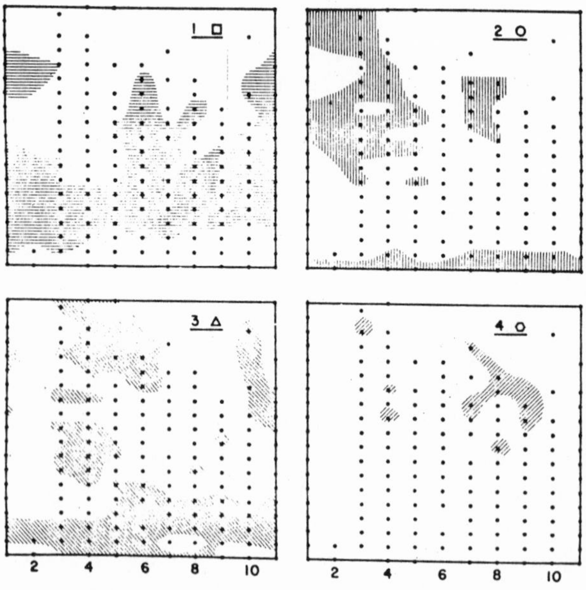

Fig.8. Distribution of factor sediment types. Shaded areas contain absolute factor loadings in excess of 0.5.

TABLE I

MEAN PARAMETERS OF FACTOR SAND TYPES

Sand types		*Mode*	*Mean*	*Standard deviation*	*Skewness*	*Kurtness*	*% fines*	*% carbonate*	*Population*		
									% coarse	*% central*	*% fine*
Factor-1	$\overline{X}$	3.15	3.35	0.30	0.08	1.10	9.15	1.93	1	11	88
	S	0.14	0.45	0.06	0.17	3.81	6.77	0.86			
Factor-2	$\overline{X}$	2.22	1.93	0.65	—0.23	1.37	1.13	1.02	25	57	18
	S	0.15	0.41	0.25	0.30	3.11	0.72	3.11			
Factor-3	$\overline{X}$	2.17	2.82	0.35	0.08	2.17	2.97	1.08	5	47	48
	S	0.19	0.13	0.11	0.29	2.94	1.85	0.74			
Factor-4	$\overline{X}$	1.03	1.10	0.81	0.07	—0.17	2.36	3.42	59	29	12
	S	0.50	0.23	0.15	0.25	0.34	2.83	2.32			

surf zone, and also in the northern offshore sector. They consist mainly of a saltation subpopulation.

Factor-3 sands are fine-grained sands of the shore face and portions of the sea floor. They consist of sub-equal amounts of saltation and suspension sub-populations.

Factor-4 sands are coarse- to medium-grained offshore sands. They are dominated by a tractive sub-population. They are the most shell rich (up to 9.0%) of the sand types, and next to the very fine-grained type 1 sands, the most enriched in fines (up to 9.5%).

INTERPRETATION OF GEOMORPHIC-TEXTURAL PROVINCES

Pleistocene substrate

The seventeen Pleistocene samples can provide only a generalized idea of the nature of the Pleistocene substrate of the study area, because: (*1*) the number of samples is small; (*2*) they were collected a considerable distance west of the study area; and (*3*) the Pleistocene outcrop pattern resulted in bias towards topographic highs (the beach and dune sands).

OAKS (1964) describes the underlying Sandbridge Formation as consisting mainly of ridges of medium- to fine-grained sand separated by bodies of fine sand and clay silt. He interprets the sequence as a prograding barrier-lagoon complex. Of seventeen samples analyzed, ten were medium-grained factor 2 sands, three were fine-grained factor 2 sands, and one was a coarse-grained factor 4 sand. The

samples differed from the modern samples in a lack of shell material (up to 21%), and in tendancy toward a high percentage of fines (up to 33%). The low shell content is undoubtedly due to post-depositional leaching, while the samples enriched in silty clay probably represent lagoonal facies not present in the modern study area.

While OAKS (1964) is more concerned with stratigraphy than with sedimentological analysis, he notes (p.162) that "fine-grained sand" is predominant in the upper member of the Sandbridge Formation. Our Sandbridge samples were predominantly medium-grained factor 2 sands; but Oaks' estimate based on subsurface core samples from sand sheets as well as surface samples of sand ridges is more reliable. Our modern inner-shelf samples are, on an equal-area basis, mainly factor 2 medium-grained sands. Apparently the prograding Pleistocene marine marginal environments filtered out coarser sediment before it reached the open coast, while the modern retreating environments do not. See later discussion.

Beach, surf zone and shore face

The hydraulic-process model. Differentiation of the heterogenous Pleistocene substrate into the well-defined textural geomorphic provinces of the shore face must be understood in terms of the hydraulic processes characteristic of the zones of shoaling and breaking waves. These processes can be conveniently divided into two groups; processes which move sediment parallel to the beach as littoral drift, and processes which move sediment onshore or offshore.

Littoral drift is the result of longshore currents generated by obliquely-breaking waves. LANGFELDER et al. (1968) have attempted to obtain quantitative data on littoral drift between Cape Hatteras and the North Carolina line by means of wave-refraction studies. Wave-height, period, direction, and duration data were collected from published studies of the U.S. Corps of Engineers, and nearshore bathymetric data were tabulated from Coast and Geodetic Survey charts. Wave-refraction patterns were calculated by computer from this data, and the directions and rates of littoral drift for successive sectors of the coast were determined. Langfelder et al. point out that "although a limited number of wave data are available, sufficient statistical data from which significant wave heights and periods could be obtained with confidence do not exist... their limitations should be kept in mind in evaluating the results...". Directions of littoral drift in the study area as determined by Langfelder et al. are presented in Fig.9. There is a major node at Oregon Inlet; with drift trending to the north and, with some minor reversals, to the south. Thus, littoral drift on Currituck Spit trends from its distal toward its proximal or mainland-tied end. OAKS (1964) has suggested that the Late Pleistocene Sandbridge beach ridges were built by such northward drift, since they bifurcate to the north.

Before investigating the grain-size data for the shore face, it is worth reviewing

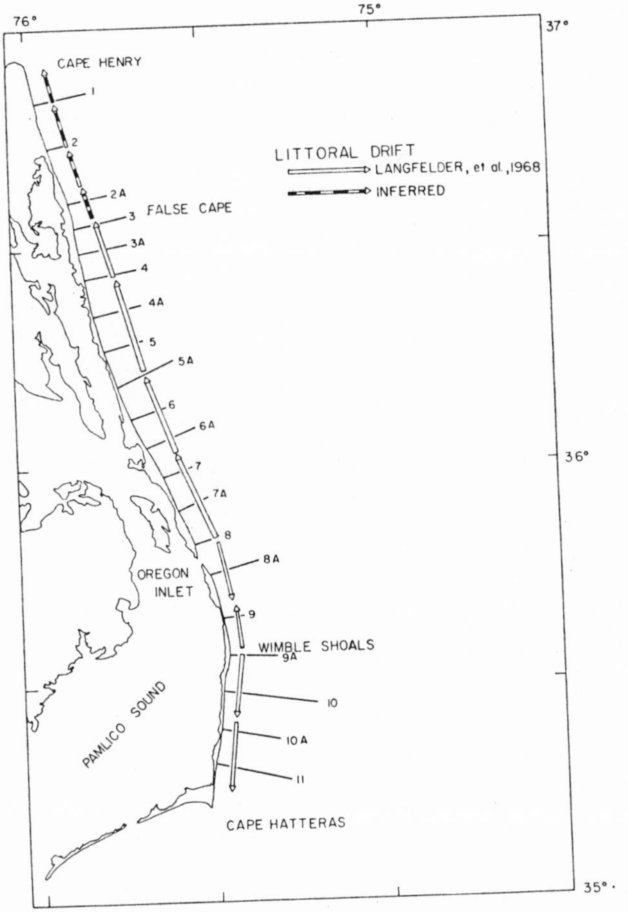

Fig.9. Littoral drift along the study area. Inferred drift based on shoaling next to wrecks and groins.

the less obvious but equally important hydraulic mechanisms for onshore–offshore transport. In the beach and surf zone, the direction of onshore–offshore transport depends on the shape and height of waves. During fair weather the long-period, flat swells which approach the beach have an extremely asymmetrical pattern of surges; landward crestal surges are much stronger than seaward trough surges (INMAN and NASU, 1956). Coarse sand moves landward as bedload onto the plunge-point bar and across it to the beach.

Fine sand, thrown into suspension as a wave crest passes, tends to be moved seaward by the mid-depth return flow associated with mass transport (IPPEN and EAGLESON, 1955), or by channelized rip currents. The fine sand settles out on the shore face seaward of the breaker as a transient, fair-weather veneer.

Here, in the zone of shoaling waves, size fractionation continues. Some

secondary sorting mechanisms may be important in this sector. KEULEGAN (1948) has pointed out that as the crestal surge of a surface wave sweeps over a rippled bottom, coarser sand is moved landward as bedload, while finer sand is suspended in the horizontal roller that develops in ripple troughs. These rise during the course of the surge; when the weaker reverse flow starts, the roller "explodes", and the finer sand is swept seaward. SCOTT (1954) has attempted a quantitative approach to the resulting size fractionation of the sand.

Another sorting mechanism is slope sorting, or the null line mechanism. This concept states that each bedload particle seeks an equilibrium position on the shore face, where the landward-directed net wave surge is balanced by the down-slope component of gravitational force. As smaller and smaller particles are considered, their mass and the corresponding gravitational effect diminish more rapidly than do surface area and the corresponding hydraulic drag. Hence, smaller particles must seek an equilibrium position further down the shore face than larger ones. See JOHNSON and EAGLESON (1966), MURRAY (1967) and SWIFT (1969, 1970) for evaluations of this mechanism.

COOK (1969) believes that sequential settling of suspended sand out of rip current heads is the dominant process, and that the sand of the resulting blanket is so well sorted that it "migrates as a unit", rendering ineffective the above secondary sorting mechanisms. While further work remains to be done before the significance of these mechanisms can be resolved, it is clear that a fair-weather swell regime serves as a fractionating mechanism for shore-face sediments.

During storms, the steep-sided waves have a more symmetrical surge pattern (INMAN and NASU, 1956), hence their ability to transport sand shoreward is not necessarily enhanced, even though they are higher and contain more energy. But because maximum orbital velocities are higher, more sand goes into suspension, and the grain-size boundary between suspensive and tractive loads is shifted toward the coarser end of the distribution, so as to include more sediment. To the extent that storm wave period tends to be relatively short, there is little change for suspended sand to settle to the bottom. The result is that a greater portion of the available ranges of sizes, and absolutely a greater amount of sand is transported seaward by the mid-depth return flow during storms. During intense storms the beach may be deprived of its modern sands. HARRISON and WAGNER (1964) report that portions of the northern part of the study area were stripped down to a substrate composed of "cypress stumps rooted in silt" during the Ash Wednesday storm of 1962. Erosion of such substrates during storms releases sediment which serves as grist for the "fractionation mill" of the fair weather swell regime, and the cycle begins again.

Sediment response in the beach and surf zones. The previously cited diagrams indicate that there is a systematic distribution of grain parameters on the shore face, which reflect the hydraulic processes described above. The first pattern to

be considered, although not necessarily the easiest to explain, is grain-size variation down the beach. Fig.4 indicates that when the data is smoothed, grain size on the berm decreases from 1.77 phi at Cape Henry to 2.73 at transect 3, then increases to 1.85 at Cape Hatteras. Fig.8 shows that in the area of the regional grain-size minimum, the fine factor 3 sands normally found just seaward of the breakers are present on the beach as well. Superimposed on this regional trend is a pronounced coarse anomaly at transects 5, 6, and 7, the Nags Head area of Currituck Spit. The anomaly on the unsmoothed curve of Fig.4 is as coarse as very coarse sand. Patches of very fine to fine gravel, rare elsewhere, are abundant on this sector of the beach.

The anomaly is the most readily resolved portion of the curve. FISHER (1967) has interpreted a relict beach ridge pattern in this area as indicative of a former inlet and points out that it is sited directly opposite the mouth of Albemarle Sound (Fig.2). We infer that when this inlet was closed, the sediment that closed it was mainly excavated by the surf from the immediate substrate, and that relatively little far-traveled sediment was supplied by littoral drift. We further infer that the substrate so mined was the gravelly channel of the former Albemarle River. If so, the inlet may have originated as a detached river mouth, and Currituck Spit as a detached mainland beach, after the manner described by HOYT (1967).

The reason for the regional trend of grain-size coarsening towards both ends of the study area must be sought in the interaction of lateral and onshore–offshore transport mechanisms, and their effect on sediment characteristics inherited from the Pleistocene substrates. Lateral transport (littoral drift) seems to have the least to do with the grain-size pattern. There is no symmetry of grain sizes about the Oregon Inlet node. The increase in grain size from the grain-size minimum at transect 3 towards Cape Hatteras may correlate with a southward increase in average wave height. Although the Corps of Engineers maintains 3 wave gages in or near the study area, they are so few in number, and are placed in such heterogenous environments that they are not able to substantiate this hypothesis. The inference is based on the fact that Cape Hatteras is a regional promontory and thus would presumably concentrate wave energy, while Cape Henry is toward the head of the Middle Atlantic Bight.

If this explanation is valid for the southward increase in berm grain size through the south half of the study area, it fails for the northward coarsening northern half. It is possible that grain size in this sector is dominated by sizes inherited with little change from the Pleistocene substrate. This sector is primarily mainland beach, and OAK's (1964) profiles suggest that Pleistocene outcrops much higher on the beach face of the mainland beach, than it does on the beach face of Currituck Spit (Fig.10). As previously indicated, back-beach peat and tree stumps are exposed on the shore face of this sector, indicating appreciable erosion.

Sediment response on the shore face. The shore face consists of a uniform blanket

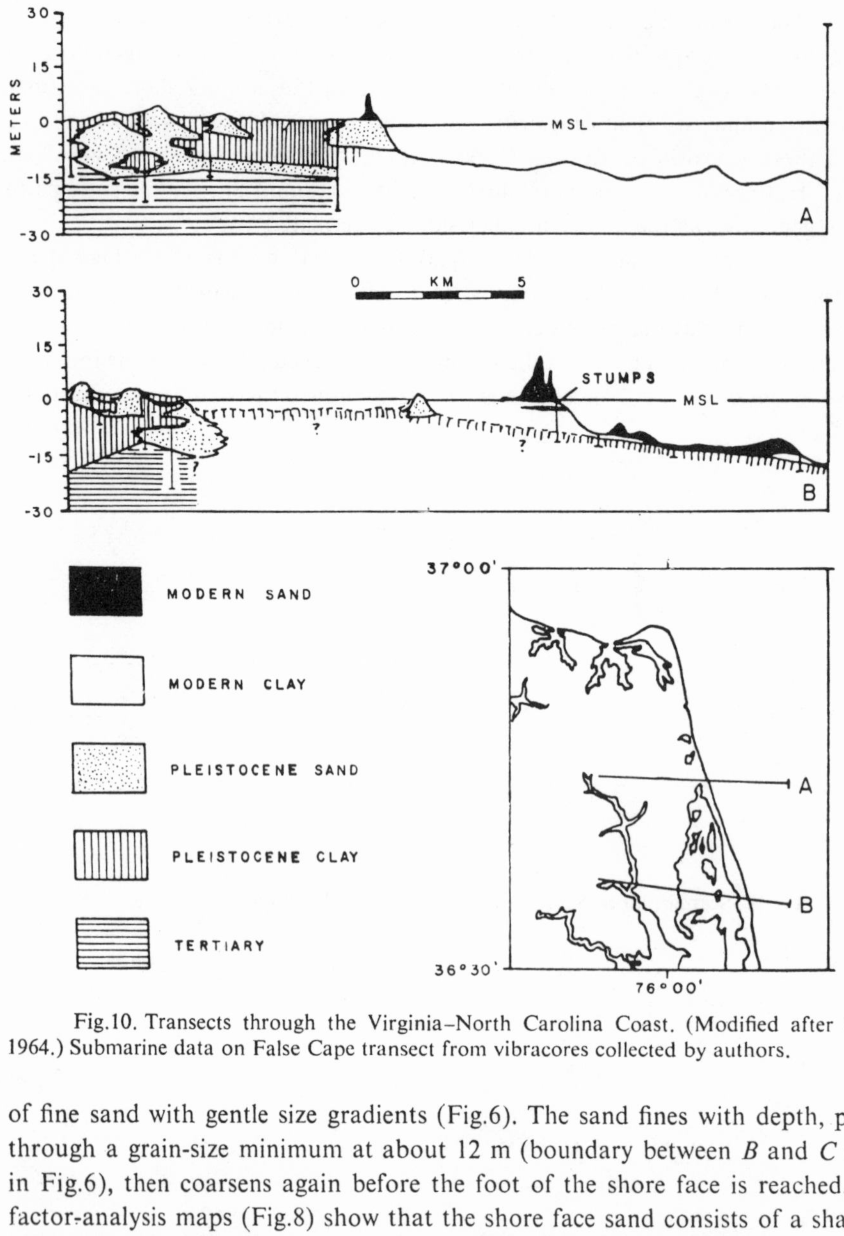

Fig.10. Transects through the Virginia–North Carolina Coast. (Modified after OAKS, 1964.) Submarine data on False Cape transect from vibracores collected by authors.

of fine sand with gentle size gradients (Fig.6). The sand fines with depth, passes through a grain-size minimum at about 12 m (boundary between *B* and *C* areas in Fig.6), then coarsens again before the foot of the shore face is reached. The factor-analysis maps (Fig.8) show that the shore face sand consists of a shallow-water ribbon of fine factor-3 sand, and a deeper ribbon of fine to very fine factor-1 sand. Both belts broaden southward, and the factor-1 zone swings seaward at transect 10 to become the north flank of Diamond Shoals. The broadening (and thickening?) of the shore face bands of fine sand towards the south may be a corrollary of the southward coarsening of the berm. With wave height increasing to the south, the surf would transport more fine sediment seaward as suspended

load. There is a suggestion (the shoaler nature of transects *10* and *11*, Fig.3), that towards Cape Hatteras the lower shore face is actually aggrading, due presumably to a southern flow of fine sand from the eroding shore face further north (see analysis of littoral drift, Fig.9). Such aggradation would be in accord with the suggestion of TANNER (1960), and EL-ASHRY and WANLESS (1968) that Cape Hatteras is a terminus for littoral drift. Diamond Shoals has perhaps thus been generated as the coast on either side has retreated.

Transects normal to shore (Fig.3) show that on the shore face the fine factor-2 and factor-3 sands become increasingly finer seaward. Fig.11 and 12 show that modal diameter decreases and percent fines decreases as exponential functions of depth. The hydraulic mechanisms believed responsible for this sorting have been discussed. Percent fines increases down the shore face presumably

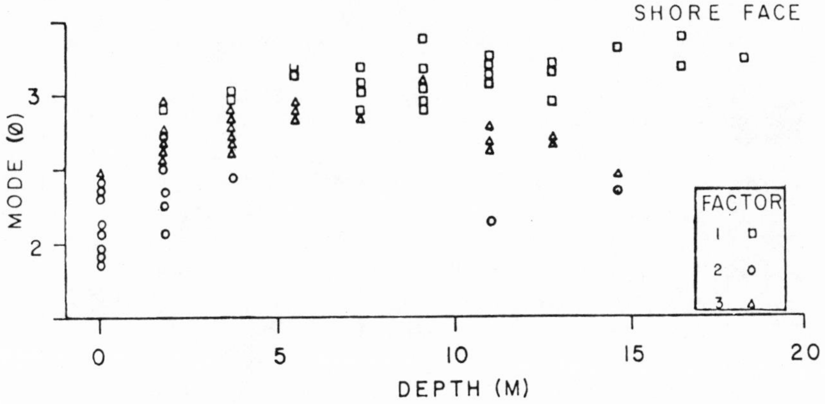

Fig.11. Scatter plot of phi modal diameter versus depth on the shore face.

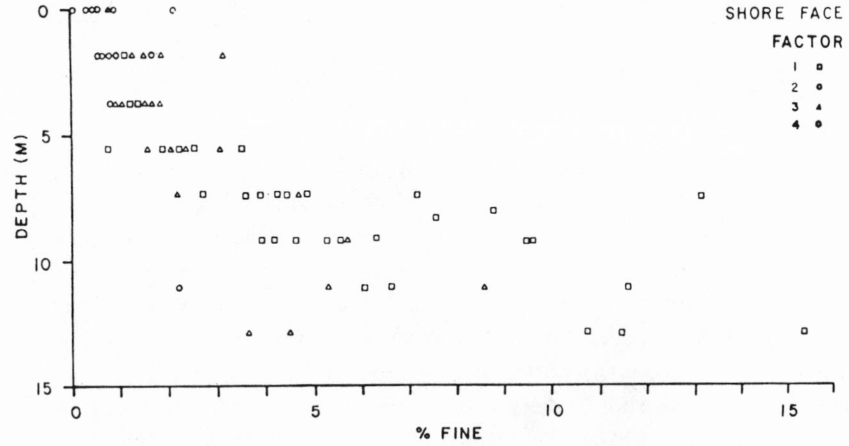

Fig.12. Scatter plot of percent fines versus depth on the shore face.

because as wave energy and the rate of coarse sediment input decreases, burrowing invertebrates are better able to work fine sediment into the substrate.

The most cryptic feature of the shore face is the grain-size minimum. Fig.13 shows in more detail the sequences of grain-size distribution down the shore face. At 10 m the fine sand belt has generally attained its maximum sorting and fineness. At 12 m, still part of the topographic shore face, the sand tends to be slightly coarser, and more poorly sorted. At 14 m a secondary coarse mode tends to supplement the mode seen at the preceding station, suggesting a mixture of sediment types. Grain size at the 16-m station, generally the last station that can be referred to the more steeply sloping shore face, is highly variable.

Clearly the fair-weather fine sand mantle of the upper shore face wedges out at about 10 m. The coarse material below may be a residuum resulting from storm erosion of the Early Holocene or Pleistocene substrate, may reflect a storm rather than a fair-weather hydraulic regime, or may be a result of both factors.

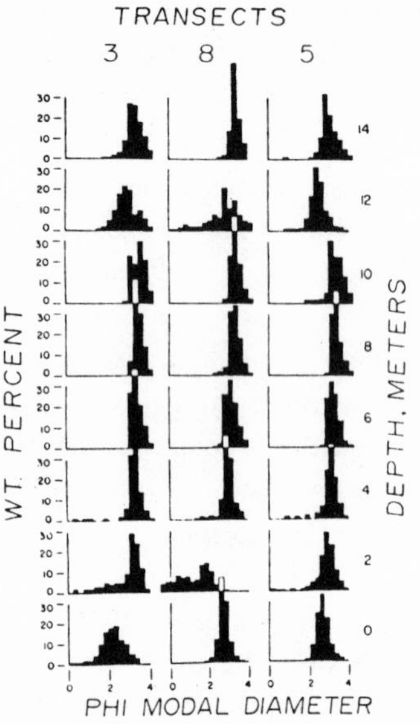

Fig.13. Size distributions of three transects.

The sea floor

Hydrographic data. While scattered hydrographic data is available for the sea floor of study area, it is insufficient to construct a coherent hydraulic process model for the sea floor. Waves are the most obvious energy input. Summer waves are predominently from the southeast; winter waves are predominantly from the northeast (unpublished data, Norfolk district Corps of Engineers). On a yearly basis, waves from the northeast predominate (18%). Of this percentage, 4% of

the waves are over 4 m high, 18% are 2–4 m high, 59% are 0–2 m high, and calm prevails 19% of the time. Wave-length figures are not available, hence it is difficult to predict the effect of these waves on the bottom. However, during Scuba dives we have observed active, rippled bottoms, out to 10 m when surface waves were less than 1 m high. Bottoms in the deeper parts of the study area (up to 20 m) were also rippled, but these deeper ripples were generally covered with a brown algal film. Presumably the sea floor in all portions of the study area mobilized during the 22% of the time when waves are over 2 m high.

Tidal currents are the next most obvious source of hydraulic energy input. In the course of ongoing hydraulic studies, we have recorded mid-tide velocities integrated from 5–15 cm off the bottom, of 5–25 cm/sec. The velocity fields that we measured probably include components of permanent coastal current. Storm surge might conceivably increase these values by 2- or 4-fold.

Sediment response on the sea floor. Fig.5 indicates that the sea floor is relatively shoal at both ends of the study area, and is deepest along transects *6* and *7*. The southern shoal has already been referred to as the north flank of Diamond Shoals. The Cape Henry Shoal may be associated with a clockwise residual current gyre (HARRISON et al., 1964), resulting from the interaction of the tidal jet of Chesapeake Bay with the shelf tide. PAYNE (1970) has referred to this shoal area as the southern flank of a Chesapeake Bay Mouth, submarine, "tidal delta". Fisher (1968b; personal communication) has suggested that this shoal sector is a drowned "False Cape headland".

The factor-loading maps (Fig.8) show the sea-floor sands as divisable into a heterogenous northern area with medium-grained factor-2 sand dominant, a central tongue of very fine factor-1 sand, a southern area with the coarse calcareous, poorly-sorted factor-4 sand dominant, and finally the fine factor-1 and factor-3 sands of the north flank of Diamond Shoals. The key to the interpretation of this gross distribution of sediment and topography is the tongue of very fine factor-1 sand which extends seaward from the shore face into a deeper basin of the sea-floor province along transects *5*, *6*, and *7*. Since it is opposite the "Albemarle Inlet" described on a previous page, it is most easily explained as the drowned former Albemarle River Valley which is presently receiving fine sand winnowed from surrounding highs and the shore face. This hypothesis supports Fisher's contention that the shoal area to the north Fig.1 is a drowned former headland, although the sector may still have been modified by the strong tidal currents associated with the mouth of Chesapeake Bay, as suggested by PAYNE (1970). Likewise the sea floor high in the southern part of the study area may be a drowned interfluve on which Diamond Shoals has been built by modern hydraulic processes.

Relief on the sea floor consists of a ridge and swale topography with an amplitude of up to 6 m (Fig.14). The system is trochoidal in that sharply rounded ridges are generally about 0.5 km wide; broad swales are more typically 2–4 km wide.

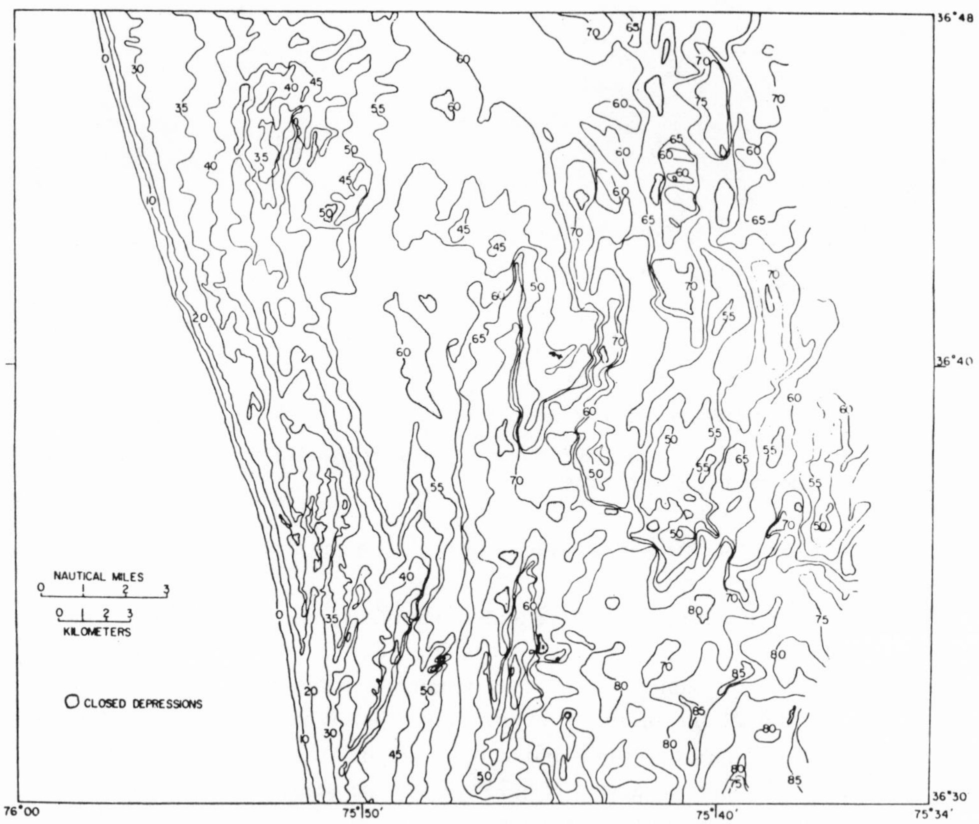

Fig.14. Bathymetry of the northern portion of the study area, from 1922 smooth of coast and geodetic survey chart. The three beach-tied ridges and the fourth offshore ridge at False Cape may represent an evolutionary sequence, with the offshore ridge being the oldest. Note that the three beach-tied ridges constitute a promontory of the shore face, which is here defined by the 50-ft. contour. This map covers transects *1*, *2*, and *3* of Fig.1.

The ridges are generally shore-parallel, but locally converge with the beach. They tie to the shore face in groups of 3 or more at 5–15 m depth, at False Cape and Wimble Shoals. The offshore ridge and swale topography has been considered a relict strand plain (FISHER, 1968b), and the False Cape system a sequence of relict beach ridges (SANDERS, 1962). However, the False Cape ridges appear to have moved shoreward up to 200 m since the 1922 Coast and Geodetic Survey (SWIFT et al., 1971). The ridges may be traced to within 100 m of the beach, in areas where, at a conservative estimate, the beach must have retreated over 1 km since its formation. Here the ridges lie in the zone swept by the retreating barrier and must be hydraulic rather than relict in origin. This hypothesis is strengthened by vibracores which reveal a watery mud of probable lagoonal origin between the modern sand sheet and a substrate of presumed Pleistocene age (Fig.10). The

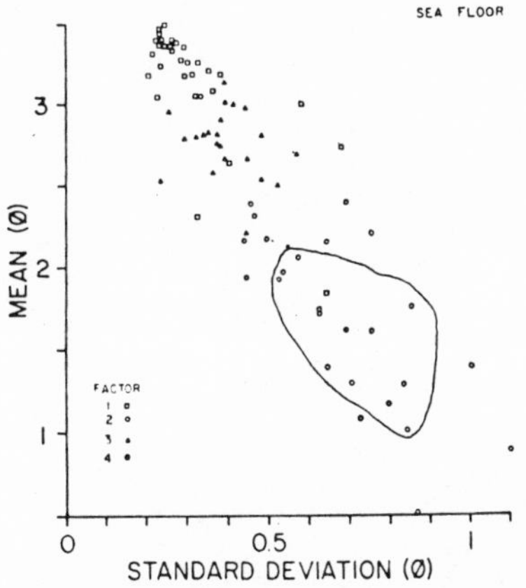

Fig.15. Phi moment mean versus phi moment standard deviation of samples. Samples from topographic highs are circled.

problem of the origin of the ridge and swale topography will be considered in more detail in a later paper.

The heterogenous nature and apparently random distribution of grain sizes on either side of the presumed Albemarle Channel is a result of the fact that the interval of the sample net is much greater than the wave length of topographic relief, and of the resulting textural variation. When mean diameter is plotted against standard deviation, and samples taken from crests of the ridge and swale topography are so labeled (Fig.15), it can be seen that crests are consistently coarser than the intervening swales.

A MODEL FOR TEXTURAL FRACTIONATION ON A RETREATING BARRIER COAST

The data presented on preceding pages, plus published studies of nearshore hydraulic processes permit us to suggest a process-response model for textural fractionation on the shore face of a low, unconsolidated coast with a low sediment

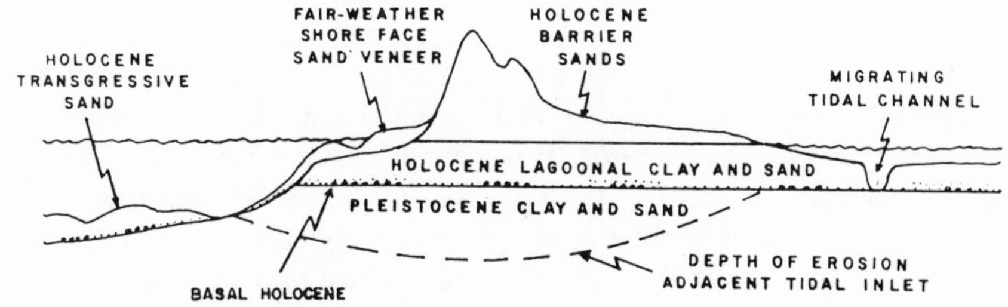

Fig.16. Stratigraphic response model for an unconsolidated coast undergoing erosional retreat during a period of rising sea-level. Note that erosion in tidal inlets and in tidal creeks supplements shore face erosion, in production of Pleistocene–Holocene disconformity.

input, undergoing erosional retreat. The process model is presented in Fig.16. Two key ideas are DILLON's (1970) concept of continuous barrier retreat, and "Bruun's rule" (BRUUN, 1962; SCHWARTZ, 1965, 1967, 1968) of sea-level rise and concomitant shore-face erosion. Dillon reviews a model for coastal retreat during a period of sea-level rise in which barriers are stabilized by a high rate of sand input, generally of fluvial origin. Their upward growth may thus keep pace with the rate of sea-level rise. However, this is a "self-defeating operation" (DILLON, 1970), since as their submarine surface area increases, more and more sand is needed to keep pace, and eventually the supply, if unchanged, must become inadequate to maintain them. CURRAY (1964, p.199) suggests that any change in the rate of supply will probably be a decrease since, as sea level rises, the back-barrier lagoon becomes broader, and an increasingly efficient trap for fluvial sand. Eventually the starving barrier must be overstepped, and the process must be repeated inshore.

Dillon believes this model to be inapplicable to the Rhode Island barrier he has studied. As in the case of the Middle Atlantic Bight barriers, there is no significant fluvial sand supply. Consequently the barrier has migrated more or less continuously landward by a process of shore-face erosion (FISCHER, 1961; SWIFT, 1968) eolian action, and storm washover. BRUUN (1962) and SCHWARTZ (1965, 1967, 1968) have considered the nature of shore-face erosion during a period of sea-level rise. "Bruun's rule" as proposed by Schwartz says that a rise in sea level will result in erosion of the shore face and an equal-volume aggradation of the adjacent sea floor. DILLON (1970, p.105) notes that this rule must be modified to the extent that the upper shore face loses sand to the barrier flat by eolian activity and storm washover. Still, there are two noteworthy studies of nearshore processes in the Middle Atlantic Bight which suggest that Bruun's rule is a valid approximation for this area. HARRIS (1954) conducted precision navigation surveys of the Long Branch, N.J., dredge-dump site for a period of 4 years. He noted that the entire shore face underwent erosion while most of the adjacent sea floor aggraded during this period. MOODY (1964, pp.142–147) compared the results of his own survey with U.S. Coast and Geodetic Surveys of the Delaware Coast dating back to 1919, and concludes that erosion of the barriers was balanced by aggradation of the adjacent sea floor during this period. Both authors appear to have reached their conclusions independently of Bruun and Schwartz.

A synthesis of Dillon's and Bruun's concepts, together with published studies on the nearshore hydraulic regime and our data for the Virginia–Carolina Coast suggests the following model for sediment fractionation during the retreat of a barrier coast (Fig.16).

The superstructure of the barrier retreats in cyclic fashion, by shore-face erosion, storm washover, burial, and re-emergence on the upper-shore face. The lower-shore face is effected mainly by the steep-sided waves of occasional storms. These appear to move sediment mainly seaward, aided perhaps by the

return flow currents associated with storm surge. Here it is added to the leading edge of the Holocene transgressive sand sheet, which becomes in this model a nearshore rather than a truly littoral facies.

The topography and grain-size distribution of the inner sea floor are problematic. A further sediment fractionation appears to be transpiring whereby fine sand is being winnowed out of ridge crests, and is accumulating in swales. The transfer of fines from ridge crests to swales is most easily attributed to the fair-weather wave regime that has been observed by divers. If so, this basically degrading regime could not build the ridges; they are more probably a response to storm surge currents, as suggested by UCHUPI (1968). MOODY (1964) notes that during the Ash Wednesday storm of March 1962, nearshore ridges of the Delaware Coast shifted as much as 70 m.

The scale of the winnowing process is not clear. As a whole, the sea-floor province (exclusive of Diamond Shoals) appears to be deficient in fine sand relative to the other modern facies, even after the distortion of Fig.8 is taken into account. This is a further suggestion that this province, with its ridge and swale topography is not simply a drowned strand plain. As noted by PAYNE (1970), it is also coarser than the subjacent Pleistocene source. Possibly the missing fines are to be sought on Diamond Shoals, in which case there is significant lateral transport and winnowing on a regional scale.

There is a second explanation for the anomalously coarse nature of the sea-floor province. It requires addition of coarse sediment, rather than subtraction of fines. It does not necessarily conflict with the preceding one, but unlike the latter, has hard data to support it. We suggest that as the belt of lagoons and estuaries retreated across the coastal plain, they buried a patchy band of fluvial sand and fine gravel, deposited where streams lost competence. The emergence of this thin wash of Early Holocene fluvial sand and gravel at the foot of the shore face below the lagoonal mud may contribute heavily to the grain-size reversal noted on the shore face, and to the generally coarse nature of the sea floor sands. The Albemarle channel sands would belong to this horizon, as would the basal gravels of the Holocene sand sheet cored by POWERS and KINSMAN (1953) and ourselves (SWIFT et al., 1971) off Chesapeake Bay mouth.

We conclude that the division of modern sea-floor facies into "nearshore modern sands" and "offshore relict sands" is valid generalization for our area, but somewhat restrictive and simplistic. All sediment in our area is relict, in the sense that it is ultimately derived by reworking of the pre-Recent substrate. The "modern" beach, surf and shore face sands are a narrow strip in our sector, much narrower than has been described from such areas as the Georgia Coast (PILKEY and FRANKENBERG, 1964). They are "modern" only in that their relict characteristics have been obliterated by the high-energy nearshore environment. The negative correlation that we have observed between phi grain size and depth suggests that the "relict" sea-floor sediments are also adjusting to the modern hydraulic regime.

albeit more slowly and subtly. If our model for sediment fractionation is valid, then the "relict" sea floor sands in our area may be more specifically referred to as "the Holocene transgressive sand sheet", and the modern sands as "the Holocene barrier prism".

ACKNOWLEDGEMENTS

Our study was supported by the Coastal Engineering Research Center, contract DACW-72-69-C-0016, and by the Office of Marine Geology, U.S. Geological Survey. Our study has profited from discussion of its substantive problems with David Duane of the Coastal Engineering Research Center, John Schlee of the U.S. Geological Survey, K. O. Emery and Elazar Uchupi of The Woods Hole Oceanographic Institution, John Sanders of Barnard College, Jack Pierce of the Smithsonian Institution, Orrin Pilkey of Duke University, Don Colquhoun of the University of South Carolina, and John C. Ludwick of the Old Dominion University Institute of Oceanography.

REFERENCES

BRUUN, P., 1962. Sea-level rise as a cause of shore erosion. *J. Waterways Harbors Div., Am. Soc. Civil Engrs.*, 88:117–130.

COLQUHOUN, D. J., 1969. *Geomorphology of the Lower Coastal Plain of South Carolina.* Div. Geol., State Develop. Board, 36 pp.

COOK, D. O., 1969. *Sand Transport by Shoaling Waves.* Thesis. Univ. Southern California, 148 pp. (unpublished).

CURRAY, J. R., 1964. Transgressions and regressions. In: R. L. MILLER (Editor), *Papers in Marine Geology.* MacMillan, New York, N.Y., pp.175–203.

CURRAY, J. R., 1965. Late Quaternary history, continental shelves of the United States. In: H. E. WRIGHT JR. and D. G. FREY (Editors), *Quaternary of the United States.* Princeton University Press, Princeton, New Jersey, pp.723–736.

CURRAY, J. R., 1969. Shore zone sand bodies: barriers, cheniers and beach ridges. In: D. J. STANLEY (Editor), *The New Concepts of Continental Margin Sedimentation.* American Geological Institute, Washington, D.C., pp. JC-2-1–JC-2-19.

DILLON, W. P., 1970. Submergence effects on a Rhode Island barrier and lagoon and inferences on migration of barriers. *J. Geol.*, 78:94–106.

DOLAN, R., 1970. Dune reddening along the Outer Banks of North Carolina. *J. Sediment. Petrol.*, 40:765.

EL-ASHRY, M. T. and WANLESS, H. R., 1968. Photo interpretation of shoreline changes between Capes Hatteras and Fear (North Carolina). *Marine Geol.*, 6:347–379.

EMERY, K. O., 1952. Continental shelf sediments off southern California. *Geol. Soc. Am. Bull.*, 63:1105–1108.

EMERY, K. O., 1968. Relict sediments on continental shelves of world. *Am. Assoc. Petrol. Geologists, Bull.*, 52:445–464.

FISCHER, A. G., 1961. Stratigraphic record of transgressing seas in the light of sedimentation on the Atlantic coast of New Jersey. *Bull. Am. Assoc. Petrol. Geologists*, 45:1656–1660.

FISHER, J. J., 1967. *Development Pattern of Relict Beach Ridges, Outer Banks Barrier Chain, North Carolina.* Thesis, Univ. North Carolina, 250 pp. (unpublished).

FISHER, J. J., 1968a. Origin of barrier island chain shorelines: Middle Atlantic States. Geological Society of American meetings for 1967. *Geol. Soc. Am., Spec. Paper*, 115:66–67.

FISHER, J. J., 1968b. Preliminary quantitative analysis of surface morphology of inner continental shelf surface–Cape Henry, Virginia to Cape Fear, North Carolina. In: A. E. MARGOLIS and R. C. STEERE (Editors), *Transactions National Symposium on Ocean Sciences and Engineering of the Atlantic Shelf.* Marine Technological Society, Washington, D.C., pp.143–149.

HARRIS, R. L., 1954. Restudy of test-shore nourishment by offshore deposition of sand, Long Branch, New Jersey. *Beach Erosion Board, Tech. Mem.*, 62:1–18.

HARRISON, W., BREHMER, M. L. and STONE, R. B., 1964. Nearshore tidal and nontidal currents, Virginia Beach, Virginia. *Coastal Eng. Res. Center, Tech. Rep.*, 5:1–20.

HARRISON, W. and WAGNER, K. A., 1964. Beach changes at Virginia Beach. *Coastal Eng. Res. Center, Misc. Papers.*, 6(64):1–25.

HOYT, J., 1967. Barrier island formation. *Bull. Geol. Soc. Am.*, 78:1125–1136.

INMAN, D. L. and NASU, N., 1956. Orbital velocity associated with wave action near the breaker zone. *Beach Erosion Board, Tech. Mem.*, 79:1–43.

IPPEN, A. T. and EAGLESON, D. S., 1955. A study of sediment sorting by waves shoaling on a plane beach: *Mass. Inst. Tech. Hydrodynamics Lab., Tech. Rept.*, 18:1–36.

JOHNSON, J. W. and EAGLESON, P. S., 1966. Coastal Processes. In: A. T. IPPEN (Editor), *Estuary and Coastline Hydrodynamics.* McGraw-Hill, New York, N.Y., pp.404–492.

KEULEGAN, G. H., 1948. An experimental study of submarine sand bars. *Beach Erosion Board, Tech. Rept.*, 3:1–42.

KLOVAN, J. E., 1966. The use of factor analysis in determining depositional environments from grain size distributions. *J. Sediment. Petrol.* 36:115–125.

LANGFELDER, J., STAFFORD, D. and AMEIN, M., 1968. *A Reconnaissance of Coastal Erosion in North Carolina.* Department of Civil Engineering, North Carolina State University, Raleigh, N.C., 127 pp.

McMASTER, R. L., 1954. *Petrography and Genesis of the New Jersey Beach Sands.* New Jersey Bureau of Geology and Topography, Trenton, N.J., 239 pp.

MOODY, D. W., 1964. *Coastal Morphology and Processes in Relation to the Development of Submarine Sand Ridges off Bethany Beach, Delaware.* Thesis, Johns Hopkins Univ., 167 pp. (unpublished).

MURRAY, S. P., 1967. Control of grain dispersion by particle size and wave state. *J. Geol.*, 75: 612–634.

NEWMAN, W. S. and MUNSART, C. A., 1968. Holocene geology of the Wachapreague Lagoon, eastern shore peninsula, Virginia. *Marine Geol.*, 6:81–105.

OAKS JR., R. Q., 1964. Post-Miocene stratigraphy and morphology, outer coastal plain, southeastern Virginia. *Office Naval Res., Tech. Rept.*, 5:1–240.

OAKS JR., R. Q. and COCH, N. K., 1963. Pleistocene sea levels, southeastern Virginia. *Science*, 140:979–983.

PAYNE, L. H., 1970. *Sediments and Morphology of the Continental Shelf off Southeast Virginia.* Thesis, Columbia Univ., 70 pp. (unpublished).

PIERCE, J. W. and COLQUHOUN, D. J., 1969. Evolution of barrier islands. In: *Abstracts with Programs for 1969.* Geol. Soc. Am., New York, N.Y., 7:178.

PIERCE, J. W. and COLQUHOUN, D. J., 1970. Configuration of Holocene primary barrier chain, Outer Banks, North Carolina: *Southeastern Geology*, 11:231–236.

PILKEY, O. H. and FRANKENBERG, D., 1964. The relict–recent sediment boundary on the Georgia continental shelf. *Georgia Acad. Sci., Bull.*, 22:1–4.

POWERS, M. C. and KINSMAN, B., 1953. Shell accumulations in underwater sediments and their relationship to the thickness of the traction zone. *J. Sediment. Petrol.*, 23:229–234.

SANDERS, J. F., 1962. North-south trending submarine ridge composed of coarse sand off False Cape, Virginia (abstr.). *Bull. Am. Assoc. Petrol. Geologists*, 46:278.

SANFORD, R. B. and SWIFT, D. J. P., 1971. Comparison of sieving and settling techniques for size analysis, using a Benthos rapid sediment analyser. *Sedimentology*, in preparation.

SCHLEE, J., 1966. A Modified Woods Hole Rapid Sediment Analyser. *J. Sediment. Petrol.*, 36:403–413.

SCHWARTZ, M. L., 1965. Laboratory study of sea-level rise as a cause of shore erosion. *J. Geol.*, 75:528–534.

SCHWARTZ, M. L., 1967. The Bruun theory of sea-level rise as a cause of shore erosion. *J. Geol.*, 75:76–92.

SCHWARTZ, M. L., 1968. The scale of shore erosion. *J. Geol.*, 76:508–517.

SCOTT, T., 1954. Sand movement by waves. *Beach Erosion Board, Tech. Mem.*, 48:1–37.

SWIFT, D. J. P., 1968. Shore-face erosion and transgressive stratigraphy. *J. Geol.*, 39:18–33.

SWIFT, D. J. P., 1969. Processes and products on the inner shelf. In: D. J. STANLEY (Editor), *The New Concepts of Continental Margin Sedimentation*. American Geological Institute, Washington, D.C., pp.DS-4-1–DS-4-46.

SWIFT, D. J. P., 1970. Quaternary shelves and the return to grade. *Marine Geol.*, 8:5–30.

SWIFT, D. J. P., SHIDELER, G. L., AVIGNONE, N., HOLLIDAY, B. W. and DILL JR., C. E., 1971. Quaternary sedimentation on the inner Atlantic Shelf between Cape Henry and Cape Hatteras: a preliminary report. *Maritime Sediments*, in press.

SWIFT, D. J. P., STANLEY, D. J. and CURRAY, J. R., 1971. Relict sediments: a reconsideration. *J. Geol.*, in press.

TANNER, W. F., 1960. Florida coastal classification. *Trans. Gulf Assoc. Geol. Soc.*, 10:259–266.

UCHUPI, E., 1968. Atlantic continental shelf and slope of the United States—physiography. *U.S. Geol. Surv., Profess. Papers*, 529-C: 1–30.

VISHER, G. S., 1969. Grain-size distributions and depositional processes. *J. Sediment. Petrol.*, 39:1074–1106.

ZENKOVITCH, V. P., 1967. *Processes of Coastal Development*. Wiley, New York, N.Y., 738 pp.

APPENDIX

SIZE DISTRIBUTION OF SAMPLES

Sample	Site[1]	Depth (m)	% Weight of 0.25-ϕ classes																				% Fines	% Carbonate
			−0.75	—0.50	—0.25	0.00	0.25	0.50	0.75	1.00	1.25	1.50	1.75	2.00	2.25	2.50	2.75	3.00	3.25	3.50	3.75	4.00		
1-A	Berm	0						1.0	2.0	2.0	6.0	9.0	20.0	23.0	17.0	15.0	3.0	2.0					0.01	1.10
1-B	Surf	2									0.5	0.5	0.5	1.5	3.0	7.0	17.0	30.0	23.0	13.0	3.0	1.0	1.28	1.12
1-C	SW	4															5.0	33.0	31.0	23.0	6.5	1.5	1.89	1.89
1-D	SW	6															1.0	16.0	40.0	26.0	14.0	3.0	3.98	3.09
1-E	SF	8											1.0	0.0	0.0	1.0	1.5	9.5	38.0	29.0	14.5	5.5	13.12	4.03
1-F	SF	9																15.0	36.0	27.0	11.5	10.5	7.60	3.41
1-G	SF	9																	39.0	31.0	22.5	7.5	8.74	3.14
1-H	SF	9																	36.0	31.0	26.0	7.0	8.84	3.28
1-I	SF	10																0.5	37.5	37.0	18.0	7.0	10.43	3.51
1-J	SF	11										0.5	2.0	9.5	38.0	26.0	12.0	6.5	3.5	1.5			3.54	2.11
1-K	SF	12											1.0	2.0	8.0	34.0	24.0	11.0	9.0	7.0	3.0	1.0	4.44	2.68
1-L	SF	13					1.0	0.5	3.5	3.0	4.5	6.5	10.0	18.0	18.0	13.0	8.0	3.0	4.5	5.0	0.5	1.0	2.58	2.99
1-M	SF	15												3.0	0.5	3.0	7.5	17.0	31.0	20.5	12.5	5.0	9.15	4.31
1-N	SF	15													1.0	4.0	16.0	24.0	25.0	17.0	13.0		9.09	4.17
1-O	SF	16													1.5	3.5	12.0	24.0	26.0	18.0	10.0	5.0	8.34	4.55
1-P	SF	18					1.0	0.0	1.0	1.0	1.0	1.0	3.0	4.0	5.5	11.5	21.0	14.0	9.0	13.5	11.5	2.0	7.68	3.28
1-Q	SF	17			0.5	1.0	1.5	1.0	1.5	2.0	3.5	4.5	13.5	26.0	22.5	11.0	3.5	3.0	1.0	1.5	0.5	1.0	2.93	1.48
1-R	SF	17								0.5	0.5	1.0	6.0	26.0	29.0	19.0	8.0	2.5	2.5	3.0	1.5	0.5	3.02	1.88
1-S	SF	17							0.5	1.5	3.0	5.5	14.5	22.0	20.0	10.0	5.0	4.0	4.0	5.0	3.0	2.0	3.98	2.77
3-A	Berm	0				0.5	0.5	0.5	1.5	1.5	4.0	9.5	16.0	18.0	19.0	13.0	9.5	4.5	2.0				0.87	0.38
3-B	Surf	2							1.0	1.0	2.0	2.0	5.0	4.0	5.0	5.0	9.0	29.0	24.0	12.0	1.0		1.16	0.75
3-C	SW	4												0.0	0.0	2.0	7.5	35.0	39.0	14.0	5.0	1.0	1.73	1.06
3-D	SW	6																27.0	39.0	21.0	10.5	2.5	0.76	4.01
3-E	SW	8																7.0	46.0	30.0	14.0	3.0	3.60	1.67
3-F	SW	10															0.5	23.0	19.0	30.0	21.0	6.0	9.51	1.35
3-G	SW	12										1.0	2.0	4.0	10.0	17.0	21.0	19.0	8.0	10.0	6.0	2.0	8.58	0.94
3-H	SW	13											0.5	0.5	1.5	3.0	8.5	27.0	26.0	18.0	11.0	9.0	10.58	2.30
3-I	SF	15						1.0	2.0	2.0	3.0	6.0	12.5	27.5	23.0	10.0	6.0	2.0	2.5	1.5	0.5	0.5	0.64	1.03

3-J	SF	14									1.0	2.0	3.5	9.0	22.0	30.5	7.0	3.0	1.0				3.78	0.82
3-K	SF	10							0.5	3.5	8.0	10.0	12.0	16.0	25.0	14.0	6.0	1.0	2.0	1.0	1.0		0.86	0.54
3-L	SF	14													1.5	4.0	30.0	42.5	16.0	3.0	3.0		3.86	1.93
3-M	SF	16									1.0	3.0	10.0	24.0	26.0	10.0	7.0	7.0	2.0	2.0	1.5	1.5	5.05	0.97
3-N	SF	16								1.5	0.5	3.5	9.5	17.5	31.5	15.5	12.0	4.5	2.0	1.0	1.0		1.81	2.15
3-O	SF	20													2.0	3.0	34.5	13.3	15.5	16.5	10.0	5.0	21.06	3.21
3-P	SF	20							1.0	1.0	4.0	8.0	17.5	25.5	22.0	10.0	6.0	3.0	1.0	1.0			3.02	1.88
3-Q	SF	22	1.0	2.0	2.0	3.5	4.0	5.5	6.5	7.5	9.0	13.0	18.0	12.0	8.0	2.0	2.0	0.5	2.5	0.5	0.5		3.32	6.29
3-R	SF	22											1.0	5.0	8.5	28.5	28.5	15.0	7.0	3.0	1.5	1.5	3.57	0.83
4-A	Berm	0											1.5	3.5	10.0	36.0	30.0	16.0	3.0				0.79	0.73
4-B	Surf	2							0.5	0.5	1.0	2.0	3.0	5.0	20.0	33.0	20.0	11.0	3.0	1.0			1.64	0.51
4-C	SW	4														2.0	10.0	44.0	31.0	10.0	3.0		1.37	0.61
4-D	SW	6														2.0	4.0	27.0	34.0	20.0	6.0		1.74	1.20
4-E	SW	8												1.0	1.0	3.0	5.0	25.0	29.0	20.0	12.0	4.0	4.80	1.70
4-F	SW	10								0.5	0.5	1.0	1.0	1.0	4.0	8.0	17.0	31.0	17.0	10.0	6.0	3.0	5.68	2.31
4-G	SW	12								0.5	1.0	1.0	1.0	2.0	6.0	9.0	24.0	26.5	17.0	7.0	5.0		5.27	2.37
4-H	SW	14											1.0	2.0	5.0	12.0	28.0	20.0	18.0	9.0	4.0	1.0	3.66	1.36
4-I	SF	16											1.0	2.5	8.5	20.0	17.0	16.0	15.0	9.0	9.0	2.0	2.48	1.35
4-J	SF	17												3.0	4.5	12.5	31.0	28.0	12.0	4.0	2.0	3.0	3.71	0.51
4-K	SF	18				0.5	1.5	1.0	4.0	5.0	8.0	12.0	17.5	18.5	12.0	7.0	7.5	3.5	1.5	0.5			1.28	3.13
4-L	SF	19												1.0	4.5	14.5	13.0	46.0	16.0	4.0	0.5	0.5	2.60	0.63
4-M	SF	19		1.0	0.5	0.5	2.0	3.0	5.0	7.0	8.0	11.0	13.0	15.0	16.0	10.0	5.0	2.0	1.0				0.74	1.44
4-N	SF	21											1.0	2.0	7.0	37.0	32.5	19.5	1.0				1.62	0.45
4-O	SF	24								0.5	1.0	1.0	3.0	3.0	11.0	28.5	25.5	13.0	5.0	3.0	2.0	2.0	43.44	1.85
4-P	SF	22												2.0	4.0	14.0	43.0	22.0	10.0	3.0	1.5	5.0	2.40	0.32
4-Q	SF	22										0.5	1.0	1.5	3.5	18.5	35.0	23.0	10.0	3.0	3.0	1.0	3.36	0.53
4-R	SF	22									1.0	0.0	0.5	3.0	5.0	12.0	29.5	25.5	16.5	6.0	1.5	0.5	1.53	1.05
4-S	SF	22								0.5	0.5	1.0	1.0	2.0	5.0	10.0	27.0	26.0	18.5	6.5	1.5	0.5	2.37	0.64
5-A	Berm	0											1.0	8.0	24.0	33.0	21.0	8.5	3.5	1.0			2.16	0.11
5-B	Surf	2								1.0	0.0	0.5	2.0	4.5	6.0	18.0	40.0	23.0	10.5	4.0	0.5		3.14	0.61
5-C	SW	4							1.0	0.0	1.0	0.0	2.0	0.0	3.0	5.5	23.5	36.0	20.0	7.0	1.0		1.03	0.44
5-D	SW	6														1.0	24.0	32.0	26.0	13.0	3.0	1.0	1.57	0.64
5-E	SW	8													0.5	0.5	1.0	31.0	42.5	15.5	7.0	2.0	4.25	0.78
5-F	SW	10											1.0	1.0	1.0	2.0	5.0	29.0	22.0	22.0	12.5	4.5	9.51	1.23
5-G	SW	12							1.0	1.0	1.0	1.5	6.0	17.5	31.0	19.0	10.0	7.0	2.5	2.0	0.5		2.23	0.74
5-H	SW	14											1.0	1.0	3.5	10.5	32.0	22.0	14.0	10.0	5.0	2.0	4.51	0.90

[1] SW zone of shoaling waves; SF = sea floor.

(continued)

APPENDIX *(continued)*

SIZE DISTRIBUTION OF SAMPLES

Sample	Site[1]	Depth (m)	% Weight of 0.25-φ classes																				% Fines	% Carbonate
			−0.75	−0.50	−0.25	0.00	0.25	0.50	0.75	1.00	1.25	1.50	1.75	2.00	2.25	2.50	2.75	3.00	3.25	3.50	3.75	4.00		
5-I	SW	16							1.0	0.5	1.5	2.0	4.5	9.5	23.0	25.0	21.0	7.0	2.0	1.0	1.0	1.0	1.47	0.91
5-J	SF	22															2.0	8.0	34.0	25.0	18.0	13.0	3.71	0.51
5-K	SF	20	3.0	1.0	3.0	4.0	8.0	9.0	12.0	9.0	11.0	11.0	8.5	7.5	6.0	3.0	2.0	1.0	1.0				1.21	3.20
5-L	SF	18	1.0	1.0	1.5	2.5	3.0	3.5	4.0	2.5	3.5	3.0	6.5	13.5	25.0	16.5	7.0	4.5	1.5				2.1	3.85
5-M	SF	20	1.0	2.0	3.0	3.0	5.0	6.0	10.0	9.5	13.5	15.0	17.0	8.0	3.0	1.0	2.0	1.0					0.91	4.51
5-N	SF	20											1.0	2.0	3.0	7.5	25.5	38.0	16.0	5.0	1.0	1.0	3.64	1.02
5-O	SF	23											1.0	0.5	3.0	6.0	39.5	20.0	16.5	9.0	3.5	1.0	5.95	0.97
6-A	Berm	0	4.0	8.0	9.0	17.0	14.0	8.5	6.0	5.0	3.5	4.0	5.0	5.0	5.0	3.5	1.0	1.5					0.44	2.29
6-B	Surf	2													2.0	4.0	52.0	25.0	12.0	5.0			1.53	0.66
6-C	SW	4														2.0	11.0	47.0	24.0	12.5	3.5		1.48	0.65
6-D	No Sample																							
6-E	SW	6															2.0	46.5	16.5	20.5	12.0	3.0	4.44	1.42
6-F	SW	8														2.0	6.0	37.0	28.5	15.5	8.5	2.5	5.28	1.01
6-G	SW	10															2.0	14.0	43.0	23.0	13.0	5.0	11.58	1.10
6-H	SF	12															1.0	15.0	35.0	25.5	16.0	7.5	15.31	1.78
6-I	SF	14															1.0	1.5	30.5	35.0	24.0	8.0	0.64	2.61
6-J	SF	16															2.0	4.0	24.0	33.0	26.0	11.0	29.69	2.12
6-K	SF	18														0.5	1.5	3.0	30.5	29.5	23.0	12.0	10.23	2.96
6-L	SF	18																1.0	21.0	27.0	33.0	18.0	16.19	2.79
6-M	SF	19											1.0	1.0	2.0	4.0	10.0	36.5	20.5	12.0	8.0	5.0	7.68	2.70
6-N	SF	19										1.0	1.0	1.0	1.5	3.5	6.0	38.0	28.0	12.5	5.5	3.0	6.16	2.98
6-O	SF	19												1.5	4.0	21.5	27.0	21.0	13.0	8.0	3.0	1.0	3.10	0.81
7-A	Berm	0							0.5	0.5	0.0	2.0	2.0	7.0	16.0	34.0	25.0	10.0	3.0				0.66	0.58
7-B	Surf	2		0.5	0.5	1.0	3.0	3.0	3.0	3.0	2.5	2.5	3.0	7.0	25.0	25.5	9.5	6.0	4.0	1.0			0.63	1.18
7-C	SW	4													1.0	0.5	26.5	36.0	24.0	9.0	2.0	1.0	1.43	0.71
7-D	SW	6											0.5	0.0	0.5	1.0	3.0	16.0	45.0	22.0	9.0	3.0	2.17	0.94
7-E	SW	8													1.0	2.0	7.0	26.0	28.0	21.0	11.0	4.0	2.69	1.29
7-F	SW	10													0.5	0.5	3.0	23.0	30.0	20.5	16.5	6.0	4.17	1.57
7-G	SW	12															2.0	11.0	40.0	22.5	15.5	9.0	6.05	2.11

7-H	SW	14											1.0	0.5	0.5	2.0	6.0	23.0	18.5	21.0	18.0	9.5	11.46	2.31
7-J	SF	18																1.0	29.0	35.5	24.5	10.0	10.32	2.75
7-K	SF	18		1.0	1.0	1.0	3.5	3.5	5.0	6.0	7.0	8.5	13.0	14.5	18.0	9.0	5.0	2.0	2.0				2.31	0.96
7-L	SF	18	0.5	1.5	1.0	2.0	4.0	4.0	9.0	9.0	12.0	11.5	15.5	15.0	9.0	3.5	1.5	1.0					1.61	1.38
7-M	SF	19				0.5	0.5	0.5	1.0	1.5	3.0	5.0	7.0	11.5	25.5	27.5	13.5	2.0	0.5	0.5			1.45	0.66
7-N	SF	19			0.5	0.5	1.0	1.0	5.0	4.0	6.0	9.0	17.0	20.0	12.0	4.0	6.0	9.0					1.04	0.44
8-A	Berm	0											0.5	2.5	11.5	42.5	30.0	9.5	3.0	0.5			0.45	0.36
8-B	Surf	2		4.0	5.0	5.0	8.0	7.0	7.5	4.5	6.0	5.5	7.0	12.5	14.0	9.5	3.5	0.5	0.5				0.66	1.38
8-C	SW	4									0.5	1.5	1.0	2.0	1.5	10.0	41.0	30.5	8.5	3.5			1.32	0.34
8-D	SW	6													1.0	3.0	26.0	33.0	22.5	10.5	4.0		2.13	1.02
8-E	SW	8													1.0	2.0	3.5	33.0	35.0	18.5	6.0	1.0	3.87	1.45
8-F	SW	10															2.0	23.0	47.0	19.0	7.0	2.0	3.91	1.38
8-G	SW	12				0.5	0.5	1.5	3.5	2.0	2.0	2.0	4.0	5.0	8.0	7.0	21.0	11.0	15.0	8.0	5.0	4.0	69.07	4.54
8-H	SW	14														1.0	3.0	16.0	49.0	22.0	9.0		4.92	1.49
8-I	SF	14	0.5	0.0	1.5	1.5	5.5	8.0	11.0	10.0	14.0	17.0	14.0	6.0	2.0	1.5	1.5	2.5	2.5	1.0			4.23	0.60
8-J	SF	15										0.5	1.0	0.5	1.5	12.0	50.5	22.0	7.0	2.0	2.0	1.0	2.53	1.05
8-K	SF	15															2.0	7.0	36.0	28.0	19.0	8.0	12.43	3.18
8-L	SF	16																1.0	27.0	29.5	29.5	13.0	17.64	3.33
8-M	SF	17		0.5	1.0	1.5	4.0	5.0	7.0	5.0	9.0	13.0	24.0	16.0	9.5	1.5	2.0	0.5	0.5				1.26	2.67
8-N	SF	18	1.0	0.5	2.5	2.0	5.0	7.0	14.0	8.5	10.5	8.0	6.0	4.0	7.0	8.0	5.0	4.5	2.5	1.5	1.5	1.0	6.81	5.79
9-A	Berm	0						1.0	4.0	3.0	7.0	7.5	13.5	23.0	20.0	11.0	7.0	2.0	0.5	0.5			0.81	0.59
9-B	Surf	2								1.0	1.0	3.0	3.5	7.5	23.0	31.0	19.0	10.0	3.0				0.94	0.68
9-C	SW	4										2.0	0.0	2.5	9.5	41.0	28.0	12.0	2.0	2.0			1.40	1.04
9-D	SW	6										0.5	1.0	1.5	1.5	2.0	18.5	32.5	21.5	13.0	6.0	2.0	3.10	1.65
9-E	SW	8												1.0	1.0	4.0	13.0	40.0	24.0	11.0	5.0	1.0	2.21	1.55
9-F	SW	10															1.0	15.0	43.0	25.0	12.0	4.0	5.63	2.02
9-G	SW	12																5.0	43.0	34.0	12.0	6.0	6.61	1.53
9-H	SW	13															1.5	3.5	44.0	26.0	18.0	7.0	10.71	1.61
9-I	SF	14														2.5	6.0	36.0	27.0	16.0	8.5	2.5	4.76	1.87
9-J	SF	15	2.0	3.0	6.0	6.0	10.0	14.0	23.0	8.0	3.0	1.0	2.0	3.0	3.0	5.5	2.5	0.0	1.0	3.0	2.0	2.0	9.50	9.03
9-K	SF	15	2.0	2.0	3.5	4.5	5.0	4.0	6.0	5.0	8.0	16.0	23.0	12.0	5.0	2.0	1.0	0.5	0.5				1.23	5.57
9-L	SF	16	7.0	7.0	7.0	9.0	11.0	11.0	11.0	8.0	7.0	6.5	7.0	3.5	2.5	0.5	1.0	0.0	0.5	0.5			3.07	7.48
10-A	Berm	0	0.5	0.5	1.0	2.0	3.5	3.5	4.0	3.0	3.5	6.5	1.0	17.0	25.0	13.5	6.0						0.33	1.53
10-B	Surf	2				0.5	0.5	0.5	1.0	1.0	0.0	1.5	3.5	14.0	29.0	33.5	11.5	2.5	1.0				1.04	0.49
10-C	SW	4											0.5	0.5	1.0	7.0	40.0	32.0	14.0	4.0	1.0		0.95	0.67
10-D	SW	6															22.0	43.0	24.0	7.0	3.0	1.0	2.30	0.77
10-E	SW	8																3.0	49.0	26.0	15.5	6.5	7.14	1.37

(continued)

APPENDIX (*continued*)

SIZE DISTRIBUTION OF SAMPLES

Sample	*Site*[1]	*Depth (m)*	*% Weight of 0.25-ϕ classes* —0.75	—0.50	—0.25	0.00	0.25	0.50	0.75	1.00	1.25	1.50	1.75	2.00	2.25	2.50	2.75	3.00	3.25	3.50	3.75	4.00	*% Fines*	*% Carbonate*
10-F	SW	10																4.0	45.0	26.0	17.0	8.0	6.29	1.62
10-G	SW	12																	36.5	28.5	24.0	11.0	7.37	1.53
10-H	SF	14																3.0	37.8	30.0	20.0	10.0	16.83	2.07
10-I	SF	16															9.0	4.5	36.5	27.0	15.0	8.0	14.70	2.14
10-J	SF	17												2.0	1.0	2.0	2.0	12.0	34.0	27.0	14.0	6.0	10.90	2.84
10-K	SF	18											1.5	1.5	1.5	2.0	4.5	30.0	23.0	19.0	12.0	5.0	11.37	3.16
10-L	SF	19															2.0	35.0	25.5	19.5	12.0	6.0	2.28	2.46
10-M	SF	20															4.5	44.5	33.0	12.0	4.5	1.5	4.30	2.00
10-N	SF	21															23.0	40.0	22.5	10.5	3.0	1.0	5.02	2.03
11-A	Berm	0							1.5	2.0	4.5	17.0	7.0	28.0	22.0	10.0	5.5	2.5					0.5	1.92
11-B	Surf	2									0.5	0.0	0.5	2.5	4.0	11.5	46.5	19.5	10.5	3.5	1.0		1.89	0.97
11-C	SW	4						1.0	0.5	1.5	2.5	4.0	5.5	7.5	12.5	22.5	19.0	14.0	5.5	3.5	1.0		0.84	3.69
11-D	SW	6															4.0	27.0	39.0	17.5	8.5	4.0	2.24	1.49
11-E	SW	8														3.5	29.0	37.0	17.0	9.0	4.0		4.71	1.82
11-F	SW	10															1.0	21.0	35.0	23.0	14.0	6.0	4.62	2.05
11-G	SW	12																8.0	33.5	28.0	21.0	9.5	7.35	2.17
11-H	SW	13											1.0	1.0	1.0	1.5	2.5	21.0	29.0	20.0	15.5	7.5	9.28	3.16
11-I	SF	14															2.0	28.0	21.0	21.0	19.0	9.0	9.46	2.78
11-J	SF	15															3.0	35.0	23.0	18.0	14.0	7.0	8.44	2.81
11-K	SF	15												0.5	1.5	3.0	8.0	35.0	27.0	14.0	7.0	4.0	4.96	2.92
11-L	SF	15												1.0	4.0	7.0	24.0	27.0	18.0	11.0	5.5	2.5	3.96	3.04
11-M	SF	16																2.0	47.0	30.0	15.0	6.0	11.49	3.32
11-N	SF	17															2.0	10.0	44.0	29.0	11.0	4.0	7.37	3.13
11-O	SF	18																18.0	50.0	22.0	8.0	2.0	6.08	2.84

12

Reprinted, courtesy of the Geological Society of America, from *Geol. Soc. Am. Bull.*
85:1319–1328 (1974)

Holocene Sedimentary Environment of the Atlantic Inner Shelf off Delaware

R. E. SHERIDAN
C. E. DILL, JR.*
J. C. KRAFT
Department of Geology, University of Delaware, Newark, Delaware 19711

ABSTRACT

High-resolution subbottom profiles have been used to identify and map the pre-Holocene erosional surface in Delaware Bay and in the Atlantic Ocean off the Delaware coast. Vibracoring has revealed the nature and age of the various seismic layers. Age identifications based on radiocarbon dating, on oxidation zones, and on lithologic characteristics confirm the partial retention of Holocene coastal environments under the inner shelf. The Holocene transgression across the ancient Rehoboth and Indian River lagoons has involved erosion through the Holocene lagoonal sediments in some areas, leaving pre-Holocene ridges exposed on the inner-shelf sea floor where once there may have been headlands. Where the ancient lagoonal deposits are thickest over depressions in the pre-Holocene surface, Holocene coastal sediments were retained. The presence of marsh peat, dated as ~7,500 yr B.P., and overlying very fine lagoonal mud proves that lagoonal conditions existed east of Delaware even then. The position of the barrier complex related to these earlier Holocene lagoons is projected on a paleographic reconstruction to be at least 12 km (7 n mi) east of the present coast. This yields a minimum average coastline retreat rate of 1.6 horizontal m per yr. (6 ft per yr).

This study of the shallow subsurface geology also revealed that the bottom morphological features of the inner shelf off Delaware are related to dynamic molding of hydraulic bedforms in equilibrium with the present marine environment. Although one ridge form was found to be a subcrop of Pleistocene sand and gravel, most of the ridge and channel topography was found to be formed within the upper Holocene sediments. This is especially true of the shoreface-connected ridges off Rehoboth Bay and Bethany Beach, the inlet-associated shoals near the mouth of Delaware Bay, and the Delaware shelf-valley flood channel. *Key words: marine geology, continental shelf.*

* Present address: Alpine Geophysical Associates, Norwood, New Jersey 07648.

INTRODUCTION

The marine areas immediately off Delaware's Atlantic coast were investigated to determine the geology of the shallow subsurface (Fig. 1). During 1970 and 1971, high-resolution 3.5-kHz and 7.0-kHz subbottom reflection profiles were made and vibracores were taken at 16 selected sites. The seismic profiling aboard R/V *Eastward* and R/V *Skimmer* and vibracores 1 through 13 taken from *Eastward* were navigated with loran A. The profiles and vibracores 14 to 16 taken from R/V *Wando River* were made with a Motorola radar transponder navigational system accurate to ±3 m (9 ft).

The areas where most of the coring data were concentrated offered an opportunity to extend seaward the interpretations of the Holocene sedimentary environment on land (Kraft, 1971). Good subsurface control was available in drill holes made on several coastal-type environments, such as (1) the spit complex of Cape Henlopen, DH–2–71 (Fig. 1), (2) the barrier island across Rehoboth Bay, R–41–1, (3) the Holocene deposits between ancient Pleistocene beach ridges of the headland at Bethany Beach, 8–DH–70, and (4) the headland held up by oxidized Pleistocene gravel at Rehoboth Beach.

The extensive study of these land drill holes led Kraft (1971) to interpret the Holocene evolution of the Delaware coast as one of relatively rapid westward transgression of the straight barrier-island shoreline across and above the lagoonal com-

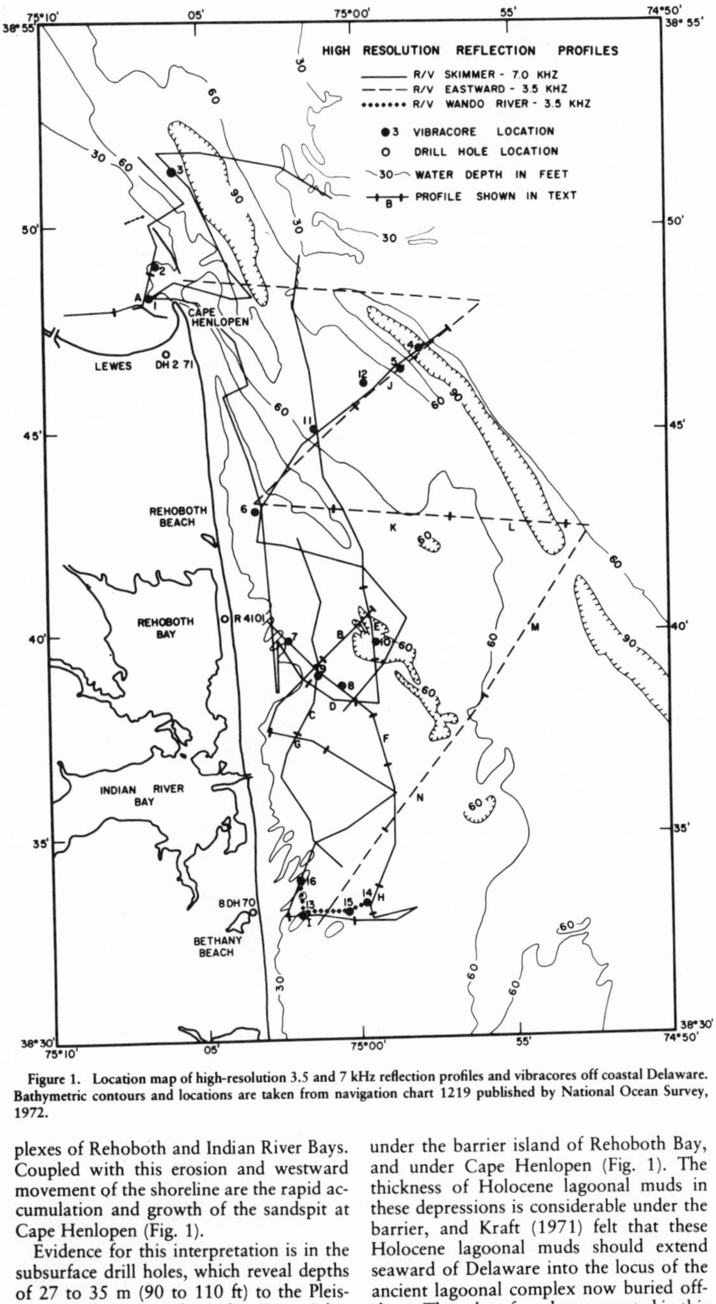

Figure 1. Location map of high-resolution 3.5 and 7 kHz reflection profiles and vibracores off coastal Delaware. Bathymetric contours and locations are taken from navigation chart 1219 published by National Ocean Survey, 1972.

plexes of Rehoboth and Indian River Bays. Coupled with this erosion and westward movement of the shoreline are the rapid accumulation and growth of the sandspit at Cape Henlopen (Fig. 1).

Evidence for this interpretation is in the subsurface drill holes, which reveal depths of 27 to 35 m (90 to 110 ft) to the Pleistocene sediments under Indian River inlet, under the barrier island of Rehoboth Bay, and under Cape Henlopen (Fig. 1). The thickness of Holocene lagoonal muds in these depressions is considerable under the barrier, and Kraft (1971) felt that these Holocene lagoonal muds should extend seaward of Delaware into the locus of the ancient lagoonal complex now buried offshore. The subsurface data presented in this report provide confirmation of Kraft's (1971) speculations.

Fortuitously, the reflection profiles and coring data were also taken in areas of the inner shelf where there are classically developed bottom morphological features whose origins are a matter of controversy. The Rehoboth Beach cross section includes two southeast-trending, inlet-associated shoals (Swift and others, 1972). Hen and Chickens shoal is the southeast-trending shoal, less than 10 m (30 ft) deep, seemingly attached to the tip of Cape Henlopen spit. The other parallel shoal crests at a depth of 15 m (50 ft) near core site 5 in Figure 1.

The Rehoboth Beach section crosses two linear depressions, thought by Swift and others (1972) to be mutually evasive ebb and flood channels associated with the Delaware Bay mouth. The eastern channel, outlined by the 27-m (90-ft) contour near core site 4 (Fig. 1), is continuous with the Delaware shelf valley extending to the southeast. Because of this continuation and its position to the right side of the bay mouth, it is assumed to be flood dominated, although no tidal current measurements are available to prove this. The main deep channel of the mouth of Delaware Bay is marked by the 27-m (90-ft) contour near core site 3 (Fig. 1). This depression shoals seaward; therefore, Swift and others (1972) considered it to be ebb dominated. Current measurements made across this main channel of the Delaware Bay mouth are in agreement with this ebb dominance for the upper part of the water column, but flood currents are apparently dominant at the bottom (Oostdam, 1971).

The transverse shelf valleys of the Atlantic continental shelf have been traditionally interpreted as the traces of ancient fluvial thalwegs, which have been preserved as the continental shelf has undergone passive submergence. Swift (1973), however, pointed out that this interpretation is not correct in detail because the initial fluvial landforms have been modified by processes of the inner shelf subsequent to transgression. As a consequence of this modification, the seaward-trending flood channel of the estuary mouth should be only approximately superimposed on the ancient subsurface fluvial channel. The subsurface data on the Rehoboth Beach section (Fig. 1) provide information on the validity of Swift's (1973) speculation.

Another controversial morphological feature of the inner shelf is observed in the studied profiles. This is the shoreface-connected shoal or ridge (Duane and others, 1972). Several of these shoals are well developed along the Delaware coast, as revealed by the northeast-trending fingerlike

pattern in the 10-m (30-ft) isobath near core sites 7, 9, 13, 15, and 16 (Fig. 1). The interpretation of the origin of these ridge or shoal forms has varied in the past, and the variant hypotheses are well summarized in Duane and others (1972) and Swift and others (1972). There have been three basic hypotheses applied to these topographic forms: (1) The ridges represent progressive steps of an advancing sea recorded as successive barrier beaches, each being younger nearer shore. (2) The ridges are formerly buried relict beach ridges from a Pleistocene lower sea-level stand now being excavated at the shoreface. (3) The ridges are an assemblage of now active or recently active bed forms originating at the shoreface and dynamically in equilibrium with the inner-shelf marine environment. The subsurface data of the Rehoboth Bay and Bethany Beach sections (Fig. 1) provide data that help to resolve which of these hypotheses is correct.

CROSS SECTIONS

Cape Henlopen Section

A cross section was made using data from drill hole DH–2–71 and vibracores 1, 2, and 3 (Fig. 1) and data from high-resolution reflection profiles, such as profile A (Fig. 2). A prominent reflector is observed deepening to the northeast at depths of 10 to 20 m (30 to 60 ft) on profile A. This reflector is correlated with the lithologic contrast found between the Holocene estuarine muds and the gravel layer associated with an unconformity over Pleistocene sediments (Fig. 3). The 5- to 6-m (15- to 20-ft) deep stream valleys with steep (>30° slope) walls incised into the Pleistocene sediments were apparently formed when these older sediments were exposed above sea level during a lower stand of the sea. The oxidized nature of the orange gravels found in vibracores 1 and 2 at this reflector is possibly the result of processes involved with the past subaerial exposure of these stream valleys and their sedimentary fill.

As shown on the cross section (Fig. 1), the beach and dune sands of the Cape Henlopen split complex do not extend north of this land feature. Rather, estuarine and shallow-marine silt and mud are found in cores 1 and 2. It should be noted that these vibracores were taken in two depressions along the section because of the low water-depth limitation prescribed by the coring rig. There are sands between core sites 1 and 2, however, and these sands form the shoal structure topped by the southwest-facing sand waves shown at the northeast end of profile A (Fig. 2). The sand shoal has apparently been building into this area from the northwest.

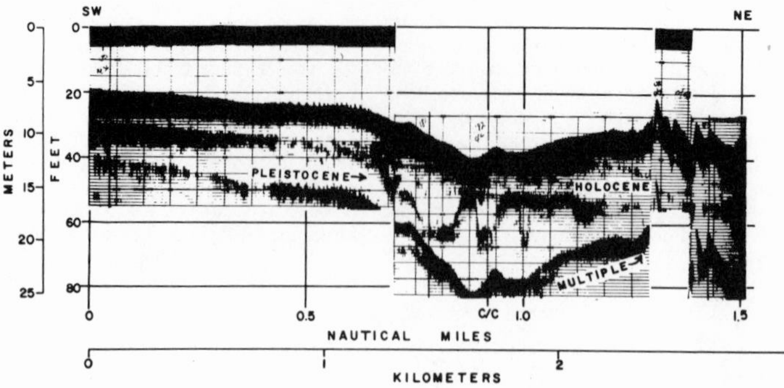

Figure 2. Original record of 7 kHz reflection profile A. Note incised valleys with steep walls (slope >30°) in Pleistocene sediments, and southwest-facing giant sand waves building sand shoal on northeast end of profile.

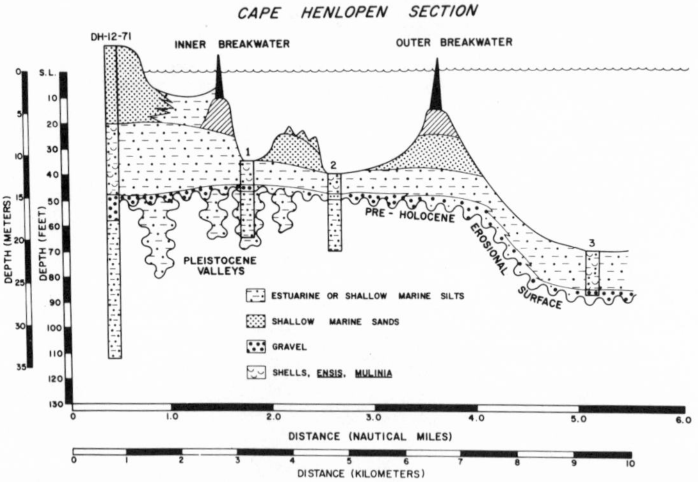

Figure 3. Diagrammatic geologic cross section north of Cape Henlopen, Delaware. Numbers 1, 2, 3 designate vibracore data; DH–2–71 designates deep drill hole on land. Solid black structures represent man-made breakwaters.

Data from this section indicate that the Cape Henlopen spit sand and gravel are locally regressive, building the land area northward over previously open estuarine or shallow-marine environments. From drilling on Cape Henlopen, Kraft (1971) found that this regression northward has been continuing for several thousand years.

Beyond the outer breakwater, the prominent reflecting surface correlated with Pleistocene sediments and with the basal Holocene unconformity is found to deepen rapidly to 25 m (85 ft) into the flanks of the main channel of the Delaware Bay mouth. Beyond this cross section, the basal Holocene unconformity is observed on high-resolution reflection profiles to be even deeper, down to about 55 m (180 ft) beneath the axis of the present channel of the bay mouth (Moody and Van Reenan, 1967).

Rehoboth Beach Section

This cross section was interpreted from the data of cores 4, 5, 6, 11, and 12 (Fig. 1) and from the high-resolution reflection profile, such as profile J (Fig. 4). The schematic interpretation for this section (Fig. 5) of the shallow subsurface geology shows that the pre-Holocene erosional surface passes beneath the ridge or shoal structures, such as Hen and Chickens shoal, without acting as a structural core for the topography. Moreover, this basal Holocene

unconformity apparently passes beneath both the ebb-dominated and flood-dominated channels without relief, although the ancient fluvial Delaware River channel is incised at a position that is slightly offset from the present topographic channels. The Delaware shelf valley extension is the deeper channel at 34 m (110 ft) near core site 4. Theoretically, this should be a flood-dominated channel, according to Swift (1973).

The pre-Holocene surface was thought to have been penetrated in cores 6 and 4, and the basal Holocene gravel was thought to be found in the bottoms of cores 11 and 12 (Fig. 5). The pre-Holocene material of core 6 is identified as an orange, oxidized quartz pebble gravel identical to the Pleistocene material outcropping 1 mi (1.6 km) away, onshore at Rehoboth Beach. Below an oxidized gravel layer at the bottom of core 4 was found a very hard, desiccated clay, suggestive of an older sediment once buried by significant overburden.

The pre-Holocene surface is therefore composite in nature, being of variable lithology and probably of variable age. Usually, however, there is a veneer of gravel associated with this erosional surface which apparently represents a Pleistocene lag deposit on this surface. Accordingly, the surface is identified ambiguously as Pleistocene and pre-Holocene in age in this discussion and on the profiles. Because the base of the Pleistocene sediments along the Delaware coast is generally slightly deeper than 37 m (120 ft; Miller, 1971), we feel that most of the pre-Holocene erosional surface mapped on these inner-shelf sections is indeed made up of Pleistocene sediments, and that only in the deeper buried channels, from 37 to 55 m (120 to 180 ft) deep, are there Tertiary coastal plain sediments immediately below this horizon.

The Holocene estuarine and shallow-

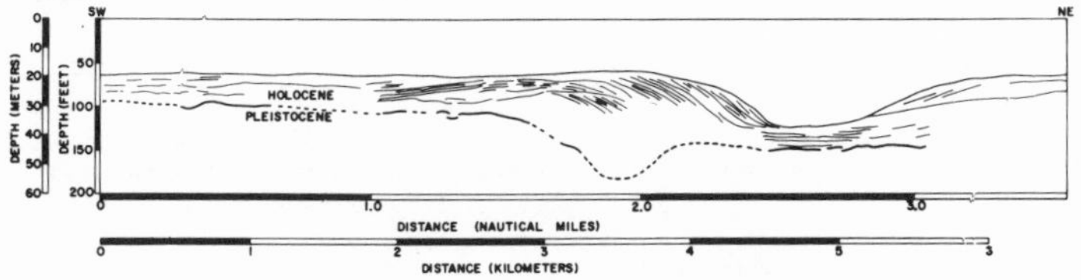

Figure 4. Line tracing of 3.5 kHz reflection profile J. Note foreset and bottomset bedding of sand shoal bordering flood-dominated channel on southwest. This bedding suggests that shoal is building northeastward to fill in channel.

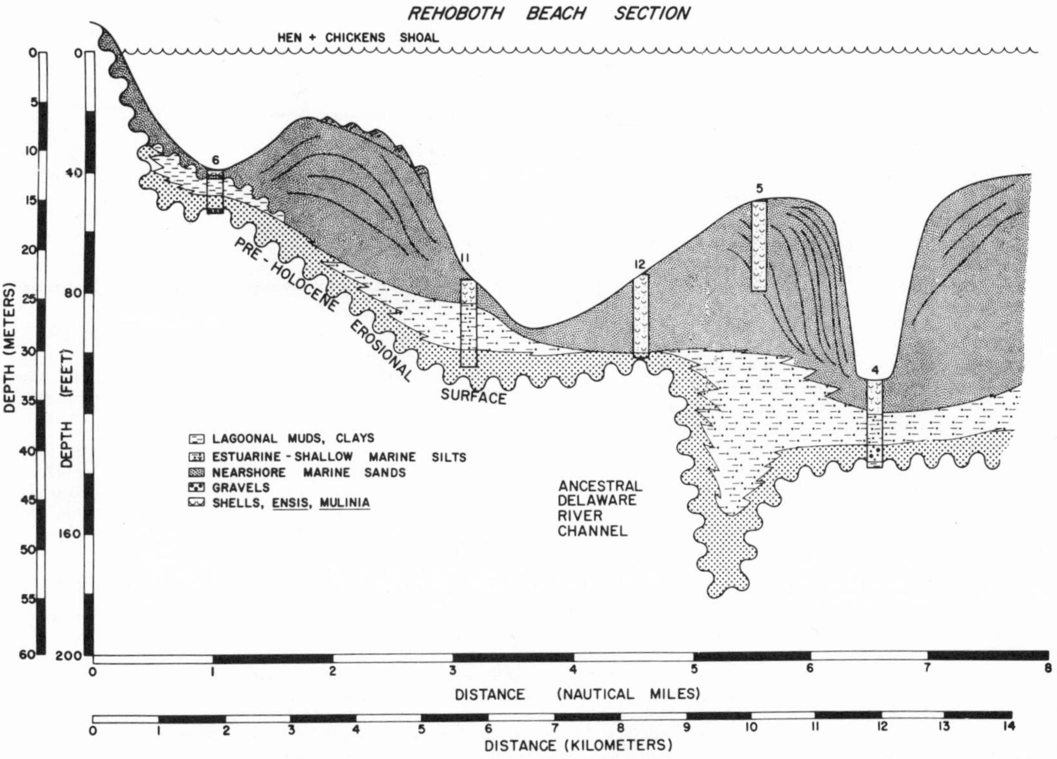

Figure 5. Diagrammatic geologic cross section northeast of Rehoboth Beach. Numbers 4, 5, 6, 11, and 12 designate vibracore data. Channel between cores 11 and 12 is thought to be ebb-dominated channel outlet of Delaware Bay; deeper channel near core site 4 is thought to be flood-dominated channel. Note that these channels are formed by build-up of Delaware Bay inlet-associated shoals, and note offset slightly from relict fluvial ancient Delaware River channel, which is now buried at depth.

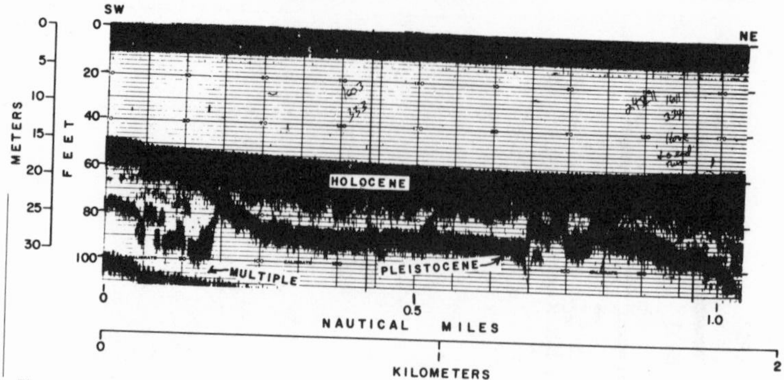

Figure 6. Original record of 7 kHz reflection profile B. Note prominent reflector with steep incised relief, which is correlated with Pleistocene sediments.

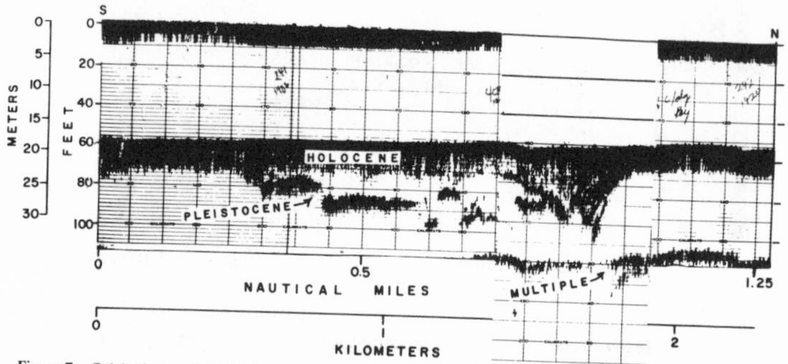

Figure 7. Original record of 7 kHz reflection profile E. Note that Pleistocene sediments are eroded into a broad valley 1 km (0.5 mi) wide. Within Holocene section, there is evidence that infilling beds, which are formed into narrower valley, have migrated to northern side of older, broader valley, which is seaward extension of ancient Rehoboth Bay lagoon.

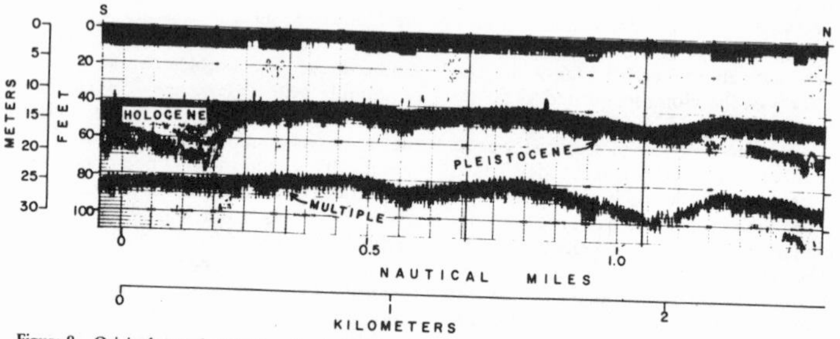

Figure 8. Original record of 7 kHz reflection profile C. Note that Pleistocene sediments rise out of deeper valleys on north and south to outcrop on sea floor in broad, gentle ridge.

marine silt recovered in cores 4, 11, and 12, is poorly sorted and more like sandy mud. This contrasts markedly with the very fine grained, blue, clay-rich sediments interpreted to be lagoonal mud in core 6 (Fig. 5). Whereas the lagoonal mud of core 6 was deposited behind a barrier in a quiet, shallow-water bay (like Rehoboth Bay), the mud in cores 11, 12, and 4 was probably deposited in an open estuary, similar to middle Delaware Bay today, subject to strong tidal flow and greater wave fetch.

As the transgression and sea-level rise progressed throughout Holocene time, these estuarine and lagoonal muds were truncated and buried by or intermixed with sand brought into the area from the marine environment. The constant daily flux of flood and ebb tides through this inlet at the mouth of Delaware Bay has piled the sand into two elongate shoals, Hen and Chickens shoal and the shoal cresting near core site 5 (Fig. 5).

Ebb currents apparently move sand, which is fed primarily by littoral drift spilling off the tip of Cape Henlopen, to accumulate in the inlet-associated Hen and Chickens shoal. Sand waves facing southwest apparently migrate up the gentle back slope of Hen and Chickens shoal to eventually deposit sand over the steeper southwest face, building the shoal upward and westward. Very well developed cross-bedding is observed in the high-resolution reflection profile of the core 5 shoal (Fig. 4). The inclination and bottomset relations of this bedding suggest that sand is migrating up and over the gentler southwest flank to be eventually deposited on the steeper northeast flank bordering the flood channel of the Delaware shelf valley. This core 5 shoal is apparently building longitudinally southeastward, as is Hen and Chickens shoal, suggesting that the sand feeding the core 5 shoal is carried out of the bay on ebb currents. Sand drifting southward off Cape May migrates as giant sand waves through what Swift and others (1972) have called the bay-mouth shoal complex to spill over into the ebb-dominated main channel of Delaware Bay. The ebb current then moves part of the sand seaward to build the core 5 shoal.

This movement of sand and the building of the core 5 shoal seem to imply that this shoal will eventually migrate northeastward to finally fill in the deep flood channel and weld with the shoal-retreat massif (Swift and others, 1972) northeast of the channel. This may be significant in the future, because this channel site has been proposed as a place for berthing Very Large Crude Oil Carriers (VLCC) of 27-m (90-ft) draft. Dredging maintenance of this channel might be required to keep its opening 30 m (100 ft) deep in the very distant future.

Rehoboth Bay Section

The Rehoboth Bay cross section was based on data from cores 7, 8, and 9 (Fig. 1) and high-resolution reflection profiles such as profiles B, C, and E (Figs. 6, 7, 8). The reflection profiles revealed a prominent reflector with rough (>30° slopes) topography incised into the surface, as if it had been subaerially eroded (Fig. 6). The overlying sediments were acoustically transparent and contained bedding suggestive of valley

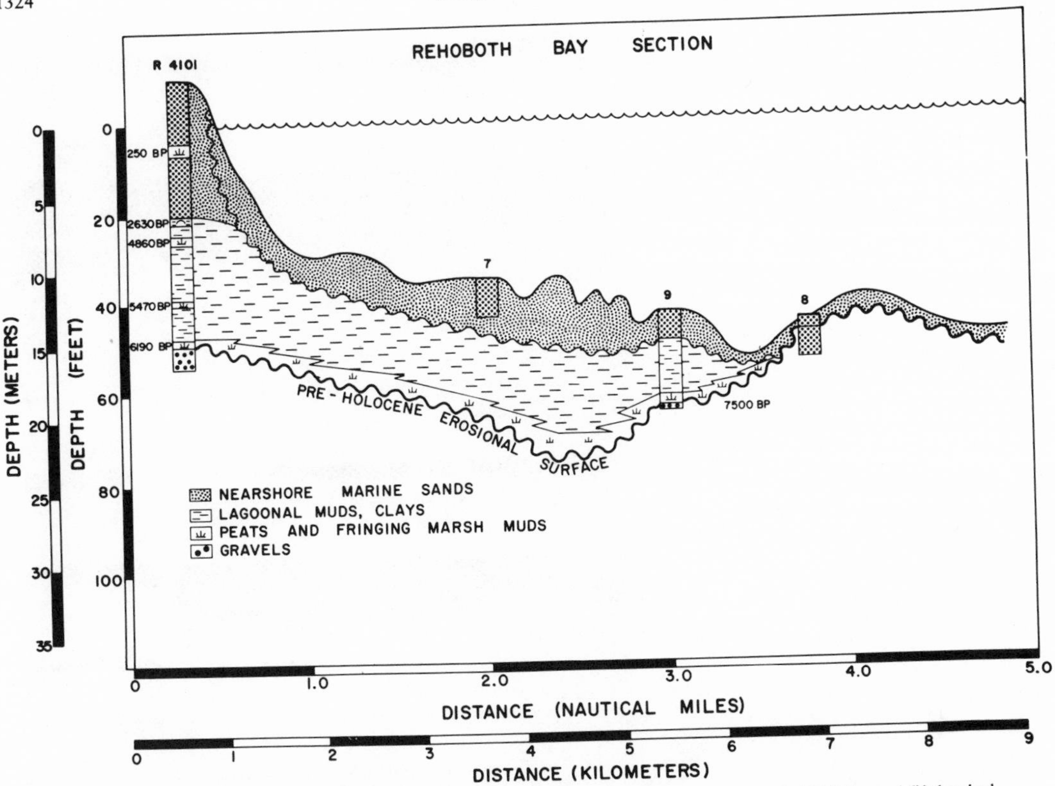

Figure 9. Diagrammatic geologic cross section southeast of Rehoboth Bay. Numbers 7, 8, and 9 designate vibracores; R–4101 designates drill hole on land.

infilling (Figs. 7, 8). The thalwegs of the larger valleys on this prominent reflector reached a depth of 30 m (100 ft), and the valleys were about 1 km (0.5 mi) wide. The surface was, therefore, sculptured by fluvial processes during the lower stand of the sea in Pleistocene time, when sea level was at least 30 m (100 ft) deeper than today. Thus, this prominent reflector is correlated with the pre-Holocene erosional unconformity underlain by Pleistocene sediments.

The reflection profiles revealed that the pre-Holocene surface formed a valley off the present Rehoboth Bay, but that this surface became shallower southward from the 30-m (100-ft) depth in this valley to about 12 m (40 ft), where it is exposed on the sea floor as a slight ridgelike topography (Fig. 8). Toward the beach, this reflecting surface was observed to deepen from the area of exposure, near core site 8, to deeper than 20 m (70 ft) beneath the sand bed forms of the shoreface-connected ridges near core site 7 (Figs. 1, 9). Unfortunately, because of the poor acoustic penetration through the sand layers of the frequencies used, the pre-Holocene surface was not well observed under these shoreface-connected sand ridges. The surface would project reasonably beneath core site 7 to be correlated with the pre-Holocene oxidized Pleistocene gravel found in drill hole R–4101 (Kraft, 1971), and these same orange, oxidized quartz-pebble sand and gravel are recovered in cores 8 and 9, lending credence to this correlation.

In core 9, above the pre-Holocene unconformity, were recovered a basal peat, radiocarbon dated as ~7,500 yr B. P., overlain by soft blue, very fine grained lagoonal mud. This is nearly an identical sequence to that found above the unconformity in drill hole R–4101 on the beach. Kraft (1971) interpreted such a sequence as being the record of a back-bay fringing marsh, now a peat, being flooded and buried by lagoonal mud as sea level gently rose during the Holocene transgression. This evidence in core 9 is proof that the sediments above the prominent reflector observed in reflection profiles, such as profiles B, C, and E are indeed the bulk of the Holocene record at those locations, and that the prominent reflector is the pre-Holocene unconformity.

The peats in core 9 and drill hole R–4101, if they truly represent back-bay fringing marshes, are the records of the past relative elevation of the sea, within a few meters. From such data, Kraft (1971) has worked out a local relative sea-level–rise curve for the past 7,500 yr. The rise had been at a rate of about 0.3 m (1 ft) per 100 yr from 7,500 yr B. P. to 3,700 yr B. P.; thereafter it decreased to about 0.15 m (0.5 ft) per 100 yr.

The nearshore marine sands in the upper part of core 9 and in core 7 are yellow and green, showing evidence of mottling, organic burrowing, and reworking, as if they were part of the dynamic bed form in equilibrium with the present nearshore marine environment. The sands are not reminiscent in structure or sorting of beach or dune sands. Therefore, of the three hypotheses discussed previously, that of Duane

and others (1972) seems to account best for the origin of these shoreface-connected ridges on the Rehoboth Bay section; that is, they are dynamic bed forms formed at the shoreface and in equilibrium with the present inner-shelf hydrodynamics.

It should be noted, however, that the somewhat broader topographic ridge near core site 8, with about 3 m (9 ft) of relief from crest to trough, is underlain by Pleistocene sand and gravel. Thus, this ridge topography is relict in nature, because it resulted from the excavation of older preexisting sediments and relief. The sediments in core 8 are strongly cross-bedded and might be interpreted as part of an ancient Pleistocene beach ridge complex formed during a lower stand of sea level. Kraft (1971) has studied these Pleistocene beach ridges in coastal Delaware, and he has found that they have a northeasterly trend at a slight angle to the present shoreline.

Bethany Beach Section

South of the exposure of the Pleistocene ridge feature near core site 8, the pre-Holocene seismic reflection surface deepens to nearly 30 m (100 ft) into the thalweg of the ancient valley of Indian River. The valley and acoustically transparent Holocene infilling are well observed on profile F (Figs. 1, 10). Again, the pre-Holocene surface rises sharply to the south out of the Indian River valley to be exposed over a broad area of the inner-shelf floor between Indian River Inlet and Bethany Beach.

Near the Bethany Beach section, the Pleistocene erosional surface deepens to from 20 to 30 m (60 to 90 ft) below steeply incised valleys with inlying transparent Holocene deposits, as seen on profile I (Fig. 11). The identification of these valley-filling sediments as Holocene lagoonal mud is made on the basis of the similarity of their acoustic appearance to that of sediments sampled in core 9 farther north. Unfortunately, core penetration was not deep enough on this cross section to penetrate to this mud (Fig. 12). Where the deeper prominent seismic reflector came close to the sea floor near core site 14, the vibracore became stuck in very stiff, hard, desiccated clay thought to be of pre-Holocene age.

The other cores, 13 and 15, taken in this cross section revealed that the two well-developed, northeast-trending, shoreface-connected ridges were underlain by at least 6 m (20 ft) of yellow, tan, and green sand similar to that of core 7 farther north. Certainly, all of the relief of these ridge features is part of the nearshore sand sheet, a dynamic bed form in equilibrium with the inner-shelf environment.

Below these ridges there is an unconformity, identifiable in the reflection

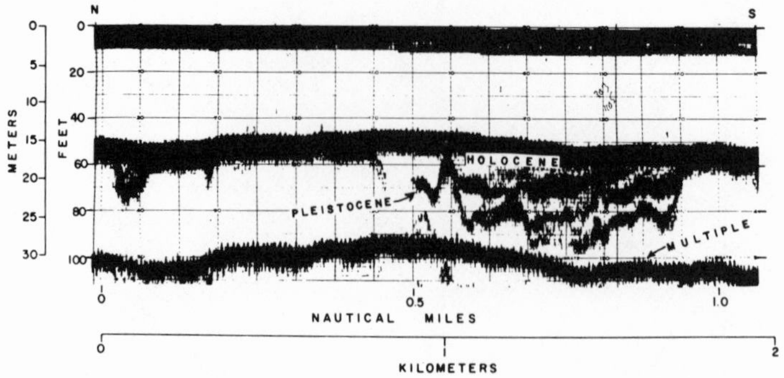

Figure 10. Original record of 7 kHz reflection profile F. Note transparent Holocene sediments filling in valley cut in Pleistocene sediments to nearly 30-m (100-ft) depth. This is seaward extension of ancient Indian River lagoon.

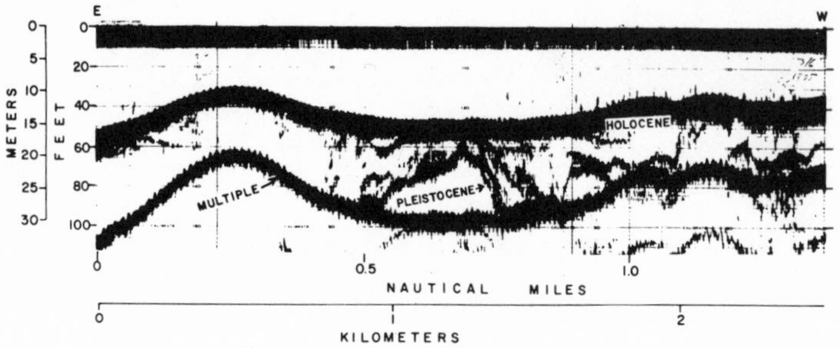

Figure 11. Original record of 7 kHz reflection profile I. Note transparent Holocene sediments filling in depressions in eroded Pleistocene sediments. Sea-floor ridges are not cored by relict sediments, but rather, near horizontal reflectors pass beneath shoals.

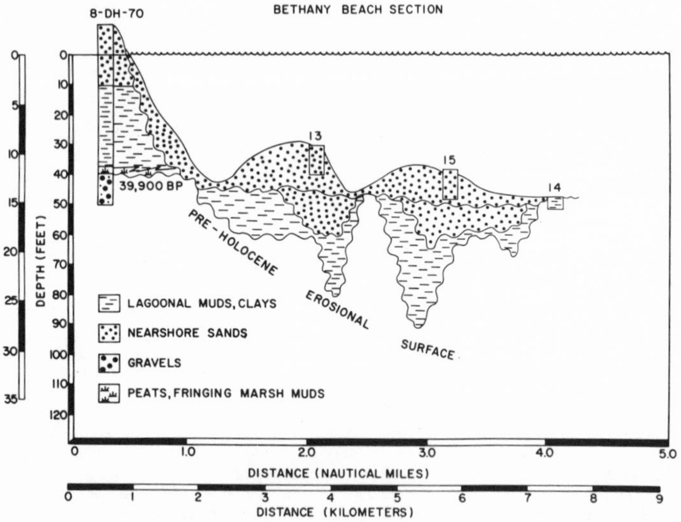

Figure 12. Diagrammatic geologic cross section east of Bethany Beach. Numbers 13, 14, and 15 designate vibracores; 8–DH–70 designates deep drill hole on land.

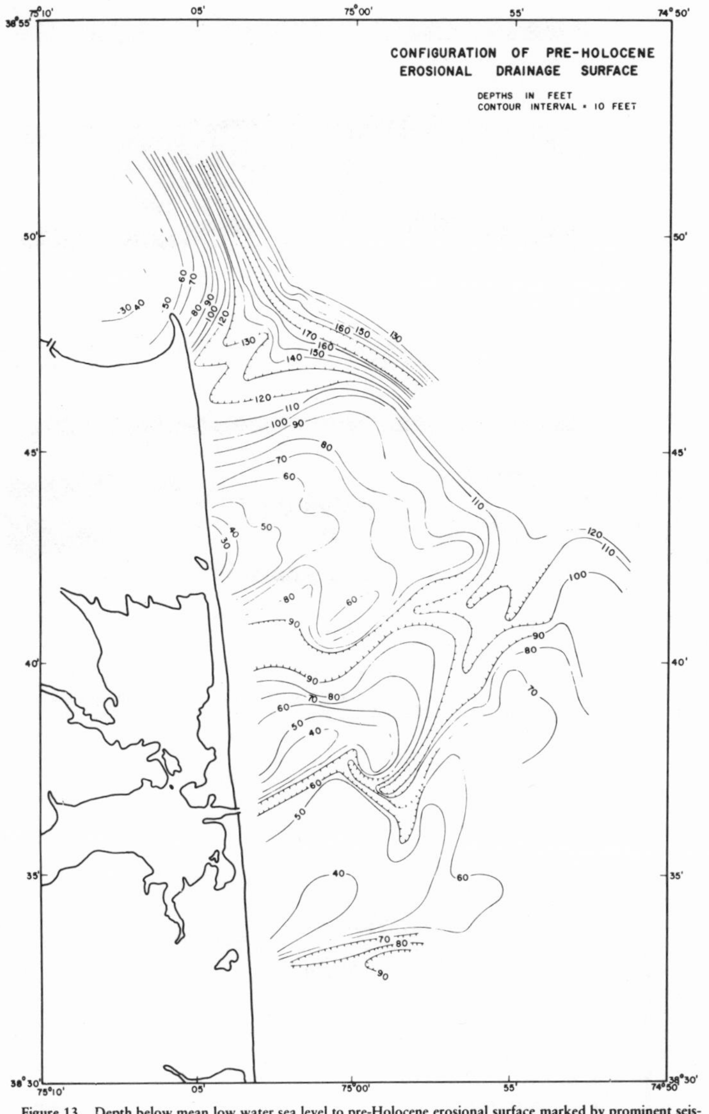

Figure 13. Depth below mean low water sea level to pre-Holocene erosional surface marked by prominent seismic reflector. Depths are in feet so that these data can be related directly to 1219 navigation chart used as base map. Chart presents in feet most accurate available soundings from study area, as will editions in next few years.

profiles, that passes nearly horizontally beneath the topography. This unconformity has resulted from the beveling of the preexisting, strong pre-Holocene relief by the shoreface erosion involved in the transgression. Certainly, the present sea-floor topography here is not cored by Pleistocene ridge structures. In fact, the pre-Holocene ridge structure observed below the beveled surface is shallowest in the swale between the ridges of the present topography (Fig. 11). This beveling effect and the nearly horizontal disconformity below the northeast-trending linear ridges is commonly observed on the inner shelf off Ocean City, Maryland, just south of Bethany Beach (Duane and others, 1972).

Drill hole 8–DH–70 (Fig. 11), made on shore near Bethany Beach, recovered a peat material at a depth of 12 m (40 ft) that has been dated as 39,900± yr B.P. Subsequent drilling has confirmed that this Pleistocene peat occurs consistently at this depth, suggesting a higher level of the sea within the Pleistocene Epoch at that time, that is, higher than the latest Pleistocene regression. The lagoonal mud found in drill hole 8–DH–70 is therefore Pleistocene in age; it formed part of the Pleistocene highland exposed at Bethany Beach. The acoustic transparency of the mud offshore, however, suggests a younger age, and this mud is correlated with the Holocene lagoonal mud seen on profiles B, C, and E.

Configuration of Pre-Holocene Erosional Drainage Surface

Based on the high-resolution reflection profiles, in which the prominent reflecting surface is identified as the pre-Holocene surface, and based on the vibracoring data, which confirmed this surface to be a true lithologic break below the Holocene sediments, depths to this surface were plotted along the profiles where the surface was apparent. These data were contoured with the interpretation in mind that the surface was probably formed initially by subaerial fluvial erosion, and that subsequently these fluvial depressions were modified into coastal lagoons as they were flooded by the sea and were trapped behind barrier islands (Fig. 13).

Essentially, the pre-Holocene surface under the Delaware inner shelf is dominated by three ancient, more or less northeast-trending, shallower interfluve areas separated by two narrow and meandering ancient channels of the Rehoboth Bay and Indian River outlets. The shallower interfluves are off Rehoboth Beach, just north of Indian River inlet and just north of Bethany Beach, all coastal areas where ancient Pleistocene headlands now exist. Off Cape Henlopen, another series of possible deep channels is contoured on the bases of the reflection profiles and of the need to match the depths to Pleistocene sediments found in the drill holes on land there. Some evidence of these deeper channels was observed in profile north of that spit (Fig. 2). The southern ancient valley bordering the shallower interfluve off Bethany Beach is poorly revealed by the profiles, but it is definitely present in profile I (Fig. 11). How this ancient valley continues to the south and east is unknown, but data from coastal drill holes (Kraft, 1971) suggest that it

should have drained the ancient counterparts of Little Assawoman Bay and been bounded on the south by another interfluve east of Fenwick Island, Maryland.

The drainage pattern of these ancient valleys, trending northeast out of Rehoboth and Indian River Bays, suggests there was a drainage at depths of 27 to 30 m (90 to 100 ft) northeastward across the main interfluve area southeast of the ancient Delaware River channel; the Delaware River's drainage, on the other hand, trended to the southeast across the shelf at depths of 50 to 55 m (170 to 180 ft). The ancient Rehoboth and Indian River channels, therefore, formed important tributaries to the ancient Delaware River drainage prior to about 12,000 yr ago. Such a paleogeography was conjectured by Kraft (1971) as a logical consequence of lowering sea level.

EVOLUTION OF LAGOONAL COMPLEX OF COASTAL DELAWARE

A logical consequence of the rising sea level of the Holocene transgression would be the flooding of these ancient Rehoboth and Indian River channels to form small estuaries at first and broad lagoons later, as the estuary mouths were closed by spit and barrier-bar development. From the core data reported here, it is apparent that calm, shallow-water conditions existed in the ancient Rehoboth Bay depression when the lagoonal mud of core 9 was deposited. Certainly, the fringing marsh peats would not have been preserved in sequence had they been exposed to shoreface erosion on the open Atlantic beach. Thus, these peats, radiocarbon dated as ~7,500 yr B.P., must have been in a back-lagoon situation, in a lagoon truly separated from the open ocean by a barrier island.

Conditions must have changed, therefore, from 12,000 yr B.P., when the Indian River and Rehoboth fluvial channels drained as tributaries into an ancient Delaware River or Bay, to the situation of 7,500 yr B.P., when these depressions had become true lagoons draining directly into the open Atlantic Ocean, probably through naturally occurring inlets in the barrier. This situation for 7,500 yr B.P. is very nearly like the hypothetical paleogeographic reconstruction of Kraft (1971) for the Dealware coast at 7,000 yr B.P.

The pre-Holocene surface map (Fig. 13) indicates that shoreface erosion has cut this pre-existing headland surface to about 10 m (30 ft) deep within a few hundred feet of the present shoreline. This is especially apparent on the Rehoboth Beach section (Fig. 5). By tracing out the 30-m (90-ft) contour on the pre-Holocene surface map, the headlands found for this depth would have been exposed when sea level was at 20 m (60 ft) deep about 7,000 yr B.P. It turns out that these headlands would have been joined by the barrier island forming the lagoons of ancient Rehoboth Bay and Indian River Bay, and this barrier island coastline, similar in appearance to Kraft's (1971) reconstruction, would have been about 12 km (7 n mi) from the present shoreline. If this were true, then the minimum average rate of Atlantic coastline retreat for the past 7,000 yr has been 1.6 horizontal m per yr (6 ft per yr).

SUMMARY AND CONCLUSIONS

The Atlantic inner shelf off Delaware is underlain by a buried, undulating pre-Holocene erosional drainage surface at depths of 12 to 55 m (40 to 180 ft). This is a composite surface of variable lithology, including oxidized sand and gravel and very stiff, desiccated clay. It is also composite in age, but where it is shallower than about 37 m (120 ft), it is mostly Pleistocene sediments. Some Tertiary sediments may form this surface where it is deeper than 37 m (120 ft). Two well-defined depressions in the surface represent the ancient fluvial drainages out of Rehoboth and Indian River Bays. These depressions, to depths of 27 to 30 m (90 to 100 ft) were filled with earlier Holocene lagoonal mud. These ancient lagoonal deposits have been subsequently beveled by shoreface erosion associated with the Holocene transgression. Consequently, only the older, deeper deposits of this lagoonal mud are retained in the deeper ancient channels on the inner shelf.

Paleogeographic reconstructions for the 7,000-yr B.P. situation of the ancient Rehoboth Bay and Indian River Bay lagoons indicate that they must have been at least 12 km (7 n mi) eastward of the present shoreline and there they were bounded on the east by a nearly straight barrier island joining the adjacent headland areas. If these paleogeographic reconstructions are true, then the westward migration of this barrier-island coast must have progressed at a minimum average rate of 1.6 horizontal m per yr (6 ft per yr).

The beveling of the pre-existing older Holocene lagoonal facies and Pleistocene headlands by shoreface erosion has left a surface at 12- to 15-m (40- to 50-ft) depths, upon which nearshore marine sands have accumulated. This surface is nearly horizontal in most cases, but occasionally a more resistant Pleistocene sedimentary facies, such as an ancient beach ridge, causes topographic ridges with as much as 3 m (9 ft) of relief to appear on this surface. These Pleistocene ridge features, although rare, might be confused with the much more common ridges formed by recent bed forms.

The northeast-trending, shoreface-connected ridges of the Delaware coast were apparently formed wholly within the recent nearshore sand deposits and are not cored by ancient Pleistocene ridges. The nearshore sands are apparently in dynamic equilibrium with the inner-shelf environment, and the ridges are therefore formed at the shoreface, rather than being formed as Holocene beach-ridge features that have subsequently submerged.

The ebb-dominated and flood-dominated channels of the Delaware Bay mouth are formed as mutually evasive pairs in response to fluctuating tidal currents. The flood channel, which is continuous with the Delaware shelf valley, is slightly offset from the deeper buried ancestral Delaware River fluvial channel. The flood channel of the Delaware Bay mouth is therefore not a relict feature but, rather, has formed in dynamic equilibrium with the nearshore marine environment subsequent to the transgression. Two inlet-associated shoals, Hen and Chickens and a parallel one in deeper water across the bay mouth, are built by the ebb transport of sand. The shoal bordering the southwest side of the flood channel entering Delaware Bay is building northeastward to close this channel.

ACKNOWLEDGMENTS

This research was supported by Office of Naval Research Contract N–000–14–69–A–0407. We thank Duke University Marine Laboratory for the use of R/V *Eastward* on cruise E–18B–70. This Cooperative Oceanographic Program is supported by National Science Foundation Grant GB–17545. We thank Al Stockel of Alpine Geophysical Associates for his assistance with the vibracoring program on *Eastward.* Ralph Shaver of Ocean Science and Engineering supervised the vibracoring from R/V *Wando River* and supplied 3.5-kHz reflection profiles of the sites. Capt. M. Willis and Party Chief Fred Kelly assisted us on *Eastward*, and Capt. M. Cooper of the R/V *Skimmer*, of the University of Delaware College of Marine Studies, assisted us during the reflection profiling.

REFERENCES CITED

Duane, D. B., Field, M. E., Meisburger, E. P., Swift, D.J.P., and Williams, S. J., 1972, Linear shoals on the Atlantic inner continental shelf, Florida to Long Island, *in* Swift, D.J.P., Duane, D. B., and Pilkey, O. H., eds., Shelf sediment transport: Process and pattern: Stroudsburg, Pa., Dowden, Hutchinson and Ross, Inc., p. 447–498.

Kraft, J. C., 1971, Sedimentary facies patterns and geologic history of a Holocene marine transgression: Geol. Soc. America Bull., v. 82, p. 2131–2158.

Miller, J. C., 1971, Ground water geology of the Delaware Atlantic seashore: Delaware Geol. Survey Rept. Inv. 17, 32 p.

Moody, D. W., and Van Reenan, E. B., 1967, High-resolution subbottom seismic profiles of the Delaware estuary and bay mouth: U.S. Geol. Survey Prof. Paper 575–D, p. D347–D352.

Oostdam, B. L., 1971, Suspended sediment transport in Delaware Bay [Ph.D. dissert.]: Newark, Univ. Delaware, p. 316.

Swift, D.J.P., 1973, Delaware shelf valley: Estuary retreat path, not drowned river valley: Geol. Soc. America Bull., v. 84, p. 2743–2748.

Swift, D.J.P., Kofoed, J. W., Saulsbury, F. P., and Sears, P., 1972, Holocene evolution of the shelf surface, central and southern Atlantic shelf of North America, *in* Swift, D.J.P., Duane, D. B., and Pilkey, O. H., eds, Shelf sediment transport: Process and pattern: Stroudsburg, Pa., Dowden, Hutchinson and Ross, Inc., p. 499–574.

Manuscript Received by the Society November 8, 1973

Revised Manuscript Received March 4, 1974

Part III

STUDIES OF FLUID MOTION

Editors' Comments on Papers 13 Through 16

13 **MURRAY**
Bottom Currents near the Coast during Hurricane Camille

14 **PALMER and WILSON**
Nearshore Current Regimes in a Linear Shoal Field, Middle Atlantic Bight, USA

15 **CASTON**
A Wind-driven Near-Bottom Current in the Southern North Sea

16 **CSANADY**
Wind-Driven and Thermohaline Circulation Over the Continental Shelves

The following series of papers dealing with long period fluid motions (longer period than wave surge) is almost completely isolated from the preceding series dealing with the concept of the equilibrium coastal profile and its behavior. Neither set of papers quotes the other. Why? Don't long period flows also entrain and transport sediment? They do indeed.

The blame for this failure in communication rests with both sedimentologists and physical oceanographers. Until very recently, most persons studying coastal sedimentation have been trained as classical terrestrial geologists. They have had little or no background in physical oceanography and fluid dynamics and little contact with physical oceanographers. The first thing that they see when standing on the land and looking seaward is the surf zone, which has been their main area of study. Their traditions are closer to the coastal engineers who study the surf zone and who do tend to have contact with sedimentologists. Consequently sedimentologists have tended to adopt the bias of coastal engineers toward waves as the dominant forcing mechanism for coastal sediment transport; they tend to see few fluid motions of interest beyond the surf until the oceanic boundary current at the shelf edge.

Sedimentologists have been aided and abetted in this attitude by physical oceanographers. The papers in this section have a common ancestor analogous to the Fenneman paper in Part I—that is, V. W. Ekman's monumental paper of 1905. We have not reproduced it because of its length, because it is in German, and because much of it is not relevant to coastal flow. Unfortunately, after this groundbreaking study, physical oceanographers paid little attention to coastal flow. It was messy and complicated—requiring frictional terms in the equation of motion—and did not seem to have the intellectual challenge of the great planetary flows of the deep ocean basins. Two studies of coastal flow during this dry period worthy of note are those of Jeffries (1925) and Baines and Knapp (1965).

Paper 13, by Murray, is the story of the current meter that was already a legend before the paper was published. Mounted on a tripod and placed in 5 m of water on the Louisiana coast shortly before Hurricane Camille struck in 1971, it recorded flows up to 160 cm/sec^{-1} until the rotor jammed, the connector was pulled out, and the sensor broken off its mount by the intensity of flow. There are several points made in this brief paper. Intense storms develop strong coastal flows that are capable of entraining and transporting sand and that last for hours or days. Their directions and intensities are not easily predicted from a knowledge of wind behavior. Winds of the proper orientation can drive offshore bottom flows, and perhaps provide the offshore transport required by the theory of erosional shoreface retreat presented in Bruun's Paper 3.

As physical oceanographers were finally turning to coastal flows, geological oceanographers were also adopting the new current meter technology to answer their own questions. Initially, they have merely raised further questions, and this preliminary statement presented as Paper 14 is no exception. Here, one of us has monitored flow over an inner shelf sand ridge—in the classic ridge topography of the Delmarva (Delaware-Maryland-Virginia) inner shelf. Offshore bottom flow again appears, together with morphologic evidence for seaward sand transport. The data, alas, are too fragmentary to establish a relationship of cause and effect, but have led to the experimental designs presently employed by several research teams that are examining fluid motion and substrate response at the time of writing of these comments.

In Paper 15, another geological oceanographer attempts to resolve the hydraulic regime in an area of complex sea floor response to flow—in the tide-induced Norfolk Banks off the Anglian Coast of Great Britain. Again the study is reconnaissance in nature and again raises more questions than it answers, but it was written

when almost any kind of near-bottom flow information in the North Sea was news—a situation that has not changed. Perhaps the major point is Caston's confirmation of the prediction of Pitt and others (Caston, p. 24): "Wind-driven currents are the most important nontidal component of the current system and over much of the central and northern parts of the North Sea can be expected to match or exceed the tidal component." For four days, the December 1967 storm increased the speed of the southerly going tide by 38 percent and reduced the northerly going tide by 25 percent.

In Paper 16, Csanady presents a review of shelf circulation especially written for the nonspecialist. We learn that wind stress overwhelms the regional pressure gradient as a forcing term for coastal flow. Patterns of wind-driven coastal circulation may generate slowly moving, long period waves, many kilometers in length, that are trapped against the coast (topographic waves). Coastal jets may develop when the inshore edge of the thermocline is deflected up by upwelling or down by downwelling so that momentum flux down through the mixed layer is modified with respect to the same flux further offshore. Long period internal waves (Kelvin waves) may develop on the thermocline that are analogous to topographic waves. Finally, we are asked to regard the coastal zone in which these phenomena occur as a coastal boundary layer, with a hydraulic climate quite distinct from that of the rest of the shelf. No attempt is made to relate coastal boundary flow to sediment transport, but it is now clear that we cannot understand the coastal sediment budget until we comprehend this relationship.

REFERENCES

Baines, W. D., and Knapp, D. G. 1965. Wind-driven water currents. *Proc. Am. Soc. Civ. Eng.* **91**(HY2):205–21.

Ekman, V. W. 1905. On the influence of the earth's rotation on ocean currents. *Arkiv fur Matematik, Astronomi och Fysik* **2**:1–53.

Jeffries, H. 1925. The effect of a steady wind on the sea level near a straight shore. *Philos. Mag.* **46**:114–25.

13

Reprinted from *J. Geophys. Res.* 75(24):4579–4582 (1970)

Bottom Currents near the Coast during Hurricane Camille

STEPHEN P. MURRAY

Coastal Studies Institute, Louisiana State University
Baton Rouge 70803

A ducted current meter, which was mounted on the bottom in 6.3 meters of water off the coast of the Florida panhandle, was operative during much of the activity of Hurricane Camille. Before the arrival of the storm an unexpected outward extension of the wave-driven longshore current was recorded. During the storm bottom current speeds ranged up to 160 cm/sec, and their direction rotated from alongshore parallel to the wind to seaward against the wind.

Hurricane Camille, which traversed the Gulf of Mexico on a northwesterly track during the period August 15–17, 1969 [*U.S. Department of Commerce*, 1969], came ashore at Gulfport, Mississippi, 160 km to the west of the present study site (Figure 1). As observational data on current fields near coasts during storms are practically nonexistent, this report presents the results from an in situ current meter that operated through most of the hurricane and a concurrent anemometer record.

A week before the storm a Marine Advisers Q16 ducted current meter (Figure 2) was installed on the bottom 360 meters offshore (90 meters seaward of the outer bar) at a depth of 6.3 meters. The meter is bidirectional: it senses the wave orbital motions but prints out the steady current speed after averaging. Calibration of this type of ducted meter was discussed by *Murray* [1969*a*]. Extensive field use has shown the operating principles to be quite sound. An anemometer mounted 10 meters above the beach (12 meters above sea level) gave a record of the local wind vector throughout the storm.

At about 1100 hours on Saturday, August 16, the first storm waves appeared on the beach, a low swell (forerunners) approaching the east-west-trending shoreline from out of the southeast. At 1700 hours on Saturday the forerunners were breaking repeatedly on the outer bar; the estimated average breaker height was 1.5–2.0 meters. Breaker height continued to increase as the storm approached.

The smoothed wind and current records are shown in Figure 3. In the figure the current speed is divided into three phases: phase A is characterized by the presence of forerunners but no storm winds; phase B, by a suddenly intensified bottom current but still no storm winds; and phase C, by the storm itself, with its wind-driven currents and waves. Before 2215 hours on Saturday (phase A) the current speed reached only 5–10 cm/sec in a direction roughly

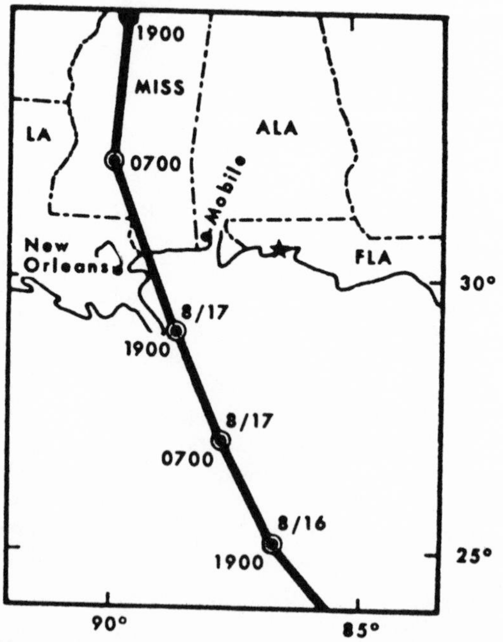

Fig. 1. The location of the eye of Hurricane Camille with respect to the study site, marked with a star, August 16–18, 1969.

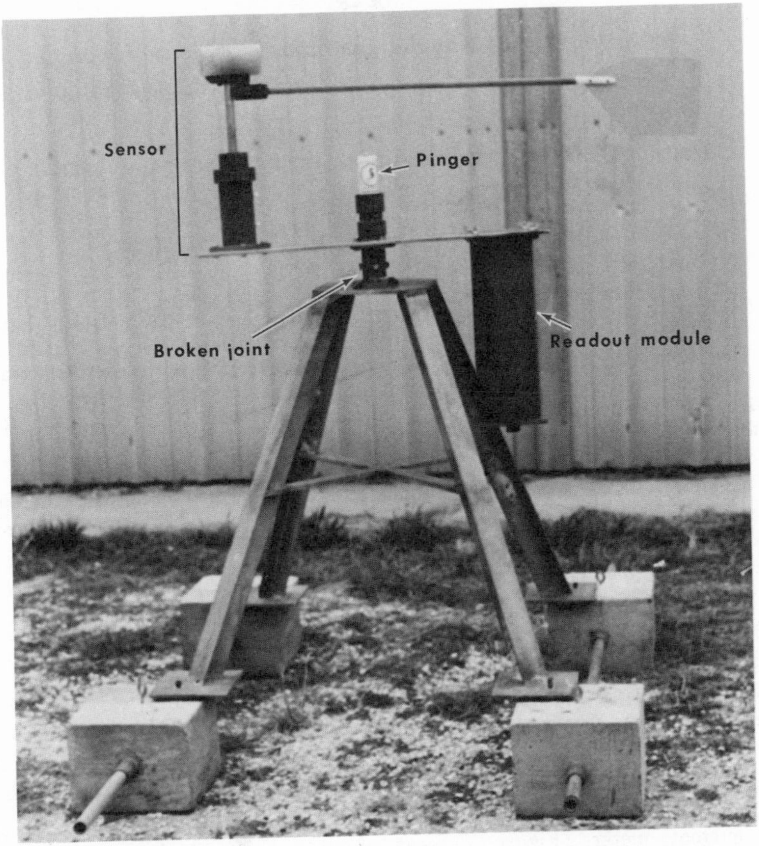

Fig. 2. The Q16 current meter and support structure. The vane shown is 90 cm long; a 2.7-meters-long vane was used in the field. The duct is 1.4 meters above the top of the blocks.

between 270° and 300°. These values are typical for moderate conditions [*Murray*, 1969*b*] in this area, and so we can conclude that the forerunners had not yet influenced the bottom current here. At the onset of phase B the current speed increased over a 3-hour period from 10 up to 35 cm/sec and remained near this level for the duration of phase B; direction became more westerly along the shore. As (*a*) the storm winds had not yet reached the study area, (*b*) prior observations have shown tidal currents here to be negligible, and (*c*) the wave drift current [*Kenyon*, 1969] was also negligible, a remaining explanation for the observed increase in the current speed during phase B is that the longshore current generated in the surf zone set in motion by lateral friction the water beyond the outer bar.

The time t required for the speed u to propagate outward by frictional stresses a distance y from a plate moving at a constant speed U in an initially motionless fluid is given implicitly by *Batchelor* [1967, p. 190] as

$$u/U = 1 - \phi[y/2(At)^{1/2}] \qquad (1)$$

where $\phi(\ )$ is the error function and A is the eddy viscosity. Although the eddy viscosity remains poorly known, a sufficiently broad range for a trial calculation here appears to be $500 < A < 1500$ g/cm/sec [*Neumann and Pierson*, 1966], corresponding to sea states 4–5 (rough to very rough seas) and Beaufort numbers 5–9 (fresh breeze to strong gale) [*Wiegel*, 1953]. If the plate speed U is identified by analogy with the longshore current over the outer bar ($\simeq$1.5 m/sec, using the equations of *Galvin* [1967], *Harrison* [1968], and *Sonu et al.* [1967]), the predicted arrival of the 35-cm/sec

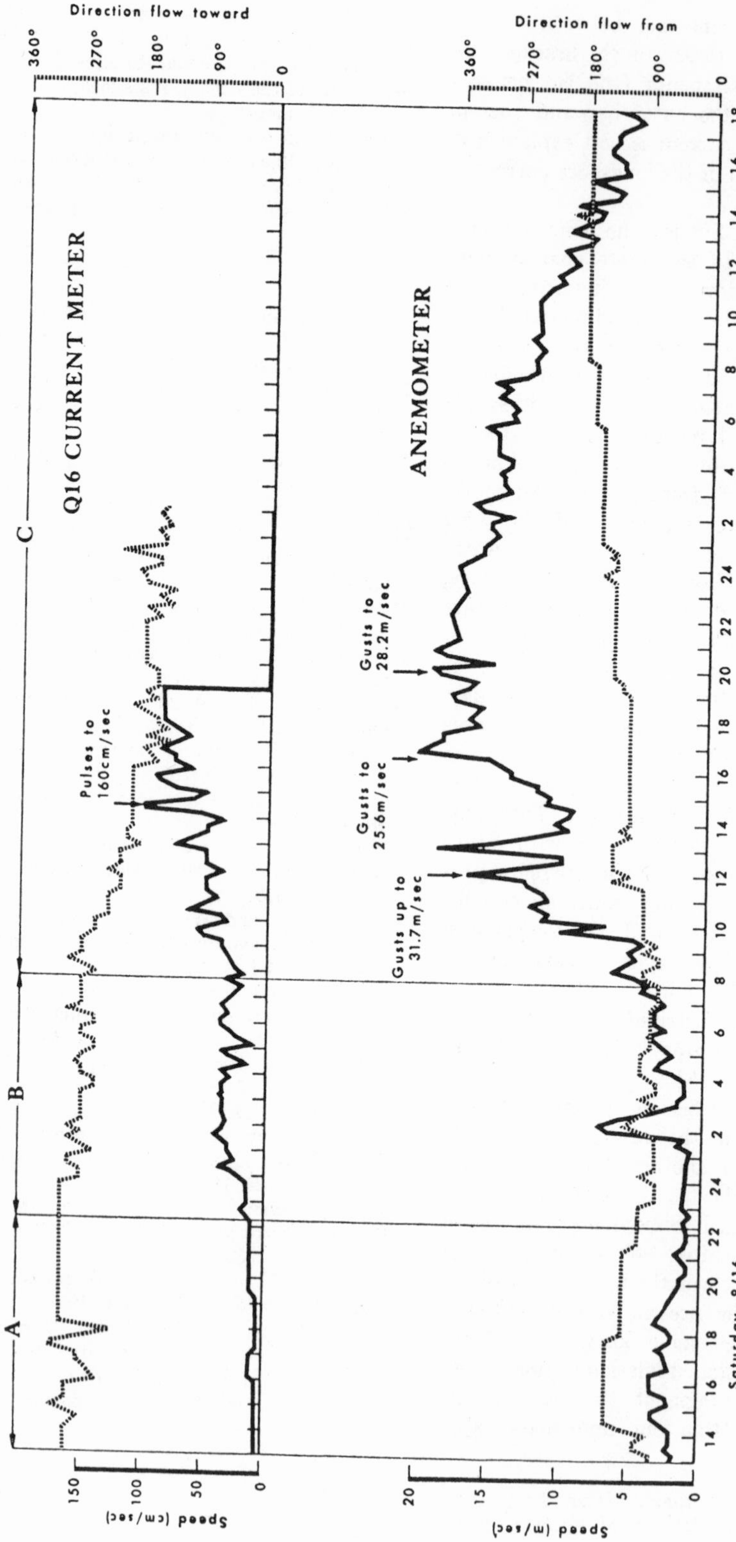

Fig. 3. The speed and direction records for the anemometer and the Q16 current meter on the same time scale.

front at the instrument falls between 5.3 and 15.9 hours. Since these times bracket the observed time of 8.5 hours (the lag between heavy breaking on the outer bar and current onset), the phase B current is best explained as a seaward outgrowth of the longshore current in the surf zone.

At 0745 hours on Sunday the wind and current speeds began to accelerate sharply with the arrival of the storm front, initiating phase C. At 0900 hours the current direction began a slow counterclockwise rotation from westerly flow parallel to the shore to southerly offshore flow. What precipitated this rotation is as yet unclear, as the wind remained parallel to the shore. At 1130 hours on Sunday the wind began a slow clockwise shift from east toward south. The ensuing onshore component of the wind stress likely produced a piling up of water against the coast [*Ekman*, 1905; *Van Dorn*, 1953]; this resulted in a seaward pressure gradient capable of bringing about the counterclockwise rotation of the current vector, as seen in the record after 1130 hours on Sunday.

When the wind had turned only as far as east-southeast at 1600 hours on Sunday, the current was already flowing offshore at a speed equal to about 5% of the wind speed. This value and direction at this depth agrees with *Reid*'s [1957, Figure 1] theory of the currents produced in a bounded channel under an equilibrium setup. According to Reid, a two-layer flow is generated: an upper layer moving onshore with the wind, and a lower layer flowing offshore. This velocity distribution is well documented in the laboratory [*Keulegan*, 1951; *Baines and Knapp*, 1965].

The speed impeller became jammed at 1900 hours on Sunday, but the direction sensor continued operating for another 7 hours; at this time the record indicates the underwater connector was unplugged. Subsequently the joint marked in Figure 3 broke off; a pinger locator was later used to recover the meter assembly.

In conclusion, the present record, although incomplete, provides some knowledge of the structure of the coastal circulation during storms and provides encouragment that a complete understanding is within our immediate capability.

Acknowledgments. I thank Marshall Cartledge and Colonel Jack G. Coblenz of Eglin AFB for their personal cooperation and the use of Proving Grounds facilities. Norwood Rector, Research Associate of the Coastal Studies Institute, was invaluable in the design, installation, and recovery of the instruments.

This work was supported by the Geography Programs, Office of Naval Research, under contract Nonr 1575(03), NR 388 002, with the Coastal Studies Institute, Louisiana State University.

References

Baines, W., and P. Knapp, Wind driven water currents, *J. Hydraul. Div., Amer. Soc. Civil Eng., 91*(HY2), 205, 1965.

Batchelor, G., *An Introduction to Fluid Dynamics*, 290 pp., Cambridge University Press, New York, 1967.

Ekman, V. W., On the influence of the earth's rotation on ocean currents, *Arkiv Matematik, Astron. Fysik, 2*(11), 1–53, 1905.

Galvin, C. J., Longshore current velocity: a review of theory and data, *Rev. Geophys., 5*(3), 287, 1967.

Harrison, W., Empirical equation for longshore current velocity, *J. Geophys. Res., 73*, 6929, 1968.

Kenyon, K. E., Stokes drift for random gravity waves, *J. Geophys. Res., 74*, 6991, 1969.

Keulegan, G. H., Wind tides in small closed channels, *J. Res. NBS, 46*(5), 358, 1951.

Murray, S. P., Current meters in use at the Coastal Studies Institute, *Coastal Stud. Bull.*, no. 3, 1–15, Louisiana State University, Baton Rouge, 1969*a*.

Murray, S. P., Wind generated currents near the coast (abstract), *Trans. Amer. Geophys. Union, 50*, 192, 1969*b*.

Neumann, G., and W. Pierson, Jr., *Principles of Physical Oceanography,* 545 pp., Prentice-Hall, Englewood Cliffs, N. J., 1966.

Reid, R. O., Modification of the quadratic bottom-stress law for turbulent channel flow in the presence of surface wind stress, *Beach Erosion Board Tech. Mem. 93,* 33 pp., U.S. Corps of Engineers, 1957.

Sonu, C. J., J. M. McCloy, and D. S. McArthur, Longshore currents and nearshore topographies, *Proc. Tenth Conf. Coastal Eng., Amer. Soc. Civil Eng.*, 524, 1967.

U.S. Department of Commerce, ESSA, Weather Bureau, *Hurricane Camille—A Preliminary Report,* 58 pp., 1969.

Van Dorn, W. G., Wind stress on an artificial pond, *J. Mar. Res., 12*(3), 249, 1953.

Wiegel, R. L., *Waves, Tides, Currents, and Beaches: Glossary of Terms and List of Standard Symbols,* 113 pp., Council on Wave Research, The Engineering Foundation, Berkeley, 1953.

14

Reprinted from pp. 137–141 of the *IXme Congres International de Sedimentologie,* Nice, France, 1975

Nearshore Current Regimes in a Linear Shoal Field, Middle Atlantic Bight, USA

Palmer, Harold D., Manager, and Wilson, Donald G., Senior Engineer;
Westinghouse Ocean Research Laboratory
Annapolis, Maryland USA

INTRODUCTION

Many continental shelves display a distinctive surface morphology which has been described as "ridge and swale" topography. These regions are characterized by elongate ridges of unconsoldiated sands tens of km long and a few km wide. They are generally disposed in parallel sets, and their origin has been attributed to both submergence of barrier island (static, relict features preserved on the shelf surface) and to dynamic constructional processes generating ridges built and maintained by present shelf currents (dynamic features molded by the present hydraulic regimes associated with storms). On the inner continental shelf, these sand ridges may connect with the shoreface, or they may occur as isolated bathymetric highs. In water depths of less than 20 meters, wave effects can couple with inner shelf currents, and their combined effects are believed to mold and modify these shoals (Duane and others, 1972). However, field data for hydraulic regimes on the shoals are sparse and short-term. This report describes 28 days of fairweather current records in the inner Middle Atlantic Bight at two sites off the Maryland coast (Fig. 1).

SITE CHARACTERISTICS

The linear nature of the shoal field off the coasts of Delaware, Maryland and Virginia is clearly displayed in Fig. 1. The site selected for this study is a shoreface-connected ridge lying 4.5 km off Assateague Island, a major barrier island forming the coast of Maryland and Virginia. The trend of this shoal's major axis is NE-SW, and it lies at an angle of 18° to the shoreline orientation. With the exception of a band of shoreface muds, sea floor sediments in this region are medium to fine well-sorted sands. High resolution seismic reflection profiling (7 kHz) reveals a nearly flat continuous reflecting horizon beneath the sand veneer (Fig. 2). This surface has been cored, and it consists of muds believed to represent lagoonal deposits which accumulated behind a barrier during the Holocene transgression.

Two features of Fig. 2 bear upon interpretation of the hydraulic regime present at this site. First, the sea floor between the shoreface and the shore side of the ridge and that seaward of the outer base of the shoal are quite flat, displaying gentle offshore slopes of about 1:750, or a dip of about 0° 04.5'. However, the surface between the shoreface and the shoal is depressed by about two meters below that seaward of the ridge. Second, there is definite transverse asymmetry to this feature, with the southeastern (offshore) slope appearing steeper than the inshore slope. Together, these points suggest that erosion, or at least non-deposition, may be the rule in the trough between the shoreface and the ridge, and that material transported through or out of this area is deposited on the seaward ridge face and/or sea floor. Such asymmetry has been noted throughout this area (Duane and others, 1972), and similar situations occur in European shelf areas (Caston and Stride, 1970; Stride, 1974). In the latter area, current measurements and internal structure within asymmetric linear shoals

confirm a net unidirectional transport of sand in the direction in which the steeper slope faces. In the case off Assateague Island, we fail to see internal structure, but current data to be presented suggest a net offshore transport toward the southeast. Historical data provided by Duane and others (1972) imply a southeasterly displacement of nearby linear shoals which averages 2 to 6 meters/year over the 80 years charts have been kept for this region. On the basis of morphology, we suspect that these shoals are dynamic features which respond to fluctuations in the inner shelf hydraulic regime. In order to test this proposition, we deployed recording current meters on the crest of the shoal and in the inshore trough.

INSTRUMENTATION

Two General Oceanics film recording model 2010 current meters were mounted on supports which placed them 100 cm above the bottom. Meter #1 was positioned on the crest of the ridge at a depth of about 7 m, while the trough installation (#2) was placed at 11m. Frame exposures were set at 12/hr, and the period of record was 16 April - 8 May, 1974. Reading of the film was accurate to ± 1 sec. Speed readings based upon the angle of deflection could be read to the nearest 5° which is equivalent to a speed of 7 cm/sec; direction was read to ± 5°. There is evidence in the records of wave contamination at speeds less than 20 cm/sec. However, divers observed little oscillatory motion of the meters for periods of 10-20 minutes, and directional data are relatively stable. The film record for the trough meter (#2) for the period 26 April - 8 May was out of focus, and interpretation was not satisfactory. Those data are not included here.

CURRENT DATA

Speeds

Current speed data for the crest and trough sites are displayed in Fig. 3. Current speeds on the crest are consistently higher than those in the trough. Currents at 100 cm above the bottom exceeded 21 cm/sec 58% of the time, while those in the trough exceeded this speed 26% of the time. Highest speeds were toward the south-southeast.

Direction

Current directions are displayed as vector average diagrams in Fig. 4. The consistent south-southeasterly trend in fairweather flow is apparent, and only a brief departure from this vector was observed at the two sites. This occurred during a 60-hour period when the average wind increased from 3.5m/sec to 7 m/sec and changed direction from the typically variable winds from the north quadrant to a steady SSW wind. Current reversal occurred some 19 hours after this shift, but the return to northerly winds and southerly current flow was nearly synchronous. The trough flow reversal lagged behind the crest, and we attribute this to shearing within the thicker water column. As Fig. 4 reveals crest and trough flow directions are essentially parallel at any given instant.

Transport

Average daily net transport of bottom water is presented in Fig. 5. Prevailing fairweather flow at both the crest and trough stations was to the south-southeast, but trough velocities are one-tenth that at the crest. The difference

can be attributed to accelerations in the flow of water escaping from the inshore trough which narrows to the south. This effect must increase during storms when more water enters this restricted region (Swift, this volume; 1974) but our evidence suggests that it may be the predominant flow pattern throughout the year. Only the intensity varies. Part of the SSE flow can be attributed to the local nearshore current regime (described by Norcross and Stanley (1967) and Bumpus and others (1972). The only other extended surveillance of local fairweather currents in the nearshore region supports the net southerly drift of bottom waters (McClennen, 1973).

CONCLUSIONS

Fairweather current data for a nearshore linear shoal indicate that a persistant offshore flow of bottom water prevails under moderate wind conditions. Net flow speed over the crest is 10 times that in the trough, but the directions of both are essentially parallel. The prevailing offshore flow, asymmetry of the shoal profile, and depth inequality on the adjacent sea floor all suggest that sand transport is offshore, and that shoal translation, although a slow process, is toward the southeast, away from the shoreline. Such flows are consistent with storm-flow regimes proposed by Swift (this volume) and thus the net offshore flow appears to be the dominant current motion at this site.

REFERENCES

Bumpus, D.F., R.E. Lynde, and D.M. Shaw, 1972. Physical Oceanography, in: Coastal and Offshore Environmental Inventory; Cape Hatteras to Nantucket Shoals; Marine Publication Series No. 2, University of Rhode Island, p.1-1, 1-72.

Caston, V.N. and Stride, A.H., 1970. Tidal sand movement between some linear sand banks in the North Sea off north-east Norfolk. Marine Geol., V. 9, p. M38.

Duane, D.B., Field, M.E., Meisburger, E.P., Swift, D.J.P. and Williams, S.J., 1972. Linear shoals on the inner continental shelf, Florida to Long Island. in Shelf Sediment Transport: Process and Pattern, ed. D.J.P. Swift, D.B. Duane and O.H. Pilkey, Dowden, Hutchinson and Ross, Stroudsburg, Pa., p. 447-498.

McClennen, C.E., 1973. New Jersey continental shelf near bottom current meter records and recent sediment activity, J. Sed. Petrology, 43:371-380.

Norcross, J.J., and E.M. Stanley, 1967. Inferred surface and bottom drift, June 1963 through October 1964, in: Circulation of Shelf Waters off the Chesapeake Bight, Harrison, W., J.J. Norcross, N.A. Pore, and E.M. Stanley, ESSA Prof. Paper 3, Washington, D.C., p. 11-42, 82 p.

Stride, A.H., 1974. Indications of long term tidal control of net sand loss or gain by European coasts. Estuarine and Coastal Mar. Sci., V. 2, p. 27-36.

Swift, D.J.P., 1974. Continental shelf sedimentation. in The Geology of Continental Margins, ed. C.A. Burk and C.L. Drake, Springer-Verlag, New York, p. 117-135.

Uchupi, E., 1970. Atlantic continental shelf and slope of the United States - Shallow Structure: U.S. Geol. Survey Prof. Paper 529-I, 44 pp.

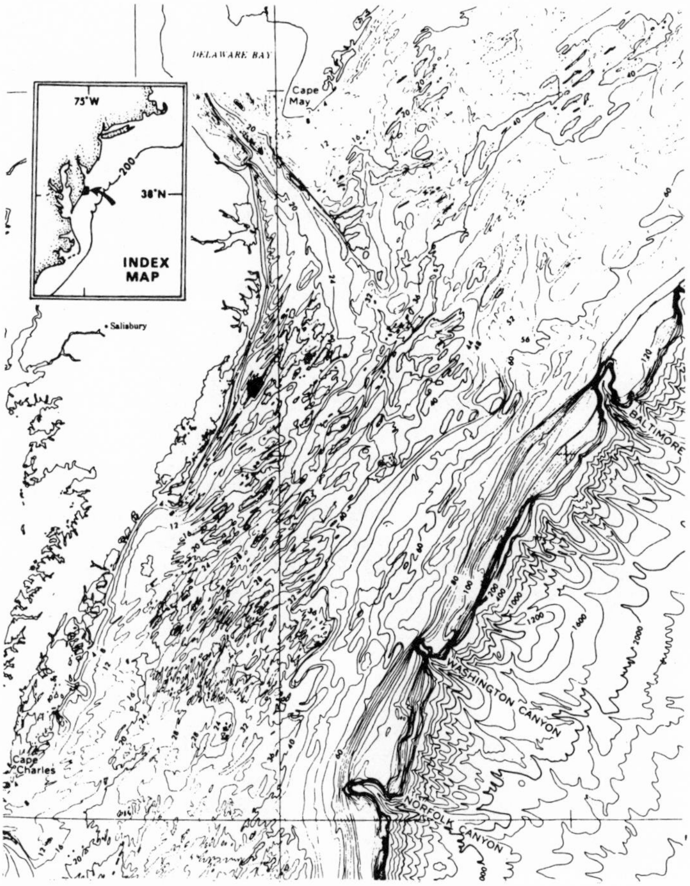

Fig. 1. Index map showing the location of the linear shoal (arrow) off the Maryland coast. Contours in meters (from Uchupi, 1970)

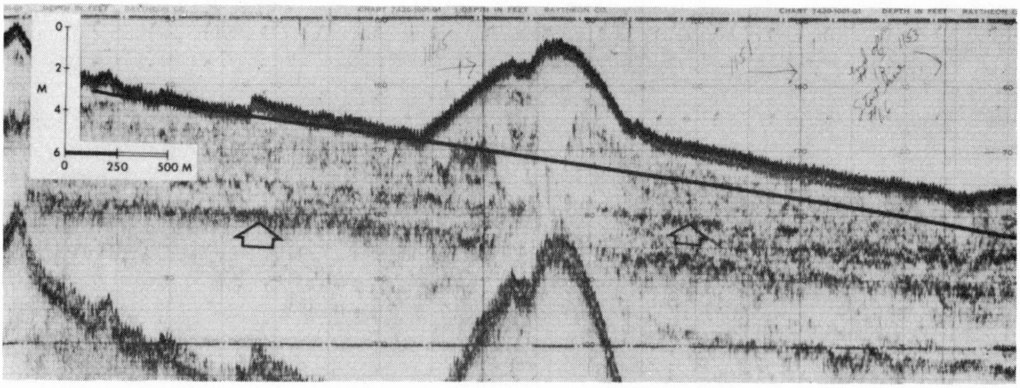

Fig. 2. 7kHz seismic profile of the linear shoal. Shore is to the left. Two smooth subsurface reflectors (arrows) are clay layers. Black line highlights depth discrepency between identical slopes shoreward (left) and seaward (right) of the crest whose water depth is 7 meters. Note asymmetry of shoal, with steeper side facing seaward.

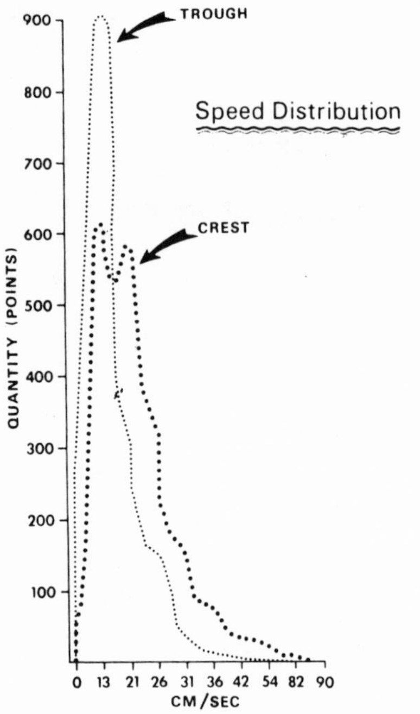

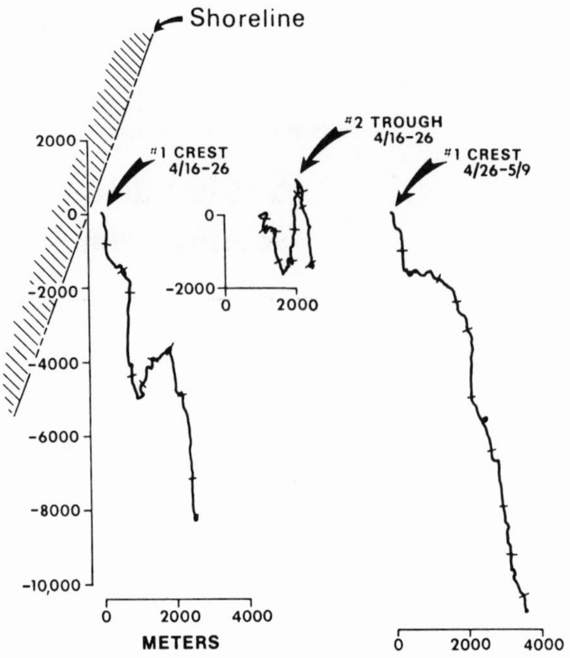

Figs. 3(left), 4(right). 3) Frequency of current speeds vs. speed. Crest speeds are consistently higher than those of the trough. Abscissa is non-linear due to variable inclination of the current meter with increasing speed and 4) Vector average diagrams for two crest records and one trough. Record at right connects to bottom of crest trace at left. Note consistency in directions between crest and trough (4/16-26) and predominance of SSE flow. Tick marks on traces are midnight (0000 hrs). Ordinate in meters.

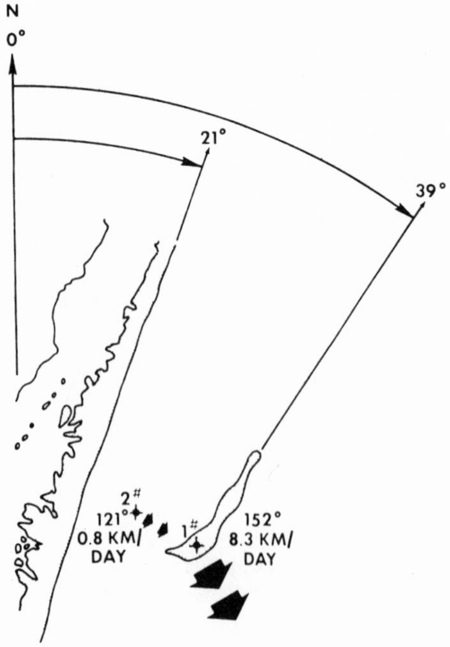

Fig. 5. Velocity summary for observations. Directional variance of 31° is due in part to lack of a second record for the trough. Inclination of ridge axis to shore (18°) is identical with the mean azimuth deviation for other shoals in this area (Duane and others, 1972).

15

Reprinted from *Estuarine and Coastal Mar. Sci.* **4**:23–32 (1976)

A Wind-driven Near-bottom Current in the Southern North Sea

V. N. D. Caston
British Petroleum Company Limited, Exploration Division, Research Centre, Sunbury-on-Thames, Middlesex, U.K.
Received 20 July 1974

Simultaneous recordings of wind and near-bottom current velocity and direction were made over the period 8 November–16 December 1967 from two production platforms situated in the south-western North Sea. The current measurements were made 4·6 m above sea bed in a total water depth of 34·7 m LAT.

Throughout the greater part of the period the recorded current velocities agreed closely with predicted values calculated on the basis of a correlation with tidal range data. Over a 125-hour period however the south-going currents were very much stronger and the north-going streams considerably less than anticipated. This corresponded with a period of high wind speeds which touched 25·7 m s^{-1} (50 kts) and exceeded 18 m s^{-1} (35 kts) for 68 hours, and which blew sub-parallel to the direction of flow of the southerly-flowing current.

The increase in velocity of the near-bottom southerly-going current averaged 23 cm s^{-1}, exceptionally 36 cm s^{-1}, and is equivalent to a figure of between 1·4 and 2% of the wind speed.

It is suggested that these anomalously high values may be related to a shift in the axes of current streams flowing between sand banks adjacent to the platform from which the current measurements were made.

Introduction

Distribution of sediments on the ocean floor is largely related to the action of currents. Although the mechanisms involved in such movements are complex and imperfectly understood, significant factors include the temperature and salinity of water masses together with such impressed forces as the action and influence of tides, winds, waves and Coriolis force.

On the continental shelves of the world the basic current regime is commonly that of the adjacent oceanic current circulation, upon which is superimposed more local effects, the relative influence of which is related to the width and depth of the shelf together with the degree of protection or isolation from deep water afforded by land barriers or submarine relief. On a world scale this leads to considerable variations in the dominant current mechanism, variations which are reflected both in the pattern of sediment transport and, not surprisingly, in the attitude and approach of those who study these phenomena. This subject is discussed in some detail by Swift *et al.* (1971).

In the semi-enclosed seas of the north-west European continental shelf, including the North, Irish and Celtic seas and the English Channel, the dominant current pattern is that related to regular tidal, i.e. lunar-induced, oscillations. A close correlation has been found between this tidal current pattern and the distribution of sediments on the sea floor (Stride,

1963). In particular it has been shown that sand-size grains in transport occur in a variety of transient bed forms, the orientation of which may be directly related to the direction of flow of the peak velocity (near-bed) tidal currents.

Despite this accordance on a regional scale between sediment distribution and tidal current velocity and direction, it is apparent, primarily from theoretical studies, that other forces may have an aperiodic and perhaps only occasional effect upon the movement of sedimentary particles. These forces are chiefly of meteorological origin, the most important being either wave-surge oscillatory bottom currents or direct wind-shear currents. However, although there has been considerable discussion concerning the magnitude and possible effects of wave surge currents (Curray, 1960; Hadley, 1964; Draper, 1967), there is as yet an almost complete lack of observational evidence and thus we have no confirmation of their presence on the scale proposed. This of course means that their effectiveness as an agent of sediment transport and dispersion (as discussed by Johnson & Stride, 1969; and McCave, 1971) is unproven.

A similar situation has prevailed in respect of direct wind-shear currents. The effect of the wind upon water movements is of continuing interest to physical oceanographers, but because their work has in general concerned near-surface water layers, usually in the open ocean, and results have commonly been reported in terms of residual water movements, their studies need have little significance to the question of sediment transport on the continental shelf.

There are nevertheless a small number of relevant studies comparing simultaneous wind and current measurements from the central and southern North Sea area. These reports have indicated that above a critical 'threshold' wind velocity, variously estimated as ranging from 5·2 m s^{-1} (10 kts) (Lawford & Veley, 1956; Hill & Ramster, 1972), to 10·3–12·9 m s^{-1} (20–25 kts) (Lee & Ramster, 1973) there is a close response of near-surface currents to wind velocities, estimated by Lawford & Veley to be 4·25% of wind speed. For instance. Lee & Ramster (1973) found that at a depth of 10 m in the eastern central North Sea sustained winds as high as 10·3 m s^{-1} (20 kts) were insufficient to markedly affect the development of tidal streams, whereas a wind speed of some 17·5 m s^{-1} (34 kts) was responsible for the generation of currents some 20 cm s^{-1} above the estimated tidal value. This response was however not found in current measurements at greater depths (36 m and 52 m), probably because of the presence of a pycnocline. A study of currents associated with a storm surge in the eastern North Sea (Gienapp, 1973) showed that bottom velocities (2 m above bottom) of up to 151 cm s^{-1} occurred in response to storm force winds exceeding 36 m s^{-1} (70 kts), but as the tidal component was not isolated the true value of the wind-generated current is not known. A review of published and unpublished current data from the North Sea area (Pitt *et al.*, 1973) concluded that: 'Wind-driven currents are the most important non-tidal component of the current system and over much of the central and northern parts of the North Sea can be expected to match or exceed the tidal component under storm conditions'. They further concluded that above a figure of *ca* 7·9 m s^{-1} (15 kts) current velocity is related to wind speed by an approximately linear relationship with a proportionality factor of about 0·02.

Interest in the sedimentological importance of wind-stress currents is particularly apparent from studies in the United States, possibly because tidal factors play a subordinate role in sediment transport in such open-shelf environments as those off Washington and Oregon (Smith & Hopkins, 1972) and the east coast (Duane *et al.*, 1972). Of particular note are a number of valuable papers included in the volume edited by Swift *et al.* (1972). Of these, the report by Sternberg & McManus (1972) concerning the analysis of some 260 days of near-bed

current data from the continental shelf off Washington represents perhaps the most significant quantitative data published to date in terms of its relevance to sediment dispersion, although once again specific evidence of the relationship and response of near-bed water movements to wind stress is lacking.

The present paper attempts to provide such evidence by illustrating the wind/current relationship observed in a part of the southern North Sea during November and December 1967. It does not pretend to provide an exhaustive treatment in theoretical terms and the relationship between tidal height and current velocity is taken only as far as is necessary to define the significant wind component.

Study information

Currents

The record in question was obtained by a Plessey self-recording current meter (Hodges, 1967), which recorded water velocity and direction of flow at 20-min intervals. The meter was hung 4·6 m (15 ft) above sea bed in a water depth of 34·7 m (114 ft) LAT off the side of the Shell/Esso platform AD in the Leman Bank field (53°05′22·8″N, 2°07′47·8″E—Figure 1) and recorded from 1612 GMT on the 8 November 1967 to 1450 GMT on the

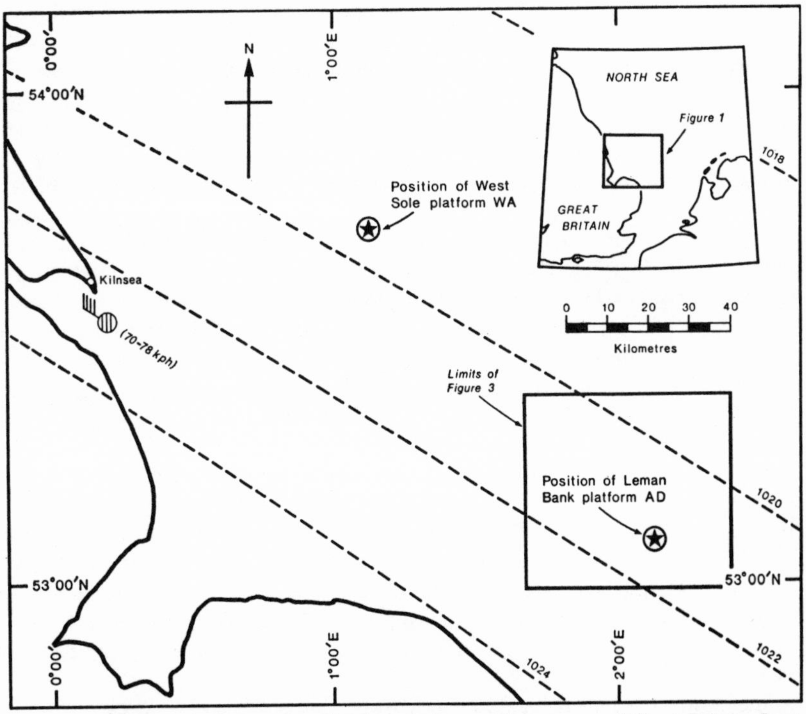

Figure 1. Location of production platforms off east coast of Great Britain. Inset shows position of Figure in relation to North Sea area. Also shown is extent of Figure 3, and meteorological information for 00 hours on 5 December 1967 reproduced from Daily Weather Report No. 38729, with the sanction of the Controller, H.M. Stationery Office and of the Director-General, Meteorological Office. Dashed lines are isobars (in mb) and wind/cloud information is that recorded at the meteorological station at Kilnsea.

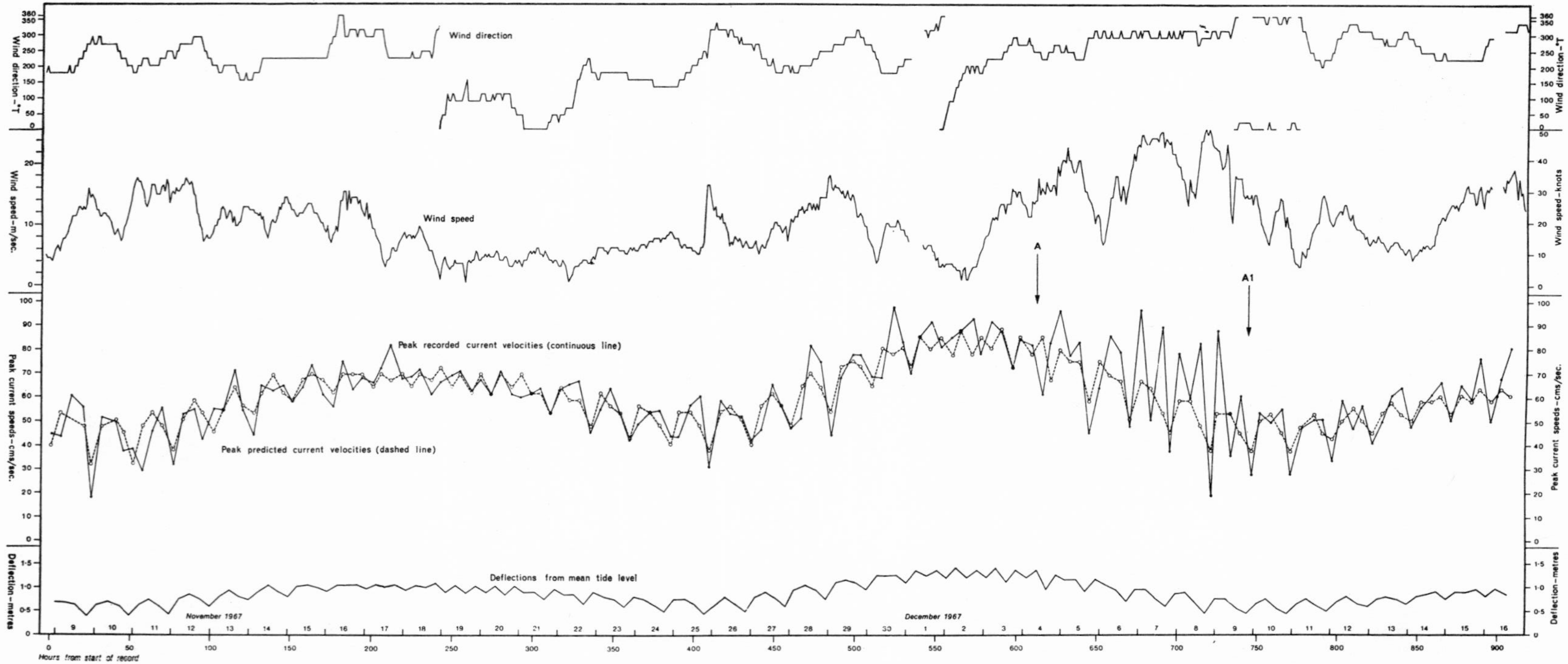

Figure 2. Wind, current and tidal data for period 8 November–16 December 1967. Wind direction consists of hourly means plotted to nearest compass point (22.5°); wind speeds are hourly mean values; peak recorded current velocities consist of a continuous line joining points representing maximum speeds recorded during each tidal cycle; peak predicted current velocities (dashed line and open circles) consist of a line joining maximum velocities predicted from formula in text for each tidal cycle; deflections from mean tide level is a continuous line joining deflection values for each tidal cycle obtained by method sketched in Figure 4. Wind data is from West Sole platform WA, current data from Leman Bank platform AD. For discussion of portion of record A–A see text.

[*facing p. 26*

16 December 1967. This record forms part of a series covering the periods 8 November to 16 December 1967 and from 13 February to 20 May 1968. The wind/current relationship described in this paper was not apparent to a comparable extent in the other records.

The record has been analysed and presented in a simplified form in Figure 2, along with the relevant wind and tidal data. Record duration is given in hours commencing at 20 00 on the 8 November 1967. The recorded current data shown consists of a continuous line joining points representing the peak velocities recorded during each northerly- and southerly-going tide, which occurred at intervals of approximately 6 h 12 min in a semidiurnal cycle. The words 'northerly-' and 'southerly-' going tide are used for convenience in this paper; in the event directions of flow at the time of maximum strength lay between 320°/325° T (between north-west and north–northwest) and 160°/175° T (between south and south–southeast). Because of the influence of bottom topography (see below) only minor changes in the directions of flow occurred throughout the period of the record and therefore current direction has not been plotted.

Platform AD is situated 48 km off the north-east coast of Norfolk within the area of the Norfolk Banks (Houbolt, 1968; Caston, 1972). The structure is located near the south-eastern end of a channel bounded by the Leman Bank (2·4 km to the south-west) and Well Bank (6·6 km to the north-east), which some 2·4 km to the north of the platform divides either side of Ower Bank. At the latitude of the platform the channel has a mean depth of about 36 m (120 ft); the Leman and Well Banks rise, respectively, to within 3·6 m and 9 m (12 and 30 ft) of chart datum (Figure 3). Tidal current flow is therefore effectively constrained along the channels between the banks.

Because of the possible effects of turbulence around underwater structural members the selection of recording position is of considerable importance in an installation of this nature. Platform AD is a fixed production platform and is rectangular in plan with its long axis aligned 063°/243°T, which is very approximately at right angles to the direction of flow of the dominant tidal currents. The meter was hung from a taut wire anchored at the seabed by a 227 kg (500 lb) weight, and attached at the top to the platform deck. This mooring was installed approximately midway between two legs on the southern side of the platform in such a position that the nearest structural members were at least 4·3 m (15 ft) away from the meter in a direction perpendicular to the axis of the instrument when aligned into the direction of flow of the principal north- or south-flowing currents.

In view of the very considerable elongation of the tidal ellipse in this area, it is therefore considered that any resultant interference would only affect the low velocity currents which change direction at slack water and which have little significance to this particular study.

Tides

Data used for this study consisted of predictions of tidal range for the position of platform AD during the months of November and December, 1967. These were computed by BP Survey Branch from harmonic constants for the River Tyne provided by the International Hydrographic Bureau (1966), together with astronomical harmonic components given in Schureman's Tables (1958).

From this data values of maximum deflection for each tidal stream at nominal intervals of *ca* 6 h 12 min have been plotted in Figure 2. The term 'maximum deflection' applies to the peak value of tidal rise or fall, in metres, either side of the mean tidal level (Figure 4). During the course of the present study it was found that this figure correlated more closely with recorded current velocity than the more commonly used value of tidal range.

Over a longer period the plot shows the expected alternations of spring and neap tides' with maxima on 18 November and 2 December separating minimum values on 11 November, 25 November and 10 December. It also draws attention to the development during neap tides of prominent 25-hourly cycles. These diminish towards the time of maximum spring tides, and are supplanted by relatively small-amplitude 12·5-hourly cycles. These cycles are reflected quite clearly in the recorded current velocities.

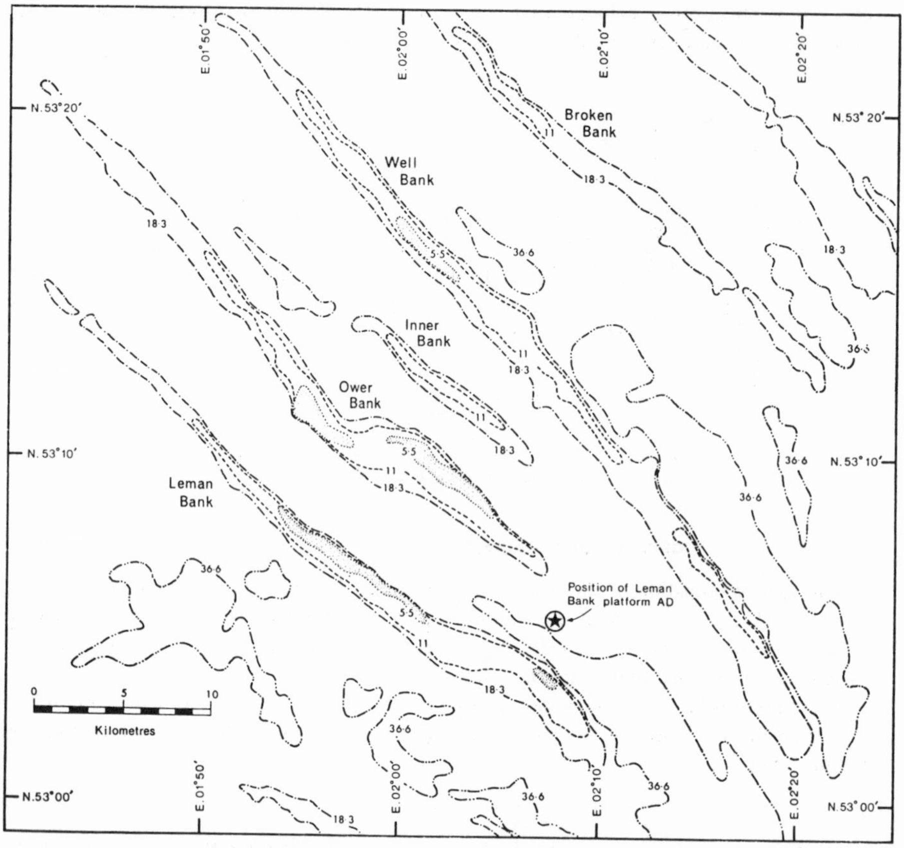

Figure 3. Bathymetry in vicinity of Leman Bank platform AD, showing Norfolk Banks. Isobaths at 5·5, 11, 18·3 and 36·6 m (18, 36, 60 and 120 ft). Reproduced from British Admiralty chart 1503 with the sanction of the Controller, H.M. Stationery Office and of the Hydrographer of the Navy.

Winds

Wind speed and direction were continuously measured throughout the period of current recording by a Munro recording anemometer installed on BP's fixed production platform WA in the West Sole Field (53°42'12"N, 01°09'00"E). Measurements were made at a height of *ca* 30·5 m (100 ft) above sea level; the velocities used in this paper are as recorded and have not been corrected to the standard exposure height of 10 m (33 ft). The wind data is shown in Figure 2.

The distance between the two platforms is 94·5 km. This separation is certainly sufficient to introduce local variations in the weather experienced in the two areas. However, study of the daily weather reports issued by the British Meteorological Office for the periods of particular interest has shown that the significant storms blew almost exclusively from a northwesterly direction and that under these conditions the same air flow would equally affect both platforms. As an example of this situation, the pressure pattern prevailing during the height of the 4–10 December storm is illustrated in Figure 1. For the purposes of this paper it is therefore concluded that the comparison of current information with wind measurements from platform WA is valid.

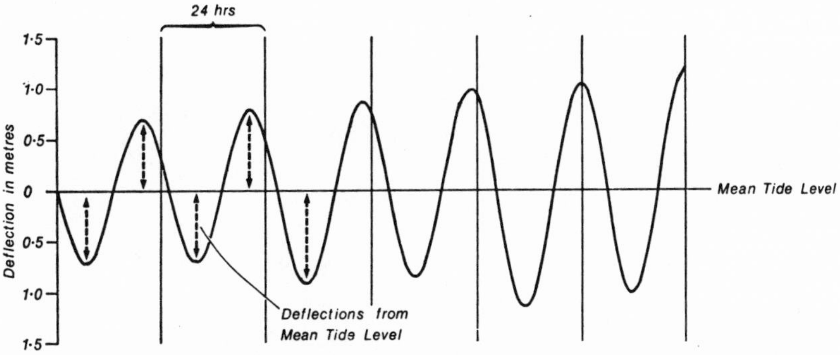

Figure 4. Diagram showing method used to obtain Deflections from Mean Tide Level. Curved line shows predicted tidal range values.

Current/tide correlation

Preliminary examination of the results suggested that throughout the greater period of the record, with the exception of the period from 4–10 December, there was a relationship between the maximum current velocity in each cycle of approximately 6 h 12 min, and the magnitude of the corresponding tidal range.

A mathematical relationship was therefore sought between these two sets of data, excluding that from the apparently anomalous period. The closest computer-derived correlation was found to be:

$$C = 49{\cdot}4\ T + 17{\cdot}4$$

where C is the maximum current velocity in cm/s^{-1} and T is the tidal deflection in metres either side (both higher and lower) from the mean tidal level. This correlation between current velocity and tidal deflection has a coefficient of 0·769; the equation is significant at the 0·1% level.

Using the above equation, the resultant current velocities calculated from the tidal predictions are plotted as a dashed line on Figure 2. It will be seen that there is a close agreement between this line and the recorded velocities throughout the record with the exception of the period A–A$_1$, the average difference between recorded and predicted velocities for the 126 observations compared being only 5·8 cm s^{-1}. It is apparent that there are a number of small-scale deviations from this general pattern, e.g. at hours 204–218, 395–425 and 469–480, and the majority of these would appear to be related to wind effects recognizable in the upper part of the Figure. However, in terms of the overall picture these

effects are insignificant, and the overall 'fit' is sufficiently close to suggest that the use of tidal data in association with the equation given above has enabled an acceptably accurate calculation to be made of the maximum tidal component of the currents flowing at the particular time, water depth and location studied. Use of this method should also enable similar estimates to be made for this location in the future.

Wind effect

The above relationship enables the anomalous portion of the record, from the 4 to the 10 December, to be isolated. Comparison in Figure 2 between the recorded and predicted velocities over this period shows that the average difference in the 21 values between 615 h and 740 h is 17 cm s^{-1}. Of these, 13 values are greater than predicted by an average of 19 cm s^{-1}, exceptionally 36 cm s^{-1}, and 8 values are lower than predicted by an average of 13 cm s^{-1}, exceptionally 24 cm s^{-1}.

The significance of velocity maxima in relation to flow directions is shown in Figure 5. In this diagram the peak northerly- and southerly-going velocities for the period from the 1 to the 14 December, taken from Figure 2, have been joined by separate lines, thereby emphasizing the relative difference in flow between the two directions. For the first 90 h covered by this diagram the strengths of the opposing currents were approximately equal, although perhaps with the northerly stream being the stronger. As from 620 h the southerly-going current clearly predominated with the exception of a single measurement at 658 h and this southerly dominance continued until the re-establishment of more equivalent velocities at 827 h. Referring back to Figure 2 it can be seen that over the period of the most extreme values, from 615 h to 740 h, 9 out of the 11 southerly-going tides flowed at an average velocity of 23 cm s^{-1} greater than predicted, or 38% greater than the average predicted value, whereas of the corresponding northerly-going currents, 6 out of 10 streams appear to have been reduced in strength by an average of 12 cm s^{-1} or 25% less than the average predicted value. In terms of relative strengths, the south-going current flowed at velocities of up to 70 cm s^{-1} in excess of the immediately preceding northerly stream.

It is considered that an explanation for this anomalous pattern is provided by study of the wind record summarized in the upper part of Figure 2. Reference to this diagram shows that the period of current velocity extremes between the 4 and 10 December corresponds almost precisely with a period of high wind speeds which exceeded 18 m s^{-1} (35 kts) for 68 h. This period included one occasion when the wind blew continuously for 32 h at greater than this velocity, during which it exceeded 23·3 m s^{-1} (45 kts) for a total of 25 h and touched 25·7 m s^{-1} (50 kts) for two separate one-hourly periods. A second important factor is that for 90 h the wind blew from between WNW (292·5°) and NW (315°), which is almost parallel to the orientation of the principal sand banks (Figure 3) and the direction of flow between the banks, as measured at Platform AD, of the southerly-going current.

The lag between wind changes and the resultant influence on the water movements is difficult to quantify because of the unknown effect of the distance between the two platforms, coupled with the unknown relative importance of wind strength and direction, but it is of interest to note that the southerly-going stream appeared to continue at a considerably higher velocity than predicted until 827 h—some 4 days after the wind had dropped to less than 18 m s^{-1} (35 kts) and during which time it had changed in direction from NW to N, to SSW and thence back to WNW again. On the other hand the drop in wind velocity over the period 640 h to 670 h appears to have been almost immediately reflected in a decrease in the velocity of the south-going current and the resurgence of the north-going stream.

Discussion and implications

Because of the particular combination of site and meteorological factors involved, it is difficult to draw any general conclusions from this data concerning the effect of wind upon near-bottom current velocities, or to attempt to extrapolate to other locations, notably a more

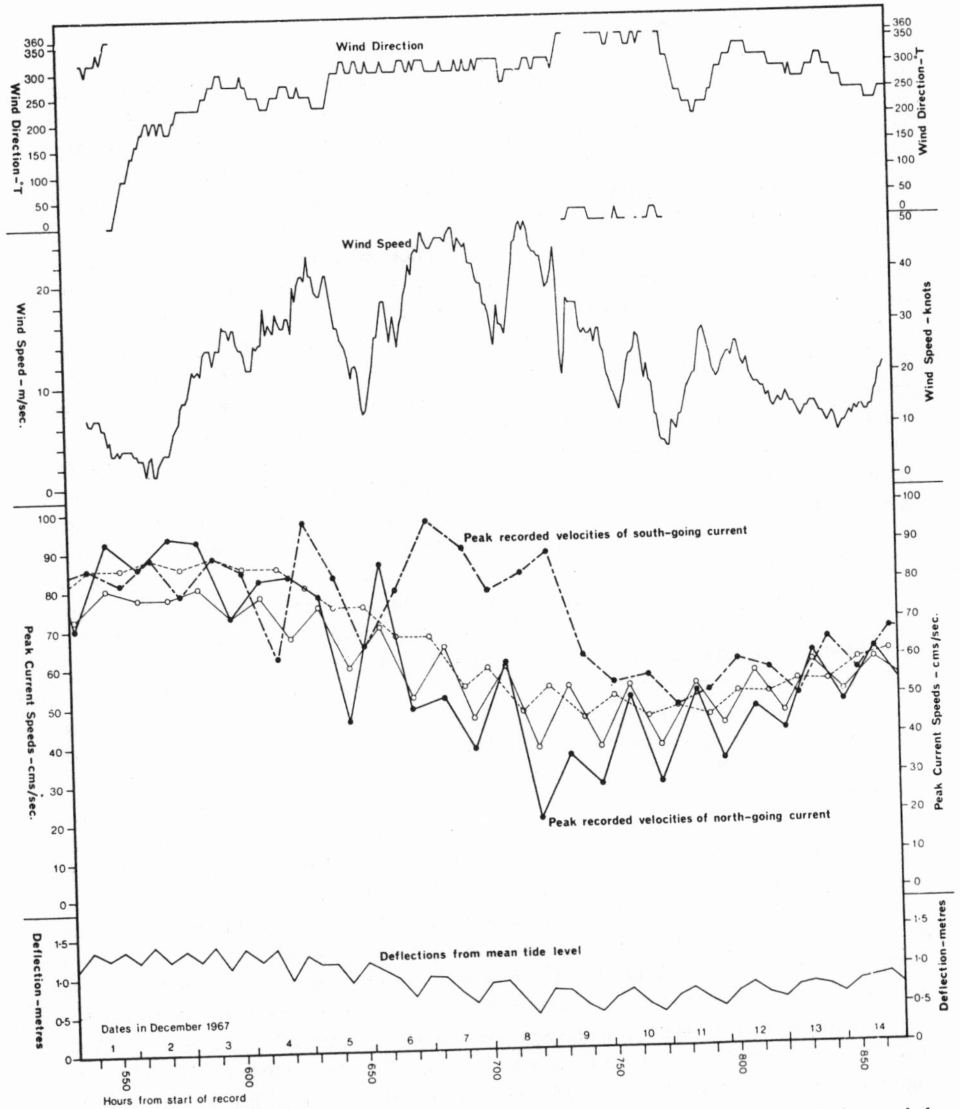

Figure 5. Wind, current and tidal data for period 1–14 December. Peak recorded values for each north- and south-going tide are joined by separate lines (thick continuous and dashed lines, respectively). Also shown are predicted north- and south-going values (thin continuous and dashed lines, respectively). Wind and tidal data from Figure 2.

open sea situation. What proportion of the wind-generated current is related to localized funnelling between the banks is also not known, but it would appear reasonable to expect a local increase in velocity resulting from a venturi-type effect. This agrees with the conclusion reached by Pitt *et al* (1973) that a local increase of current from 50–100% might be expected where a coastline or shoaling water lies to the right of the wind. Under the circumstances it therefore seems justifiable only to report that in this location a wind with a strength of between 35 and 50 kt appears to have been responsible for the generation of a water movement of up to 36 cm s^{-1} superimposed upon the semi-diurnal tidal pattern, i.e. a velocity of between 2 and 1·4% of wind speed at a depth of 30 m.

We do not know whether this represents a true increase in the overall velocity of the south-flowing stream, or whether, in view of the known complexities of the current pattern in this area (Caston & Stride, 1970) it might not have been caused by lateral movement of the axes of adjacent current streams (Pitt *et al.*, 1973). At Platform AD, predictions show that the south-going current is the stronger, although the recorded values are approximately equal (Figure 5). 400 m to the south-west, however, sand waves lie steep to the north-west, implying a predominant transport in that direction. It may be that the platform is situated approximately mid-way between opposing streams and is thus particularly susceptible to axial shifts which would temporarily bring the structure into the regime of the alternate current.

Whatever the mechanism, the record discussed in this paper presents evidence for a short-term interruption to the apparently established pattern. This interruption is significant, first, because the currents generated were clearly southerly-flowing, and second, because over an overall period of 120 h they were so much stronger than the corresponding northerly streams. Unfortunately we have no sea bed survey data from before and after the storm of the 4–10 December to provide evidence of any changes which might have taken place, but it is reasonable to suppose that considerable sediment transport must have occurred in a southerly direction over this period.

In geological terms, the significance of this event is that it must represent one of the infrequently-recorded examples in marine geology of an episodic, if not 'catastrophic' process in action, albeit on a small scale and in relation to only a single location. It is suggested that in future the simultaneous monitoring of such a phenomena together with measurement of resultant changes on the sea bed would be of major sedimentological importance.

Furthermore, in an area of the North Sea where there is activity in connection with exploration and production of hydrocarbons, a third implication is of considerable sedimentological and engineering interest. Had the wind-generated current been superimposed on top of the spring tidal velocities, rather than mid-way between springs and neaps, then the maximum velocity experienced would have been 124 cm s^{-1}, or 40% more than the calculated tidal maximum, at that time, of 88 cm s^{-1}.

Acknowledgements

Part of this work was originally undertaken on behalf of the Sand Movement Study Group. I thank the participants, including the AMOCO, AMOSEAS, BP, Burmah, Gulf, Shell, Signal and Total oil companies and the General Post Office, together with the National Institute of Oceanography (now Institute of Oceanographic Sciences), for the use of their data, and in particular Shell U.K. Exploration and Production Ltd for their co-operation and assistance and the British Petroleum Co. Ltd, as Operator for the Group, for permission to publish the results. I am indebted to my colleagues, A. Haugh and S. J. White, the former

for providing tidal prediction graphs for the platform AD position, and the latter for his help in determining the relationship between tidal range and current speed, which was based in part upon his work on this project within BP. My thanks go to G. F. Caston and J. Ramster for their advice and for critically reviewing the manuscript.

References

Caston V. N. D. 1972 Linear sand banks in the southern North Sea. *Sedimentology* **18,** 63–78.

Caston, V. N. D. & Stride, A. H. 1970 Tidal sand movement between some linear sand banks in the North Sea off northeast Norfolk. *Marine Geology* **9,** M38–M42.

Curray, J. R. 1960 Sediments and history of Holocene transgression, continental shelf, northwest Gulf of Mexico. In *Recent Sediments North-west Gulf of Mexico* pp. 221–266. (Shepard, F. P., Phleger, F. B. & Van Andel, T. H., eds) American Association of Petroleum Geologists, Tulsa, Oklahoma.

Draper L. 1967 Wave activity at the sea bed around northwestern Europe. *Marine Geology* **5,** 133–140.

Duane, D. B., Field, M. E., Meisburger, E. P., Swift, D. J. P. & Williams, S. J. 1972 Linear shoals on the Atlantic inner continental shelf, Florida to Long Island. In *Shelf Sediment Transport: Process and Pattern* pp. 447–498 (Swift, D. J. P., Duane, D. B. & Pilkey, O. H., eds) Dowden, Hutchinson & Ross, Stroudsburg, Pa.

Gienapp, H. 1973 Strömungen während der Sturmflut vom 2. November 1965 in der Deutschen Bucht und ihre Bedentung für den Sedimenttransport. *Senckenbergiana Maritima* **5,** 135–151.

Hadley, L. M. 1964 Wave-induced bottom currents in the Celtic Sea. *Marine Geology* **2,** 164–167.

Hill, H. W. & Ramster, J. W. 1972 Variability in current meter records in the Irish Sea. *Conseil international pour l'exploration de la Mer. Rapports et Procès-Verbaux.* **162,** 232–247.

Hodges, G. F. 1967 The engineering for production of a recording current meter. *International Hydrographic Review* **44,** 151–168.

Houbolt, J. J. H. C. 1968 Recent sediments in the Southern Bight of the North Sea. *Geologie en Minjbouw* **47,** 245–273.

International Hydrographic Bureau 1966 *List of Harmonic Constants.* Special Publication No. 26. Monaco, 103 pp.

Johnson, M. A. & Stride, A. H. 1969 Geological significance of North Sea sand transport rates. *Nature, London* **244,** 1016–1017.

Lawford, A. L. & Veley, V. F. C. 1956 Change in the relationship between wind and surface water movement at higher wind speeds. *Transactions, American Geophysical Union* **37,** 691–693.

Lee, A. J. & Ramster, J. W. 1973 Presentation and analysis of the current measurements made during Operation RHENO. *Conseil international pour l'exploration de la Mer. Rapports et Procès-Verbaux.***163,** 80–98.

McCave, I. N. 1971 Sand waves in the North Sea off the coast of Holland. *Marine Geology* **10,** 199–225.

Pitt, E. G., Carson, R. M. & Tucker, M. J. 1973 The current system around the British Isles as it relates to offshore structures. An assessment. *N.I.O. Internal Report No. A.*62. National Institute of Oceanography, Wormley, Surrey.

Schureman, P. 1958 *Manual of Harmonic Analysis and Prediction of Tides.* U.S. Coast and Geodetic Survey Special Publication No. 98.

Smith, J. D. & Hopkins, T. S. 1972 Sediment transport on the continental shelf off of Washington and Oregon in light of recent current measurements. In *Shelf Sediment Transport: Process and Pattern* pp. 143–180 (Swift, D. J. P., Duane, D. B. & Pilkey, O. H., eds) Dowden Hutchinson & Ross, Stroudsberg, Pa.

Sternberg, R. W. & McManus, D. A. 1972 Implications for sediment dispersal from long term bottom current measurements on the continental shelf of Washington. In *Shelf Sediment Transport: Process and Pattern* pp. 181–194 (Swift, D. J. P., Duane, D. B. & Pilkey, O. H., eds) Dowden, Hutchinson & Ross, Stroudsberg, Pa.

Stride, A. H. 1963 Current-swept sea floors near the southern half of Great Britain. *Quarterly Journal of the Geological Society of London* **119,** 175–199.

Swift, D. J. P., Duane, D. B. & Pilkey, O. H. (eds) *Shelf Sediment Transport: Process and Pattern.* Dowden, Hutchinson & Ross, Stroudsberg, Pa.

Swift, D. J. P., Stanley, D. J. & Curray, J. R. 1971 Relict sediments on continental shelves: a reconsideration. *Journal of Geology* **79,** 322–346.

16

Reprinted from pp. 31-47 of *Effects of Energy Related Activities on the Atlantic Continental Shelf*, B. Manowitz, ed., Brookhaven National Laboratories, 1976, 260 pp.

Wind-Driven and Thermohaline Circulation Over the Continental Shelves

G.T. CSANADY
Woods Hole Oceanographic Institution, Woods Hole, Massachusetts 02543

INTRODUCTION

The principal driving forces of water movements over the continental shelves are the wind stress and the horizontal pressure gradients due to density differences and to tidal waves propagating toward shore from the deep ocean. Tides with their well-defined frequencies constitute a special kind of motion that is better understood than other motions; they are treated elsewhere in this Conference.[1] Here we shall discuss the more chaotic patterns of flow induced by the winds and by the differences in temperature and salinity between different parts of the shelf, which we shall refer to as thermohaline circulation. In attempting to understand these motions many difficult dynamical problems arise, some of them requiring considerable mathematical development. Here I shall discuss the physical properties of water movements over continental shelves, omitting the mathematical arguments necessary for a more rigorous approach.

The most fundamental question in the atmospheric sciences generally is, In what sense can we expect to understand a system as complex as the atmosphere or the ocean? The human mind can appreciate simultaneously only a relatively small number of quantitative parameters. *All* the details of water motions in a given portion of the ocean cannot be described, even in principle, by a finite number of parameters, and even reasonable detail would be beyond our capacity to absorb. It is therefore necessary to distill from experience what might be called distinct phenomena, certain conspicuous features of observed motions that are continuously present or recur regularly, and attempt to understand these phenomena in isolation. It is usually possible to construct a conceptual model for the purpose of such piecemeal understanding, i.e., an imaginary ocean with simple characteristics and subject only to a limited number of external influences, in which calculations reproduce a given

distinct phenomenon qualitatively and, to a degree, quantitatively. Several examples of such conceptual models are given below. For a satisfactory degree of understanding in a given subfield of the atmospheric sciences, many conceptual models are usually required. In the case of continental shelf dynamics we are certainly far from this stage. The number of models is far too few, and what we have is poorly related to observation. Several of these models arose in connection with work on our great inland shallow seas, the North American Great Lakes, especially in the course of efforts related to the International Field Year on the Great Lakes (IFYGL). Most of these models are probably applicable to other shallow seas, especially the continental shelves, but some of the quantitative relationships may be affected by salinity differences and by interaction with tides, two influences absent in the Great Lakes. In addition, the existence of an open boundary at the edge of the shelf poses new and complex problems. Thus the discussion below is far from complete; in fact, it is only an attempt to piece together the few extant conceptual models I think important, and to speculate on a few other problems.

CHARACTER OF FORCING

At our latitudes winds are quite irregular, although westerlies generally predominate. The stress of the wind on the sea surface is proportional to the square of the wind speed and is therefore dominated by periods of strong wind. Episodes of strong winds (storms) occur with a typical frequency of once in 100 hours and last for 10 hours or so, but there is no regular periodicity involved in either their frequency or their duration. The intermittent nature of forcing has an important influence on the response of a body of shallow water to wind stress. Among other effects, intermittent forcing evokes a number of interesting large-scale wavelike responses, as might be expected on general dynamical principles.

Some aspects of this variable forcing are illustrated by a frequency distribution of wind stress, in categories of magnitude and direction. Saunders[2] has prepared wind-stress roses of this kind for the eastern seaboard, in 1° latitude-longitude squares; an example is shown in Figure 1. Concentric circles are spaced 1 dyne cm^{-2} apart, and directions are in 45° quadrants. This stress-rose applies to the sea just south of the eastern half of Long Island, for the winter (Dec.-Feb.) period. The average wind stress has components (0.61, -0.46) dynes cm^{-2}, northward and eastward respectively, while the root-mean-square deviation from the mean is 1.61 dynes cm^{-2}. The high rms-to-mean ratio at once highlights the influence of storms.

If a wind stress of 3 dynes cm^{-2} acted as a uniformly distributed force over a water column 100 m deep (a typical depth over continental shelves) for 10 hr, and if no other force acted on the water, the impulse of the wind force would generate a velocity of a little over 10 cm sec^{-1} in the water. There are many complications, but this is the correct order of magnitude of wind-driven flow velocities in shallow seas.

Horizontal density contrasts arise from insolation, especially in spring, and from the influx of fresh water at the shore. The shallowest portions of a marginal sea heat up first and thus become lighter (by $\leqslant 1$ part/1000) than the waters farther offshore. Over continental shelves this effect is usually less important than fresh-water inflow at the shore, which similarly lightens nearshore waters, by ≈ 3 parts/1000 in a typi-

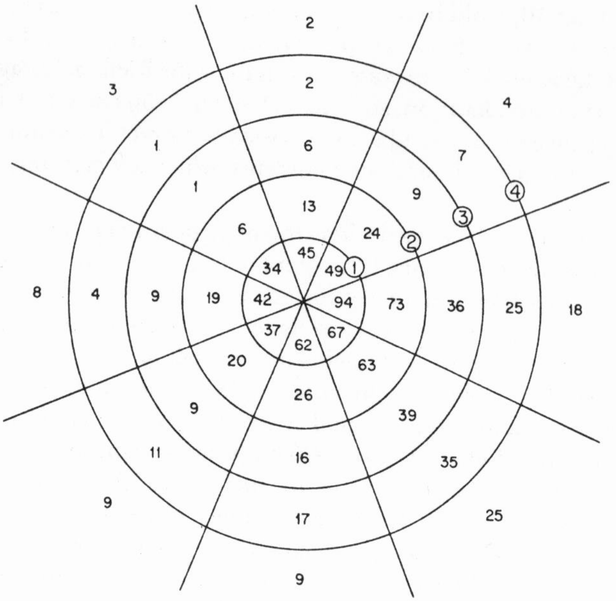

Figure 1. Wind stress frequencies between latitudes 40° to 41° and longitudes 72° to 73° by magnitude (dynes cm^{-2}) and direction (45° sectors). Top sector is stress directed *toward* true north, ±22½°. Numbers in each field give frequencies in parts per thousand. Thus, eastward (±22½°) wind stress of magnitude 0 to 1 dyne cm^{-2} occurs for 9.4% of the time.

cal case, in comparison with waters several tens of kilometers offshore. A static fluid cannot sustain such horizontal density differences, and the lighter waters tend to spread out over the top, the heavy waters, to slide in at the bottom. Given the small density differences actually existing, these motions are sufficiently slow to be deflected by the earth's rotation, so that thermohaline motions are more nearly parallel to surfaces of constant density than perpendicular. Therefore the velocity of these motions cannot be calculated in quite so elementary a way as wind-driven flow. Except near zones of concentrated density changes known as fronts, thermohaline velocities are slower than wind-driven ones, their typical magnitude being 3 cm sec^{-1}, as shown in greater detail below. Although there is also a seasonal variability in fresh-water inflow, spring runoff being greatest, thermohaline forcing varies much less drastically than wind forcing.

WIND-DRIVEN COASTAL CURRENTS

Over the continental shelves the depth of water increases gradually and more or less monotonically from zero at the shore to 100 to 200 meters at the shelf break, where it begins to increase a little more abruptly to about 2000 m. The typical bottom slope over the shelf is a few times 10^{-3}, so that within a few kilometers from shore the water depth is still only about 10 m. Our simple calculation of momentum input by a storm would therefore have shown a typical velocity there of 100 cm

sec^{-1}, rather than 10, valid in depths of about 100 m. The shores prevent perpendicular movement of water, so that this calculation is true only for the longshore component of the wind. Also, in water this shallow the frictional drag of the sea floor soon resists further acceleration, so that velocities >50 cm sec^{-1} are only rarely reached. Nevertheless, our calculation is basically correct in exposing the dominant influence of depth on wind-generated currents, which are certainly strongest in the coastal zone.

Because of surface-level variations the force of gravity exerts an important influence. Although the sea surface is horizontal in the mean, it is never so instantaneously, not even after the influence of surface waves has been filtered out by averaging over a few minutes and corrections have been made for the phase of the tide to remove any tidal wave. A wind blowing perpendicular to the coast, for example, produces a rise or fall of level sufficient to balance the wind stress, since the water cannot accelerate in a direction normal to the shore. A longshore wind would not be subject to this kind of effect if a shore were very straight and very long, which is never the case. On a real coastline, peninsulae, shoals, and bights all present obstacles to wind-driven flow, and the result is some degree of piling-up or depletion of water. Clearest is the situation in an enclosed basin where, between the upwind and downwind shores, a marked level difference, known as the setup, is established.

When the setup exactly balances the wind stress, the acceleration produced by the force of gravity equals the wind stress divided by the mass of the water column:

$$g\frac{\partial \zeta}{\partial x} = \frac{F}{h}. \qquad (1)$$

Here g is acceleration of gravity, ζ is the elevation of the free surface above its equilibrium value, and x is the coordinate axis along the direction of the wind. $F=\tau/\rho$ is the "kinematic" wind stress, i.e., the stress τ in dynes cm^{-2} divided by the density of water, the units of F being $cm^2\ sec^{-2}$. The mass of the water column is ρh per unit surface area (h=depth), so that F/h is force per unit mass, due to wind stress.

Given the typical values of $\tau=1$ dyne cm^{-2}, i.e., $F=1\ cm^2\ sec^{-2}$, $h=100\ m\equiv 10^4$ cm, and $g\simeq 10^3$ cm sec^{-2}, the surface slope, sufficient to balance the wind stress, is seen to be in the range 10^{-7}, or 1 cm in 100 km. Surface slopes of this order are indeed present in shallow seas and are so directed as to balance the wind stress. However, because of irregular variations of depth, Eq. (1) can be satisfied only in a very small fraction of a real basin. Where the depth is greater than necessary for the balance expressed by Eq. (1), the pressure gradient overcomes the wind, whereas in shallow water the wind stress dominates. The distribution of surface elevation over the entire basin is such that the *total* wind force (over the entire surface area) is more or less balanced by the setup, so that the average gradient $\partial\zeta/\partial x$ is determined by the average basin depth.

In the coastal zone, where the depth is much less than average, the wind stress completely overwhelms the pressure gradient. For example, at a depth of about 10 m, horizontal accelerations due to wind stress of 1 dyne cm^{-2} are in the range 10^{-3} cm sec^{-2}, an order of magnitude larger than $g(\partial\zeta/\partial x)$, which is in the range 10^{-4} cm sec^{-2}, given an average depth of 100 m.

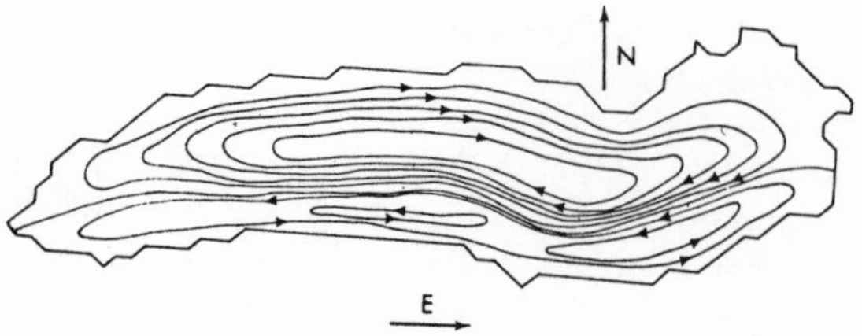

Figure 2. Water transport in Lake Ontario due to an eastward wind stress. Note downwind flow in shallow water, return flow in deep.

A simple conceptual model of an enclosed shallow sea may now be envisaged in which the wind stress is partly balanced by a certain distribution of surface level, with the difference between the wind-stress force and the gravity force accelerating the water. The wind is assumed to be "switched on" at time zero and the flow pattern to be calculated for some realistic time t, e.g., 6 to 10 hr, during which the wind is assumed to be constant and other complications absent. An arbitrary depth distribution may be handled with the aid of a computer. Patterns of depth-integrated transport (average velocity times depth) calculated in this way by Rao and Murty[3] for Lake Ontario are shown in Figure 2.

Patterns of this sort are always characterized by at least two closed gyres that are related to the depth distribution, i.e., to the topography of the sea floor, and are therefore called topographic gyres. Their downwind legs coincide with shallow water along both shores, while return flow occurs over the deepest portions of the basin. In a cross section perpendicular to the wind, flow is *with* the wind where the depth is less than the average for that section, and *against* the wind in deeper water. As already pointed out, the physical explanation is that the pressure gradient opposing the wind stress balances the latter exactly where the water depth is average, but "wins" over the wind in deep water (the gravity force being proportional to mass and therefore depth) and "loses" in shallow water, the wind stress being the same everywhere.

The transport pattern is slightly misleading because it does not show the high velocities in shallow water. The spacing of transport streamlines is proportional to transport, and, when this is roughly constant, velocities vary as h^{-1}. The details near shore also depend on friction and on the density distribution in the vertical, but it is basically true that high-speed currents occur near shore.

The application of similar conceptual models to the continental shelves presents a problem because it is difficult to predict the longshore pressure gradient. Along a straight, infinite shore no permanent longshore gradients can exist, whereas in a closed basin the gradient is determined by the average depth in a section perpendicular to the wind. Neither of these ideas is useful on a real shelf, with an open boundary seaward. In the conceptual models so far proposed for shelf circulation, the longshore gradient appears as an arbitrary external parameter, a *deus ex machina*. In the actual physical case this gradient is determined by shelf topography and by the nature of the forcing, but at present we do not know how to estimate its magnitude in a given location under given conditions.

TOPOGRAPHIC WAVES

Another important influence on water motions of reasonably large scale is the deflecting force (Coriolis force) due to the rotation of the earth. A coastal current of the kind under consideration would tend to be deflected to its right (in the Northern Hemisphere) within a few hours of its establishment. The shore prevents actual massive deflection, but a small quantity of water moves in or out from shore just enough to establish an elevation gradient in the *offshore* direction, which then balances the Coriolis force acting on the *longshore* flow. This equilibrium condition is known in meteorology and oceanography as geostrophic balance. A coastal current a few kilometers wide requires at the shore a rise (when the current leaves the shore to the right) or drop (shore to the left) in elevation of only a few centimeters for geostrophic balance. However, a broad current over a wide shelf (e.g., off the Gulf of Maine, produced by a longish storm) can lead to considerable buildup and depletion of water levels at the coast, amounting to ± 50 cm.

In a closed basin the wind-induced coastal current along the right-hand shore (looking along the wind) produces a *rise* in level as geostrophic balance is established, and a *fall* on the left-hand shore. Thus a pressure difference acts across the upwind and downwind ends of the basin. This pressure gradient may be small, but it is not opposed by the wind or any other force and gradually accelerates the water in a longshore direction. Gradually the downwind-flowing right-hand jet is extended across the downwind end of the basin, and the left-hand jet across the upwind end. In other words, the stagnation points of the flow pattern characterizing the topographic gyres move around the basin in a counterclockwise sense. This movement is quite slow, covering about 40 km day^{-1} in Lake Ontario, but nevertheless has striking observable effects. After passage of the stagnation point at some nearshore location, the coastal current spontaneously *reverses* direction. Thus, during a quiescent period following a storm, in some locations coastal currents are still moving in the direction of the wind that started them, while in other places the current is already moving in the opposite direction.

The flow structure started by a storm may be described as a wave propagating at a given speed in a given direction. Further analysis of the problem shows that depth variations are essential for the existence of this type of wave, which is therefore known as a topographic wave. A simple conceptual model of a shore zone in which such a wave may be produced is shown in Figure 3. The distribution of longshore velocity calculated for the idealized case is illustrated in Figure 4. The amplitude of this wave reduces to zero where depth variations vanish. The calculated speed of propagation agrees closely with observations in Lake Ontario. Note that significant motion is confined essentially to <15 km from shore, so that waters farther offshore are not affected by this phenomenon. The wave may be said to be trapped at the shore.

The existence of topographic waves on continental *slopes* (outside the shelf break) has been postulated by theoreticians for some time, and some evidence on slow, wavelike progression of small-amplitude variations in sea level has been interpreted as a "signature" of such waves. Some direct evidence on longshore velocities along the slopes also supports such an interpretation. However, a conclusive demon-

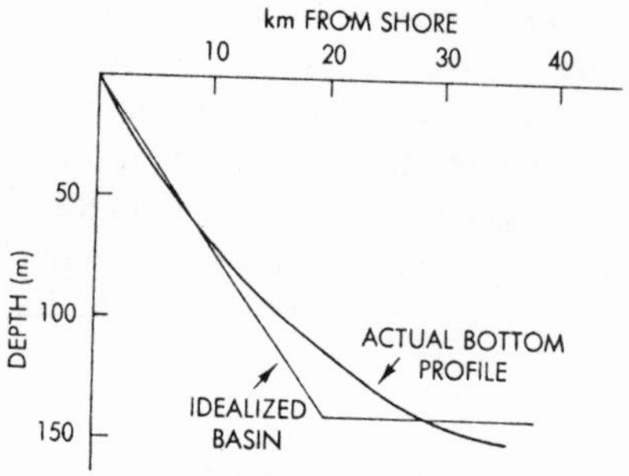

Figure 3. Idealized shore zone with beach of constant slope.

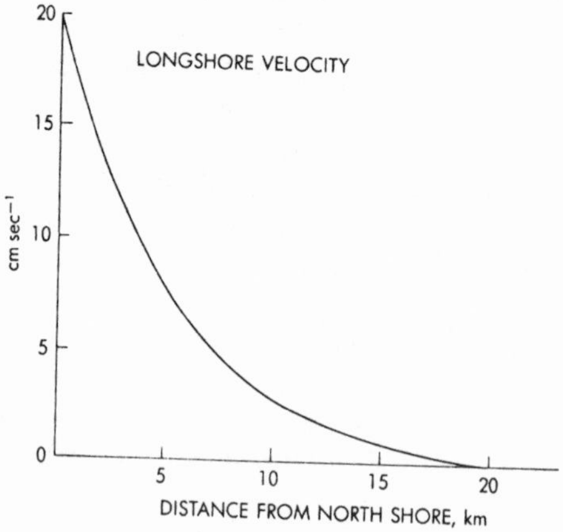

Figure 4. Distribution of longshore velocity in a topographic wave of 1-cm elevation amplitude at shore, in idealized shore zone of Figure 3.

stration that any given flow episode over the continental slope was a topographic wave has yet to be made. The principal signature of such a wave resides in its characteristic velocity field, associated surface elevations being small and subject to alternative interpretation. To observe the velocity structure over the continental slope in sufficient detail would require considerable expenditure. However, the demonstration of topographic waves in the shore zone of Lake Ontario, added to the other available evidence, makes it very likely that such waves will frequently occur over other strongly sloping parts of the continental shelf, i.e., beyond the shelf break.

The depth distribution across the continental shelf south of Long Island (at Tiana Beach, where our experimental work is proceeding) is shown in Figure 8 (discussed later). One may surmise from the character of this distribution that different natural modes of topographic waves will exist with concentrations of longshore velocity over the shore zone and over the slope, and that probably no part of the shelf will escape their influence. Flow structures of this sort are remarkable: one

finds flow velocities of up to a knot, extending over a massive body of water, which possess no obvious driving force analogous to wind stress in a wind-driven current. We have seen some apparent instances of topographic wave flow off Long Island, but our data are not yet complete enough for confirmation.

Note that although topographic waves are free modes of motion, their generation is due to wind-stress impulses. In this sense they are part of the picture of the total wind-driven circulation, with variable winds producing a complex pattern of topographic waves.

COASTAL JETS

The discussion so far has ignored any density variations in the water column. In summer a seasonal thermocline exists over the continental shelves, at a depth of some 30 m, separating a top, light layer from a bottom layer that is heavier because of higher salinity and lower temperature. The thermocline itself is a region of rapid density variation in the vertical. The density difference between top and bottom layers is a few parts in 1000, and most of this is concentrated in the few meters of the thermocline thickness. Wherever the density of a fluid decreases this strongly upward, the arrangement of fluid layers is markedly stable, sufficiently so to suppress turbulence. With turbulence absent, the force of the wind cannot be communicated directly to the bottom layer, and our ideas on wind-driven coastal currents must be modified.

A simple conceptual model is a two-layer fluid. The thermocline is compressed into a single plane, in which the density change takes place jumpwise. This interface between two fluids of slightly differing density is assumed to be frictionless. In the simplest approach, the total depth of water is taken as constant, and the shoreline as long and straight.

As a longshore wind is "switched on" over such a shore zone, the wind stress acts on the top layer alone and accelerates it downwind, which produces higher velocities than if the momentum were distributed over the entire water column. The deflecting force of the earth's rotation now produces some peculiar effects: the top layer may glide freely offshore over the frictionless thermocline (if the wind blows so as to leave the shore to the *left*); the bottom layer then must move onshore to compensate for the loss of water in the shore zone. In this manner a considerable thermocline tilt may develop near shore. If the process continues long enough, the thermocline may intersect the surface, producing upwelling of cold water at the shore. The inclination of the thermocline creates horizontal pressure gradients in the offshore direction that are in geostrophic balance with the strong current in the top layer. This current has many properties in common with the atmospheric jet stream and is therefore known as a coastal jet. A schematic illustration of coastal jet development based on the two-layer conceptual model is shown in Figure 5.

The characteristic width of coastal jets is the internal radius of deformation, $R = c_i/f$, where c_i is the speed of propagation of long waves on the thermocline and f is the Coriolis parameter. Typical magnitudes are $c_i = 50$ cm sec^{-1} and $f = 10^{-4}$ sec^{-1}, so that R is about 5 km. Within a nearshore band of this width, the wind-imparted momentum is concentrated in the top layer. Farther from shore the long-

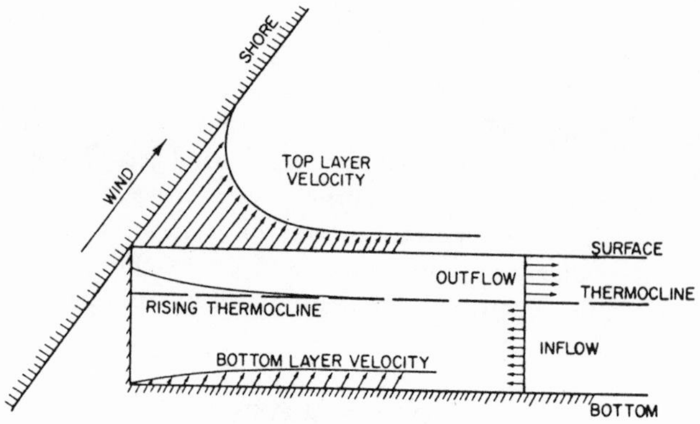

Figure 5. Schematic illustration of coastal jet flow structure in two-layer fluid of constant depth. The onshore-offshore velocities have constant amplitude, while the longshore velocity and thermocline elevation grow linearly with time.

shore momentum is more or less evenly distributed over the two layers, much as in a homogeneous water column, in spite of the isolation of the bottom layer by a frictionless thermocline. In this offshore region the Coriolis force transfers the momentum vertically downward. The equation of motion for the bottom layer, in the absence of longshore pressure gradients and friction, is

$$\frac{du}{dt} = fv \,, \tag{2}$$

i.e., acceleration equals Coriolis force per unit mass.

Integrating with respect to time we find

$$u = f\eta \,, \tag{3}$$

where η is offshore displacement of the fluid. As noted, such displacements occur in the bottom layer as the coastal jet develops. A quantitative treatment shows that they are just sufficient to equalize the longshore velocity between top and bottom layers outside a band of scale-width R.

A wind leaving the coast to the right produces a downward movement (downwelling) of the thermocline and an associated coastal jet in geostrophic equilibrium with that thermocline structure. The amplitude of thermocline movements in the shore zone is quite large, as may be inferred from Margules' equation expressing geostrophic balance across an inclined frontal surface. If $\varepsilon = \Delta\rho/\rho$ is the proportionate density defect and ζ' is the elevation of the thermocline above equilibrium, the velocity jump across the front is

$$u - u' = \frac{\varepsilon g}{f} \frac{\partial \zeta'}{\partial y} \,. \tag{4}$$

A velocity jump of 20 cm sec^{-1}, with $\varepsilon = 2 \times 10^{-3}$, thus requires a thermocline slope of 10^{-3}, or 1 m/km, or a total elevation amplitude of about 5 m (given $R = 5$

km, the scale-width of the upwelling-downwelling). As already mentioned, the thermocline may intersect the free surface after a strong wind episode, corresponding to a displacement amplitude equal to the top layer thickness of 20 to 30 m.

Episodes of opposing storms cause episodes of thermocline upwelling-downwelling near shore, accompanied by corresponding coastal jets. These are fascinating phenomena, and the massive onshore-offshore water movements that take place as one or the other develops are important in that they renew the coastal waters. Coasts subject to frequently alternating upwellings and downwellings can thus be kept remarkably clean in spite of considerable pollutant discharges, as the Chicago waterfront, for example, demonstrates.

KELVIN WAVES

The Coriolis force is again implicated in the generation of coastal jets and results in an upwelling on a left-hand shore (looking along the wind), a downwelling on the right. In a closed basin, the situation is similar to that for topographic waves: across the upwind and downwind ends of the basin thermocline elevation gradients develop, implying unbalanced longshore pressure gradients. These gradients again accelerate fluid across the original stagnation points of the initial coastal jet flow pattern, extending the right-hand jet across the downwind end, the left-hand jet across the upwind end. The analogy with topographic waves is strong; the pattern again propagates in a counterclockwise (cyclonic) sense around the basin. The speed of propagation is the velocity of long waves on the thermocline, which, as already noted, is typically 50 cm sec^{-1} (40 km day^{-1}). At a fixed location on shore, the coastal jet reverses direction as the wave passes, and the thermocline slope changes appropriately, from upwelling to downwelling or vice versa. This type of flow structure is known as a Kelvin wave.

The idealized shore zone containing a two-layer fluid is again an appropriate model. Kelvin waves of arbitrary wave form may propagate along such a shore, but always so as to leave the shore to the *right*. The main signature of the wave is now the relatively large-amplitude thermocline movement, associated with a considerable longshore velocity *difference* between top and bottom layers [given by Eq. (4)]. The amplitude of the wave decreases exponentially with distance from shore, with an *e*-folding scale of R ($R \cong 5$ km). Particle movements are everywhere alongshore, and those in the surface layer can be of quite large amplitude; long Kelvin waves could theoretically (i.e., in the absence of friction) lead to longshore excursions of about 100 km. Even if this is scaled down to, say, 30 km, it is clear that motions accompanying Kelvin waves on the thermocline make important contributions to the circulation of shallow seas. The progression of a Kelvin wave around Lake Michigan has been demonstrated by water intake temperatures.[4] Such Kelvin wave progression episodes have been documented more recently and in greater detail in Lake Ontario.[5]

In Lake Ontario the speeds of propagation of a topographic wave and an internal Kelvin wave are very close, so that the two travel almost in phase. The wave speeds are, however, determined by different physical factors (e.g., the slope of the shore zone and the density defect of the top layer) and, while they are likely to be of

the same order of magnitude in shallow seas of similar depth, in general they are easily distinguishable. The progression of an internal Kelvin wave along a seacoast does not seem to have been demonstrated so far, no doubt because so little attention has been paid to the 0 to 10-km range from shore. Note that such Kelvin waves are also "trapped" at the shore, within the first 10 km or so.

THE COASTAL BOUNDARY LAYER

We have so far discussed four distinct phenomena and corresponding simple conceptual models: wind-driven coastal currents (the experimentally verifiable part of topographic gyres), topographic waves, coastal jets, and Kelvin waves. All these affect primarily a coastal zone some 10 km in width, and their combined effect is to generate a current "climate" within this zone which differs markedly from the climate farther offshore. We may regard this difference in climate as a distinct phenomenon, perhaps belonging to a class of phenomena different from waves and jets. The physical factors leading to shore-trapped flow structures are all connected with the presence of the boundary, in close analogy with the case of the well-known boundary layers over aeroplane wings, and it is therefore appropriate to refer to the special coastal zone as the coastal boundary layer, CBL.

Experimentally, the differences between CBL and "offshore" (outside CBL) are best exhibited by current spectra, extracted from long time series of current measurements at fixed locations. The most complete evidence, which again comes from Lake Ontario, is summarized by Blanton.[6] Within the CBL, currents are usually parallel to the shore and more or less persistent. Outside the CBL, water movements in all directions occur with similar frequencies and are dominated by relatively short-period oscillations. In the Great Lakes the "offshore" current regime in the summer features predominant near-inertial oscillations known as Poincaré waves. These have not been observed over the continental shelves, but tidal oscillations are prominent and provide the main velocity signal. Density effects and the influence of storms are superimposed on this "carrier signal" almost in the manner of an amplitude-modulated radio transmission.

The width of the CBL may be determined empirically as the extent of the region within which shore-parallel flow dominates. On the north shore of Lake Ontario this was found to be 7 km. In other places around the Great Lakes it is similar, but varying distances can be expected elsewhere. Along the New Jersey coast, for example, the existence of a CBL has been demonstrated on an empirical basis, but its exact width is unknown.

In discussing the continental shelves, the question arises whether there is a second boundary layer at the shelf break, i.e., at the open boundary of the shelf. There is some evidence to suggest that there is, and that it is connected with the existence of a semipermanent front near the shelf break, an inclined surface of rapid density change, separating shelf water from slope water. Within the shelf-break zone currents would be expected to flow parallel to this frontal surface, much as they flow parallel to shore within the CBL. However, a systematic experimental study of the shelf-break zone has not yet been carried out.

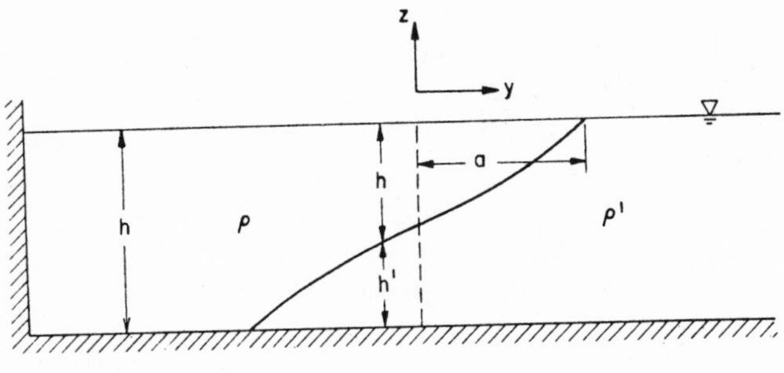

Figure 6. Adjustment of an initially vertical front (dashed line) separating light nearshore waters of density ρ from heavier offshore waters, density ρ', to an inclined position in geostrophic balance (solid curve).

FRONTAL ADJUSTMENT

We now consider some flow phenomena connected with horizontal density gradients, arising from freshening or heating of nearshore waters. As mentioned above, a semipermanent inclined surface, with a relatively rapid density variation across it, separates shelf water from slope water at the edge of the continental shelf. A similar spring thermocline separates nearshore waters in the Great Lakes from the cold central mass early in the season, after the shallow coastal waters have warmed up. Although this thermocline, with a typical lifetime of 6 weeks, is ephemeral compared with the shelf-edge front, it is clearly maintained by similar dynamical factors. The behavior of the spring thermocline was documented in considerable detail during IFYGL, and this provides some clues for gaining an understanding of the shelf-edge front.

We consider first the simple conceptual model illustrated in Figure 6. In a constant-depth shore zone (straight, infinite shore along the x-axis), at $t=0$ the waters between the shore and $y=0$ are somewhat lighter than those beyond. In an idealized experiment a thin wall coinciding with the dotted line could be suddenly removed. In a real ocean a storm might mix the water column vertically, with some freshwater addition at shore. Left to itself, the system shown in Figure 6 begins a readjustment of mass, the light fluid moving offshore at the top, the heavy fluid onshore at the bottom. The Coriolis force deflects both, so that longshore motions develop, of a velocity given by Eq. (3), that are proportional to the offshore displacement of water particles. During such an adjustment process following a sudden change, some oscillations develop at the inertial frequency f (Coriolis parameter), which proceed about an equilibrium configuration of the front separating lighter and heavier waters. Ignoring friction and any mixing of the waters across the front, the equilibrium configuration can be calculated from the postulates of geostrophic equilibrium after adjustment and the conservation of potential vorticity during it. The latter principle may be expressed by the condition

$$\frac{f - \partial u/\partial y}{h} = \text{const} , \tag{5}$$

where h is the depth of the water column. This equation is applied separately to light and heavy fluid columns. The result in a typical case is also illustrated in Figure 6: a wedge of light water moves to overlie a wedge of heavier water. The half-width a of the wedges is $R=f^{-1}\sqrt{g\varepsilon h}$, where ε is the proportionate density defect of the light water. This again ranges from a few kilometers for the nearshore zone of the Great Lakes (and other shallow seas) to 10 km or so for the shelf-edge front. The light water moves so as to leave the shore to the right, the heavy water in the opposite direction. The slope of the inclined front is h/R, so that [according to Eq. (4)] the velocity difference across the front is $\sqrt{\varepsilon gh}$, or 30 to 100 cm sec^{-1} in the two typical cases mentioned. In the coastal zone the longshore velocities generated by frontal adjustment are thus comparable with wind-driven flow velocities, whereas they dominate the latter near the shelf edge.

Given the large velocity differences across the inclined front, friction cannot be negligible even if the flow is laminar in the region of the steepest density gradient. Very little is known about boundary layer processes within such frontal zones, but the example of the sea surface (where the density contrast between air and water is extreme) suggests that a quadratic friction law prevails even in the presence of stable stratification. Friction at the interface in Figure 6 would gradually destroy geostrophic equilibrium and require further offshore movement of light water, with compensating inflow of heavy fluid, to reestablish the balance. Eventually, therefore, the inclined front should "relax" toward a more nearly horizontal position. In the Great Lakes coastal zone, a frontal zone under quiescent conditions doubles its width in a few days.

Yet it is wind-induced mixing, not such gradual decay, that usually destroys a frontal structure like that in Figure 6. Strong wind acting over shallow water breaks down the stratification of the water column and reestablishes a situation similar to that prevailing in our model *before* adjustment, in which density differences occur only in a horizontal direction. Surface cooling in a cold outbreak further enhances vertical mixing.

Weaker winds, which occur more frequently, produce less drastic effects on a frontal structure of the type shown in Figure 6. Such winds, if acting in the longshore direction, preferentially impart longshore momentum to the light fluid, which occupies more of the surface. If the wind leaves the shore to the *right*, this effect enhances the velocity contrast between light and heavy fluid and, in accordance with Eq. (4), steepens the gradient of the interface, making the angle of the wedges (light and heavy fluid) more obtuse. On the other hand, when the wind impulse opposes the direction of flow of the light fluid necessary for geostrophic balance, an offshore displacement of the light fluid must take place, much as in the case of frictional decay. Given a strong enough wind (leaving the shore to the left) the warm body of fluid might separate completely from the shore. These possibilities may be explored with the aid of our model (Figure 6), by adding an arbitrary longshore wind-stress impulse to the light fluid alone. One possible consequence is a lens-shaped thermocline (see Figure 7). In the outer portion of such a light lens, the velocity contrast [according to Eq. (4)] is like that over a wedge-shaped front, but it is reversed over the inner portion. The latter is a direct consequence of the wind-stress impulse, the former, of

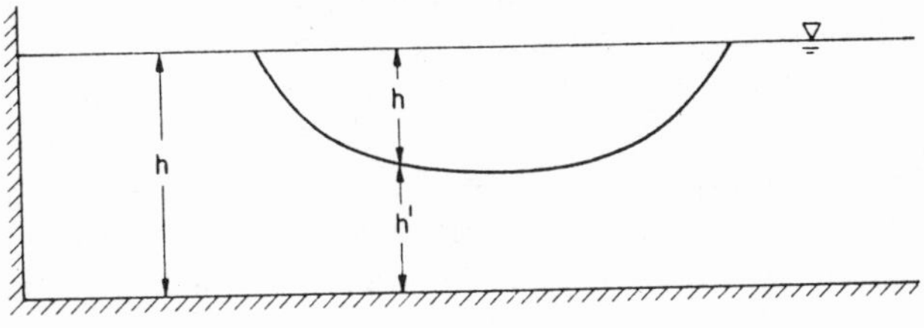

Figure 7. Lens of light water that develops when longshore wind (leaving the shore to the *left*) acts on nearshore band of light water.

an offshore displacement more than large enough to compensate for wind-imparted momentum. The heavy fluid moves in this case everywhere in the direction of the wind.

Some examples of wedge-shaped and lens-shaped thermoclines from the shore zone of Lake Ontario and associated longshore flow patterns have been given by Csanady.[7] Similar frontal structures (the wedge type frequently, separated "bubbles" occasionally) have been observed near the edge of the continental shelf, although little is known about the associated flow patterns. Given rather complex differences (subject to seasonal variations) in salinity as well as temperature between shelf water and slope water, it is certain that the behavior of the shelf-edge front is more complex than the above simple ideas would suggest. However, it is reasonable to suppose that under suitable conditions the phenomena described above will play an important role in determining the flow pattern prevailing in the shelf-break zone.

FRICTIONAL CIRCULATION

In the above discussion of conceptual models, no account has been taken of the velocity *differences* arising in a water column as the latter is subjected to wind or bottom *stress*. Although turbulent friction is subject to more complex laws than viscous friction, it is still generally true that vertical transfer of momentum requires a velocity gradient down which the momentum is transferred. Thus, as the wind stress is applied at the surface, the near-surface layers of the water column must move faster than those below, in order to effect momentum transfer. The Coriolis force acting on the faster near-surface flow also deflects it more (this force being proportional to velocity) and produces a complex velocity distribution involving variations of both horizontal velocity components in the vertical direction.

Simple conceptual models for the understanding of similar phenomena were proposed as early as 1905 by Ekman and are now well known. The simplest such model, applying to a deep ocean far from coasts, results in the celebrated Ekman spiral. In shallow water near coasts, both bottom friction and the condition of no net onshore-offshore flow must be taken into account (under steady conditions), and a more complex conceptual model is required. Suppose the wind to be parallel to a straight long coast and to blow with constant force. Longshore currents are produced, strong enough so that bottom friction exactly balances the longshore wind

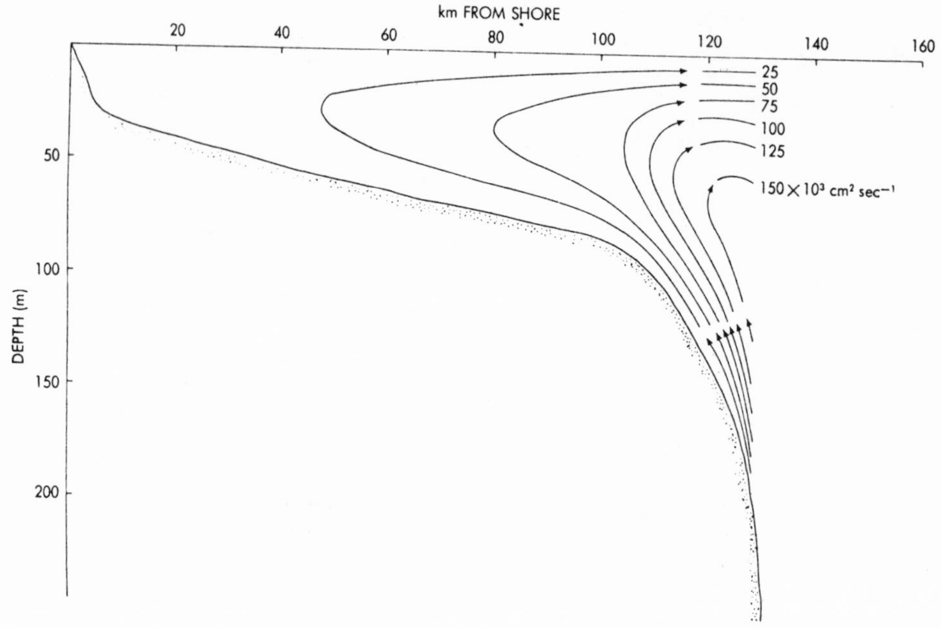

Figure 8. Calculated frictional secondary circulation in a plane normal to the coast over the continental shelf south of Long Island, caused by a strong eastward gale.

stress. A pressure gradient is established in the offshore direction in order to eliminate any net onshore-offshore flow. When the wind blows so as to leave the shore to the *left*, the top layers move offshore, the bottom layers onshore; the reverse occurs with an opposite wind. An example of such calculated frictional-equilibrium flow in a plane normal to the coast, using the depth distribution off Tiana Beach and realistic friction parameters, is shown in Figure 8. The wind was assumed to be very strong, exerting a stress of 10 dynes cm^{-2}. A much weaker wind produces a less intense circulation, similar in character but closing much nearer shore. The water column in such models is assumed to be homogeneous, and therefore similar circulations can be expected only under winter conditions.

Another classical model of this type is relevant to the thermohaline circulation in winter over the continental shelves. Although the water column is vertically well mixed, horizontal density gradients due to freshwater influx are present. Under frictionless geostrophic equilibrium conditions a southwestward current would be present in this situation, increasing in velocity linearly with vertical distance from the bottom. With vigorous turbulence present, this implies a certain rate of vertical momentum transfer. At the surface, in the absence of wind, the stress must reduce to zero, so that something akin to an Ekman spiral again forms, within which some offshore flow takes place. Return flow occurs in a bottom boundary layer. The velocity structure of such frictionally controlled thermohaline circulation is illustrated in Figure 9.

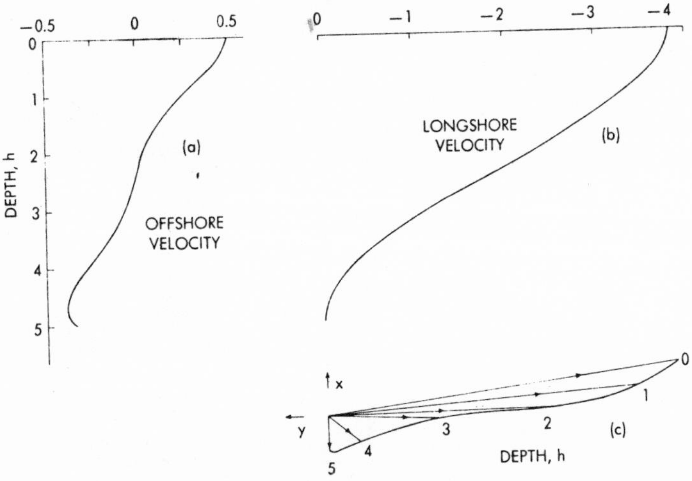

Figure 9. Velocity distribution: (*a*) offshore, (*b*) longshore component, (*c*) hodograph, in steady density driven flow due to horizontal salinity gradient over the continental shelf in winter. Depths are marked in units of Ekman depth (typically 15 m); velocities are in cm sec^{-1}, positive offshore and eastward.

Although these classical models are widely accepted as realistic, nowhere have they been satisfactorily related to observation. Part of the reason is no doubt the absence of suitable instrumentation; current meters have to be quite efficient to describe accurately the secondary-flow type of circulation in a vertical plane normal to the coast, where the dominant velocity component is longshore. Another problem is that any steady-state model is suspect, given the variable nature of forcing. Perhaps there are frictionally induced circulations such as those shown in Figures 8 and 9, but whether they can be identified as distinct phenomena is not certain, given the many transient forces and impulses affecting the coastal zone.

CONCLUSION

Even this brief and incomplete review shows the impressive complexity of flow phenomena over continental shelves. We have not even touched upon such questions as how different distinct phenomena (e.g., tides and wind-driven currents) may interact, or what a long-term (seasonal) "mean" circulation might look like. A great deal of empirical input will be needed to develop a satisfactory understanding of water motions over continental shelves. It is equally important, however, that our observational efforts be guided by realistic conceptual models; random survey work over such a vast area as the continental shelves, which are subject to such complex physical influences, is likely to be very wasteful.

To draw one specific conclusion, much of the "action" seems to take place in a coastal zone about 10 km in width, and this zone should receive special attention in any large-scale observation program. A deployment of current meters on an even grid (of, say, 20-km spacing) is likely to miss this zone almost entirely, a resolution

of 1 to 2 km being necessary near shore. Much the same goes for the frontal zone at the shelf break, where a resolution of 3 to 5 km would appear to be necessary in the cross-isobath direction. We have begun to work on these important boundary layers, but it will probably take a decade or two of solid effort to achieve a scientific understanding of continental shelf circulation.

Acknowledgment: This work was supported by Brookhaven National Laboratory under its coastal shelf transport and diffusion program, sponsored by the United States Energy Research and Development Administration.

REFERENCES

1. Garrett, J.R., See paper in this Conference.
2. Saunders, P.M., On the uncertainty of wind stress curl calculations, Unpublished, Woods Hole Oceanographic Institution, 1976.
3. Rao, D.B. and Murty, T.S., Calculation of the steady-state wind driven circulation in Lake Ontario, *Arch. Meteorol. Geophys. Bioklimatol.* **A19**, 195-210 (1970).
4. Mortimer, C.H., Frontiers in Physical Limnology With Particular Reference to Long Waves in Rotating Basins, *Great Lakes Res. Div. Univ. Mich. Publ.* **10**, 9-42 (1963).
5. Csanady, G.T. and Scott, J.T., Baroclinic coastal jets in Lake Ontario during IFYGL, *J. Phys. Oceaonogr.* **4**, 524-41 (1974).
6. Blanton, J.O., Some characteristics of nearshore currents along the north shore of Lake Ontario, *J. Phys. Oceanogr.* **4**, 415-24 (1974).
7. Csanady, G.T., Spring thermocline behavior in Lake Ontario during IFYGL, *J. Phys. Oceanogr.* **4**, 425-45 (1974).

Part IV

STUDIES OF SUBSTRATE RESPONSE

Editors' Comments on Papers 17 Through 20

17 **INMAN and RUSNAK**
Changes in Sand Level on the Beach and Shelf at La Jolla, California

18 **COOK and GORSLINE**
Field Observations of Sand Transport by Shoaling Waves

19 **LUDWICK**
Tidal Currents, Sediment Transport, and Sand Banks in Chesapeake Bay Entrance, Virginia

20 **LAVELLE et al.**
Preliminary Results of Coincident Current Meter and Sediment Transport Observations for Wintertime Conditions on the Long Island Inner Shelf

In this final section, we present a sequence of papers in which geological questions are posed and answers sought within the context of a general model of fluid dynamical process and substrate response. The authors of these relatively late papers show a heightened awareness of fluid dynamical concepts. They are concerned with observing directly the sea floor processes that operate over long periods of time and result in the geomorphic and stratigraphic phenomena described in Parts I to III.

The papers vary greatly in their goals and approaches. The ultimate goal in each is the quantification of sediment transport, and a theme explicit or implicit in each paper is the sediment continuity relationship, presented by Ludwick as:

$$\frac{\partial \eta}{\partial t} = -\epsilon \frac{\partial Q}{\partial x},$$

where η is bed elevation relative to a datum plane, t is time, ϵ is a dimensional constant related to sediment porosity, Q is weight rate of bed-sediment transport per unit width across the transport

direction, and x is distance along the transport direction. The equation states that the change in sea floor elevation with time is proportional to the change in sediment discharge along the transport pathway.

We may resolve the pattern of sediment transport in an area of study by making measurements that apply to one side or the other of this equation. In Paper 17 Inman and Rusnak devote themselves to measuring the lefthand term, the time rate of change of bottom elevation. Anyone who has undertaken a program of systematic bottom monitoring by scuba techniques will stand somewhat in awe of this paper. The authors seem to have managed quite well without such modern luxuries as pingers, pinger locaters, and precision electronic navigation. The greatest achievement, however, is simply to have maintained a diving schedule through the winter period of high winds, high seas, low underwater visibility, and near freezing temperatures.

Cook and Gorsline present more general observations of inner shelf process by means of scuba technique in Paper 18. In a sense, their approach is more limited in that their data are not cumulative between observations, and they can only report fair weather conditions. Nevertheless, Cook and Gorsline's observations lead them to propose a modified version of the generally recognized model for onshore-offshore sand budget. They conclude that onshore winds may result in offshore bottom flow, and they assign a minor role to the null point model for sediment sorting.

Ludwick's Paper 19 stresses not only the field observations (twenty-four current meter stations in Chesapeake Bay) but the model for which the data are used as input. This volume has paid scant attention to river mouths on the grounds that such a complex coastal environment should rightfully be the topic of a separate benchmark volume. Ludwick's method is of such fundamental importance, however, that the paper's inclusion in this volume is mandatory; here is a simple, yet elegant model for sediment transport, which is the most satisfactory field-tested numerical model at the time of this writing. The model gives only relative, not absolute, values and is not verified by a complimentary program of sediment transport measurements by direct observation. Its greatest value is as a demonstration of the kinds of calculations that are becoming possible as theory evolves and our field instrumentation improves.

The final paper in this section, by Lavelle and others, is similar in its objectives to Ludwick's. The authors attempt to examine fluid motion and resulting sediment transport on a straight coast between estuaries, rather than at an estuary as did Ludwick's

model. They do not attempt to resolve a spatial pattern of sediment transport, because within the study area, flow was highly coherent. Paper 20 does, however, illuminate the temporal pattern of sediment transport on a straight coast in response to wind driven flow, by direct measurement of both fluid and sediment transport. On such a coast, transport occurs in short, intense bursts. The threshhold velocity of bottom sediments is exceeded for periods of hours or of one or two days, and such periods are separated by days or weeks of quiescence. Not all events are alike. One intense storm accomplished more sediment transport than did the many mild ones during the observation period. Flow during this storm was to the west, the prevailing direction of transport indicated by bottom topography. Milder storms were to the east.

17

Reprinted from pp. 1–30 of *Beach Erosion Board TM-82*, U. S. Army, Corps of Engineers, Washington, D. C., July 1956, 58 pp.

CHANGES IN SAND LEVEL ON THE BEACH AND SHELF AT LA JOLLA, CALIFORNIA

D. L. Inman and G. A. Rusnak
University of California, Scripps Institute of Oceanography

ABSTRACT

Changes in the level of the sandy bottom were measured periodically over an interval of almost three years at stations extending from near the surf zone to a depth of 70 feet. A reference level was established at each station by forcing six rods, spaced at intervals of 10 feet, into the bottom and changes in sand level were based on the average of the differences in length of rod exposed from survey to survey. Measurements, which were performed by swimmers equipped with self-contained underwater breathing apparatus, resulted in a standard error of about ±0.05 foot per survey in the determination of net sand level.

The total range in sand level probably exceeds 2 feet at the 18-foot deep station, where changes in excess of 0.6 foot were measured before the reference rods were lost. The magnitude of change decreases with increasing depth; changes of 0.29, 0.16, and 0.15 foot were measured at stations where the depth of water was 30, 52, and 70 feet respectively. Estimates of sand level variation, made for monthly and seasonal periods, indicate that significant changes occur between monthly periods at the 30 and 70-foot depths, and that significant seasonal changes occur at the 30 and 52-foot depths. At the deepest station overall seasonal variations are relatively small in comparison with changes of shorter period.

Comparison of acoustic soundings with reference rod measurements indicated that the accuracy of the acoustic sounding method was of the order of ± $\frac{1}{2}$ foot for these operating conditions.

INTRODUCTION

The qualitative evaluation of the deposition and erosion of sediments near shore is usually based on differences obtained by comparing successive topographic surveys. In general the surveys are quite accurate for those portions of beach which are above water level, where standard leveling and positioning techniques can be employed.

*Contribution from the Scripps Institution of Oceanography, New Series No. 854.

Below water level the depth to the bottom is usually measured acoustically or with sounding lead and the bottom elevation estimated by correcting that depth for tides, waves, and various other effects. This method of hydrographic survey is relatively inaccurate as compared to the precise leveling techniques which are carried out above water.

Seasonal and long term changes in the level of nearshore sediments commonly are greatest along the narrow section of beach foreshore which falls within the intertidal zone. The magnitude of change in level decreases seaward, probably reaching a minimum on the broad flat shelves extending away from the beaches. Since the shelf areas are large in comparison with the narrow strip of beach bordering them, relatively small errors in determining sand level on the shelf may result in very large errors in estimating the total sand budget.

Attempts to evaluate the accuracy of hydrographic surveys have been made on several occasions, but the reliability of such investigations has been subject to uncertainty because of the lack of an accurate level upon which to base the surveys. The advent of self-contained breathing apparatus (SCUBA), which allows swimmers to move about freely and to make measurements and observations underwater, has provided a means of establishing underwater levels and accurately measuring changes from this level. A level is established by forcing a long rod a sufficient distance into the sandy bottom and in such a manner as to assure its remaining in a stationary position so that it does not move when material is deposited or eroded from the area. Thus, ideally, a direct comparison of the level of the exposed portion of the reference rod on two successive surveys will give the net value of erosion or deposition between surveys. A series of four stations using reference rods of this type was established on the sandy shelf off La Jolla in order to determine accurately the magnitude of erosion and deposition and to evaluate the accuracy of hydrographic surveys which use acoustical methods. Three of these stations were sampled periodically during an interval of almost three years. In addition to these stations, a series of reference rods was placed along a beach profile extending through the surf zone in order to measure small changes in sand level which occur during short intervals of time such as half-tidal cycles.

METHOD

A series of four stations were established on a range running normal to the coast line and extending from near the breaker zone out to the level portions of the sandy shelf (figure 1). Locations were selected where the depth of water was approximately 20, 30, 50, and 70 feet to allow a progressive comparison in changes of sand level from the surf zone, where fluctuations in level were known to be large, out to greater depths on the shelf. In the final analysis the exact location of the stations was determined by the proximity of good visual ranges and cross-ranges, since ease of positioning is an important factor in the economy of field operations.

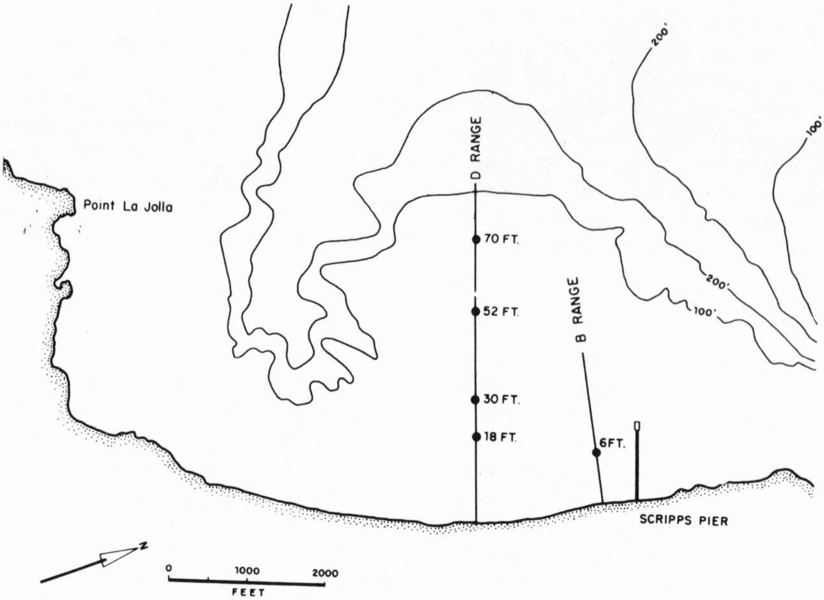

Index chart showing the location of ranges and stations. Six reference rods were placed at each station on D range. Single rods were placed at 20-ft. intervals along a profile through the surf zone on B range. Depths are in feet below mean lower low water.

FIGURE 1 · INDEX CHART

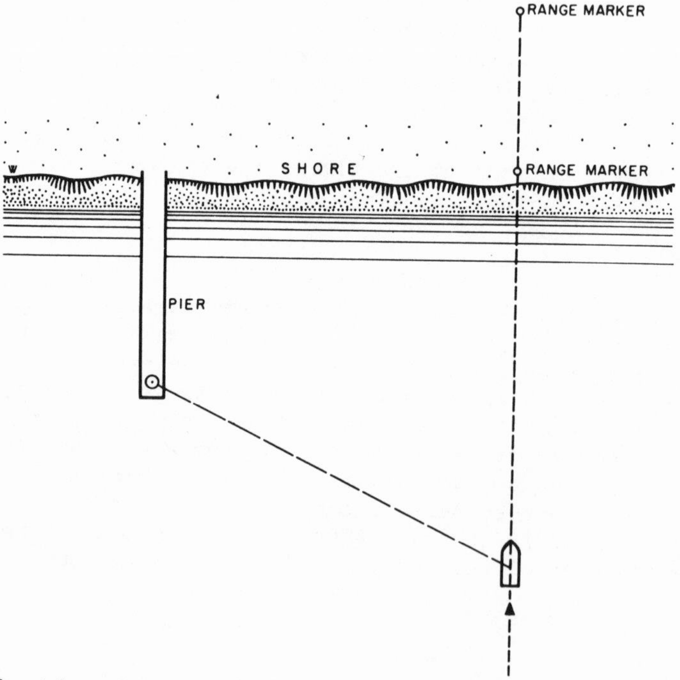

The four stations on D range were located by running shoreward along a visual range to positions determined by the angle between the range and the pier. The angle was measured by sextant from the boat.

FIGURE 2 · METHOD OF POSITIONING

It was the intention in the experimental design of the field work to increase reliability of observation by making the number of independent measurements at any one station and time as large as practical, and by sampling each station frequently. Initially six reference rods were placed at each station, and on the average they were measured from one to two times per month. Because the changes in sand level, especially in deeper water, were small, the data were subjected to various statistical analyses in order to evaluate the contributing factors. The procedures used in processing the data are described in a separate section.

POSITIONING

During surveys the station locations were determined by range and horizontal sextant angle. A range was obtained by aligning visually two range markers on shore and the small craft proceeded shoreward along this range until the appropriate angle was obtained between the range and the end of the Scripps pier, at which point a marker was dropped in the water (figure 2). The marker consisted of a small cylindrical float with line wrapped around it and a ten-pound lead attached to the free end of the line. The lead sounded rapidly, causing the cylinder to spin in the water as the line unwound. Swimmers equipped with SCUBA were then guided to the location of the station by following the line to the bottom (figure 3).

This method of positioning proved to be accurate and quick and provided the swimmers with a visual reference upon which to orientate themselves. As a measure of accuracy, the rest position of the sounding lead relative to the six reference rods is shown for the 45 observations made at the 30-foot deep station (figure 4). The position of the sounding lead reflects the effects of waves and currents as well as the error in positioning at the time the marker float was placed in the water. Even so, the sounding lead landed within a radius of 5 feet of the target (reference rod number 1) about 50% of the time and within a radius of 10 feet about 80% of the time. The radius of error was slightly greater at 52-foot and 70-foot depths.

REFERENCE RODS

The basic problem was to find a means of establishing a stationary reference level near the bottom from which fluctuations of the sandy bottom could be measured. The solution used here was to force rods into the bottom in such a manner as to assure their remaining stationary when material is eroded or deposited in the area. For this purpose, brass rods 3/8 inch in diameter and 4 feet long were pounded approximately 3 feet into the bottom with a light sledge hammer. Circular brass rings were silver-soldered around the lower portion of the rods so that once emplaced in the sand they could not easily be raised or lowered. Deformed steel reinforcing rods were also tested, but their

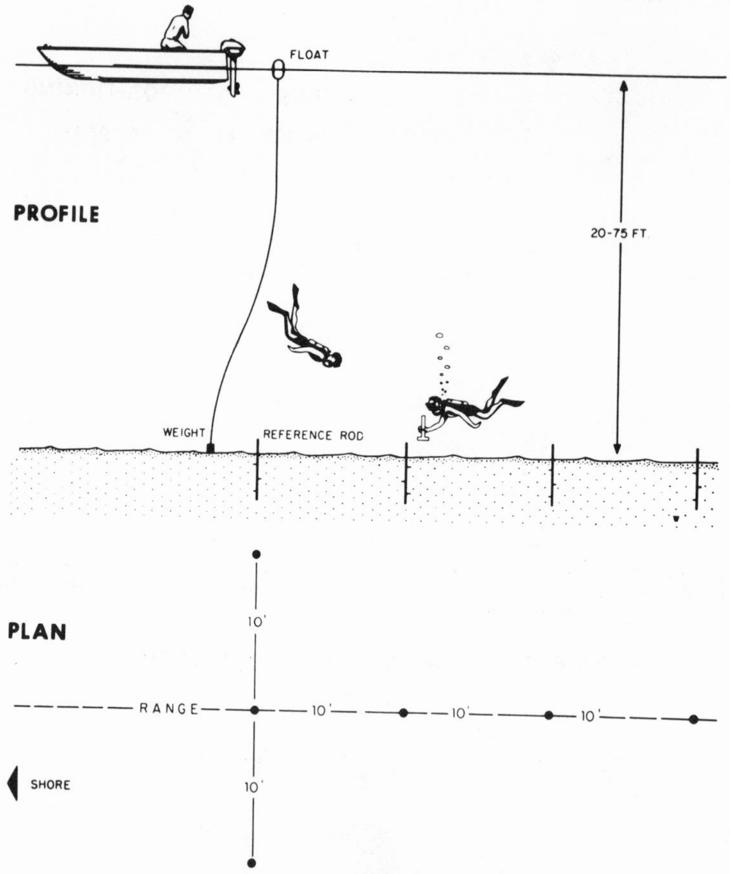

Schematic diagram illustrating the technique used in locating the station and measuring the reference rods. Six rods were placed in the sandy bottom at each of the four stations on D range; three across the range and four along the range. The station locations are shown in figure I.

FIGURE 3 · TECHNIQUE USED IN MEASURING REFERENCE RODS

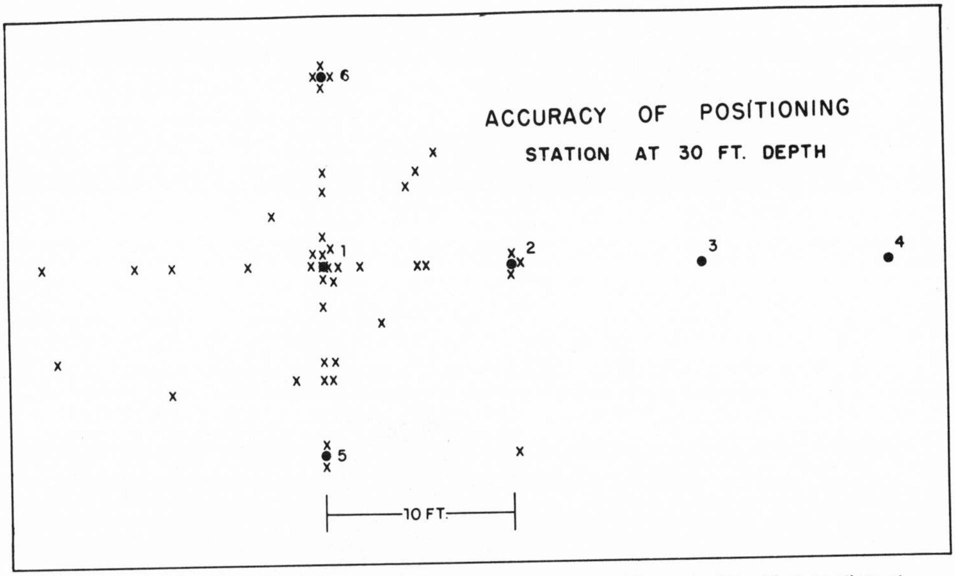

Accuracy of positioning indicated by the rest position of the sounding lead for each of the 45 observations at the 30-foot depth. The position of the sounding lead, shown by X, reflects the effect of waves and currents as well as the error in positioning the boat.

FIGURE 4 · ACCURACY OF POSITIONING

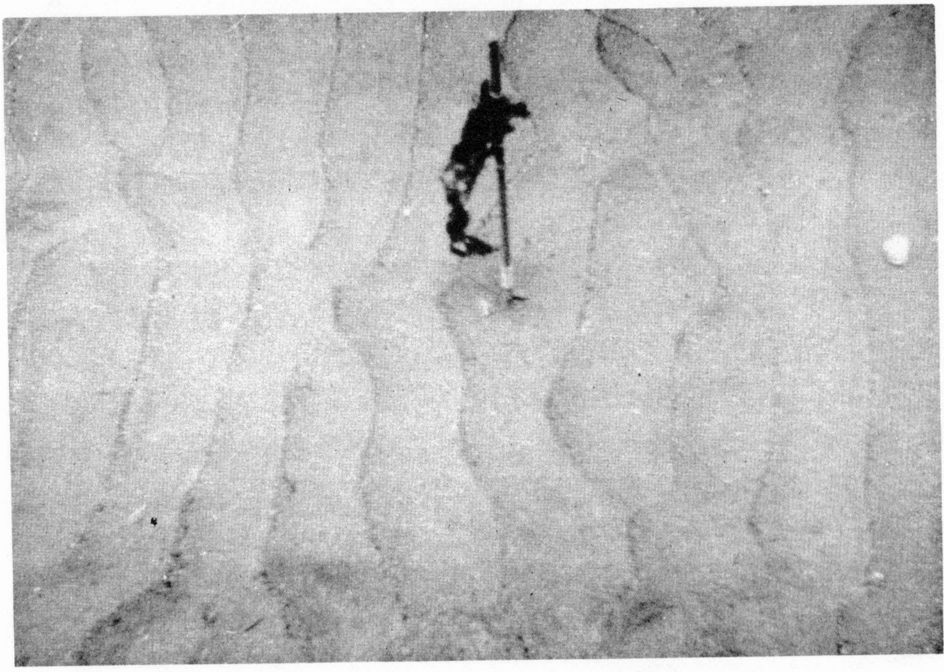

Underwater photograph of one of six reference rods at the station in 30 feet of water. The rod is 3/8-inch diameter brass; approximately I foot is exposed, and 3 feet are buried in the sand. The dark object is a plastic tag with growing hydroid. Note the slight scour depression at the base of the rod.

FIGURE 5 · UNDERWATER PHOTOGRAPH OF REFERENCE ROD

use was discontinued because it was found that hydroids and other organisms grew profusely on the steel, whereas the brass was found to be relatively free from fouling organisms. Fouling on the steel rods made measurement more difficult and increased the stress exerted on the rod by the moving water and the degree of scour at the base of the rod. In general, only a very slight scour depression formed in the sand at the base of the brass rods, as shown in figure 5.

Six rods were forced into the bottom at each of the four stations on D range. They were placed 10 feet apart in a "T" formation, with four rods along the range and three across (figure 3). The distance between rods was sufficient to eliminate the possibility of significant hydraulic interference or influence between rods.

After the rods had been in place several days and were in equilibrium with their environment, the station was again visited and the length of each rod carefully measured. This length was recorded as the reference length for that rod and is the length upon which future changes in sand level were based. For each successive survey the new length of rod was subtracted from the reference length, and the mean value of the differences for the six rods used as a measure of the new sand level. The differences from reference length for each rod during each survey, together with the mean difference and standard deviation, are listed in Appendix IA-ID. The mean difference is graphed as a function of time in figure 7. The particle size distribution analyses of the sand samples collected by the swimmers at each station are listed in Appendix IIIA-IIID.

The length of rod was measured by placing a plexiglas device upright on the sand alongside the rod and marking the device with a grease pencil. The measuring device, which consisted of a 2-foot long strip of plexiglas with an 8-inch wide by 1-inch thick strip of wood at its base, resembled a draftsman's "T" square (figure 6). The base was wide and thick so as to bridge or cover holes that might be present at the base of the rod and to avoid settling into the sand when placed beside the rod.

The lengths of each of the six rods at each station were marked on the device. When the swimmer returned to the surface, these lengths were measured to the nearest hundredth of a foot and recorded for processing as described above.

During the summer of 1953 a series of reference rods were placed at 20-foot intervals along a profile through the surf zone on B range (figure 1). The line extended from the upper portion of the beach foreshore out to a depth of approximately 6 feet below MLLW (figure 8). The section of beach above water level was surveyed by transit, and the portion under water by swimmers and reference rods in the manner outlined for the stations on D range, excepting that only one

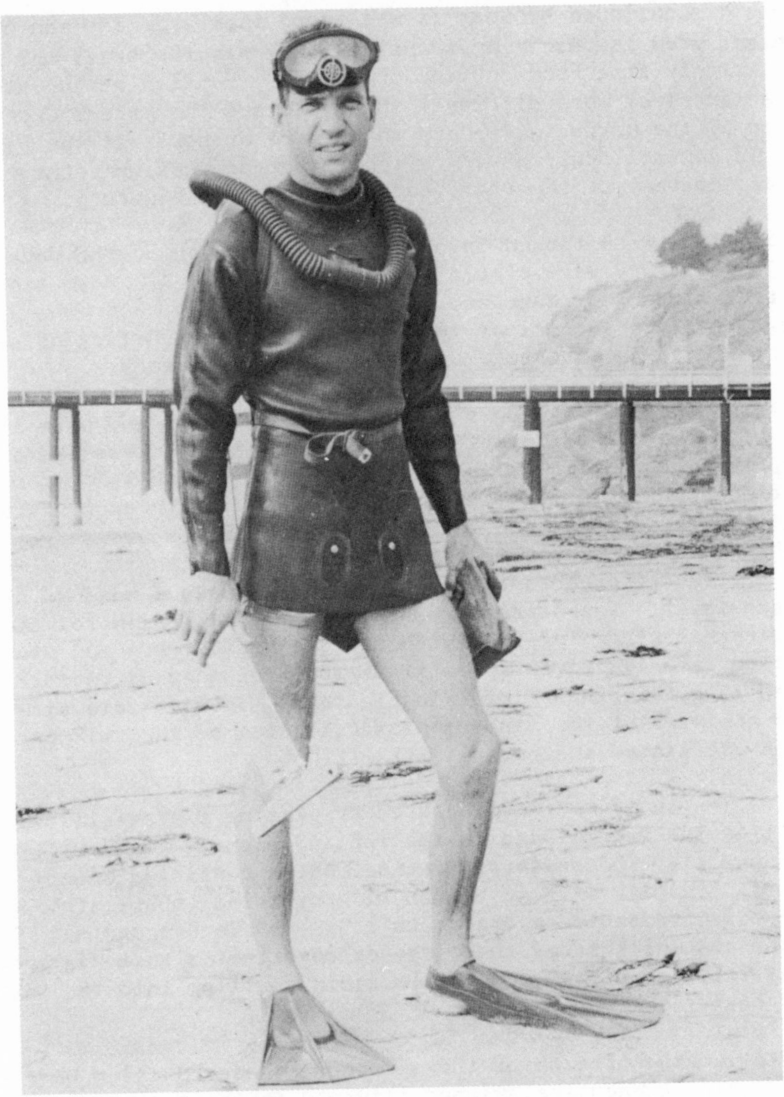

Swimmer equipped with self-contained breathing apparatus (SCUBA), exposure suit, and plexiglas device used in obtaining the length of reference rods. The length is marked on the plexiglas with grease pencil (attached to device by clip) and measured after returning to the surface.

FIGURE 6 · SWIMMER EQUIPPED WITH SCUBA AND MEASURING DEVICE

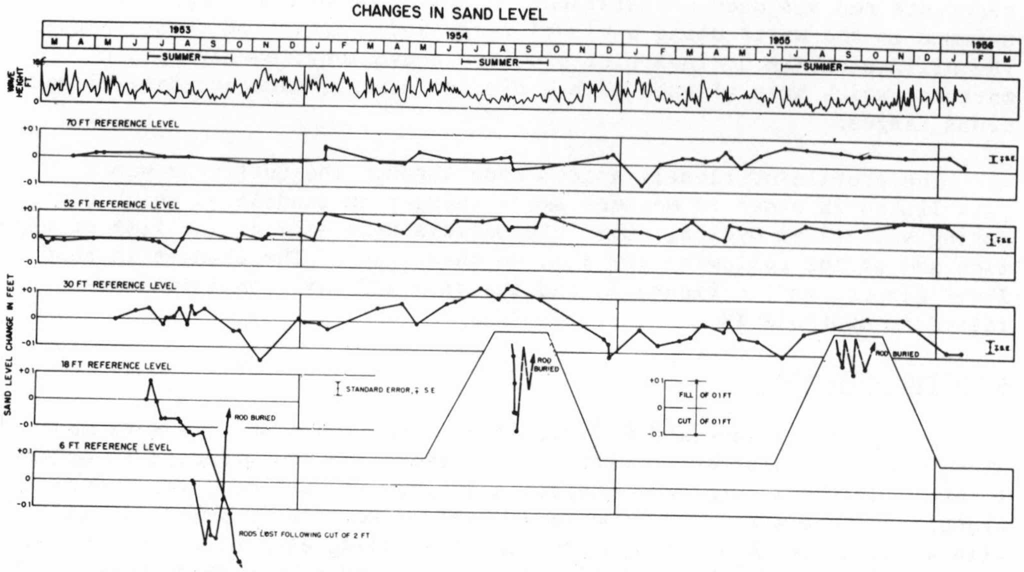

Changes in sand level with time. The stations at depths of 18, 30, 52, and 70 feet are on D range and the fluctuations in sand level are based on the mean of six reference rods at each station. The magnitude of two values of the standard error, ± S.E., is shown as a measure of reliability of the observations. The station at a depth of 6 feet was on B range and consisted of a single reference rod which was completely covered with sand during the winter. The wave heights are for significant breakers at the point of wave convergence near D range.

FIGURE 7 · CHANGES IN SAND LEVEL WITH TIME

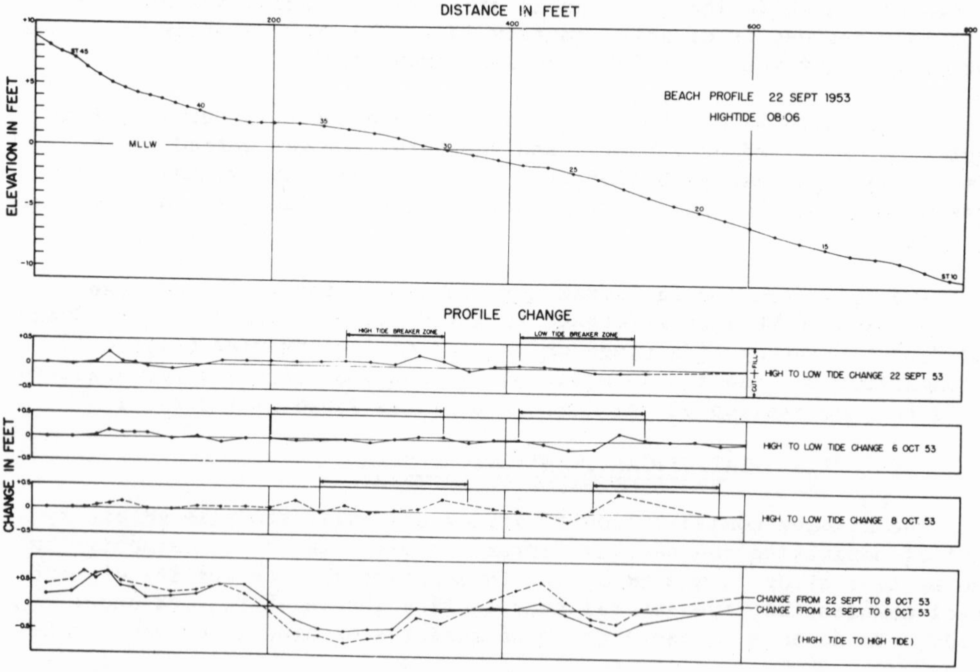

Changes in sand level on B range between time of high and low tide on three separate days. Lower graph shows the change in profile between the first day's survey and those of the succeeding days. Wave and tide conditions are listed in Appendix IV.

FIGURE 8 · CHANGES IN BEACH PROFILES DURING TIDAL CYCLES

reference rod was used per station. Positioning was performed by the swimmer in the water using a visual range and a series of cross ranges. The pilings of the Scripps pier provided convenient, evenly spaced markers, which when aligned with a distant marker, made accurate visual cross ranges.

The profile of closely spaced rods through the surf zone was established in order to measure small changes in sand level which occur during half-tidal cycles. Complete surveys were made at the time of high tide and at the following low tide on three days. The changes in sand level are graphed in figure 8, and the tide and wave conditions are listed in Appendix IV.

ACOUSTIC SOUNDING

During the summer and fall of 1953 hydrographic surveys were made along D range concurrently with the reference rod measurements in order to evaluate the accuracy of the acoustic sounding method. The hydrographic surveys were made from an amphibious vehicle (DUKW) equipped with a Bludworth NK - 2 fathometer and a sounding element rigidly mounted to the side of the DUKW approximately a foot and a half below water level. This fathometer is equipped with gain, frequency, and power controls, marker bar, and has a record scale of 6½ inches equal to 60 feet in depth. The fathometer was calibrated with a lead line which was graduated in tenths of a foot and held vertically in the water beside the sounding element. Approximately six checks were made at each of two depths while the DUKW was in still water. Calibrations were made in water depths of approximately 15 and 30 feet in areas where the bottom was flat and sandy (Shepard and Inman, 1951).

Positioning was performed by visual range and horizontal sextant angles as described previously, and the fathogram was marked at the time of passage over each station. Approximately five sounding runs were made along the range just prior to measuring the reference rods at each station.

The data from the fathogram were corrected for depth and reduced to the datum of MLLW by reference to a tide record from the U. S. Coast and Geodetic Survey tide gauge on the end of Scripps pier (figure 1). A comparison between the acoustic survey and the reference rod measurements for the station at the 30-foot depth is shown in figure 9.

STATISTICAL TREATMENT OF DATA

The primary consideration of any experimental study is generally that of separating the several factors involved in the experiment. In the present study it was important to separate the variability of sand level changes into the several contributing groups of factors under study. To do this, a large group of measures of sand level were collected

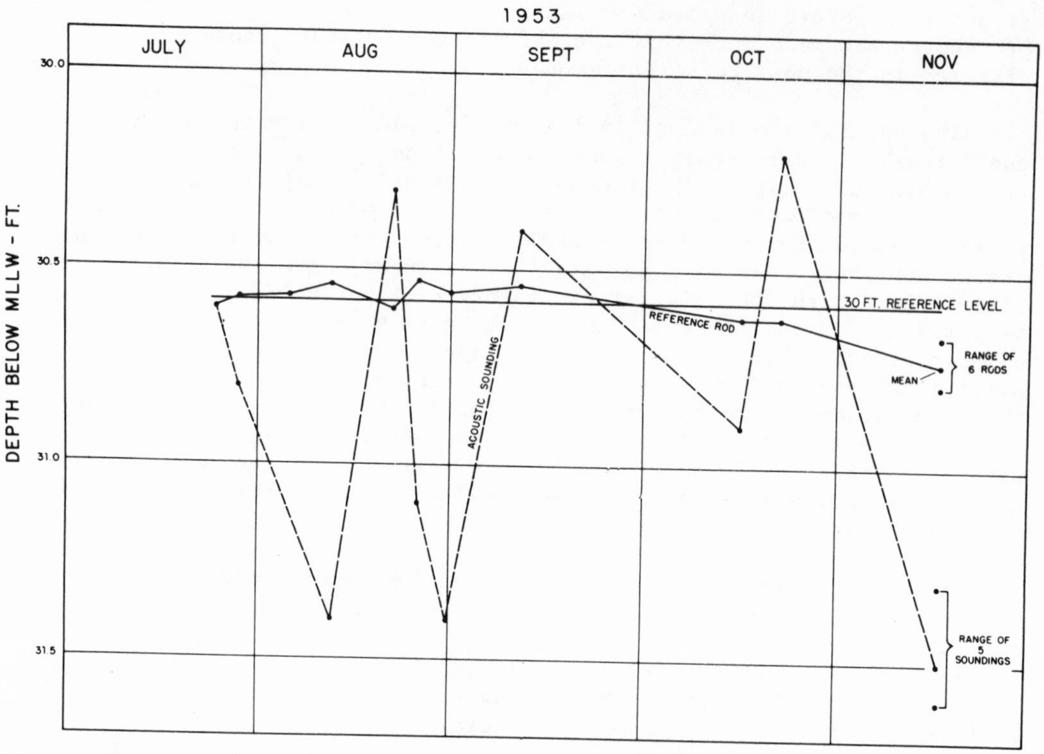

Comparison between sand levels obtained by acoustic sounding and reference rods at the 30-foot deep station on D range. Each point represents the mean of approximately 5 acoustic sounding runs or 6 reference rod measurements. Within a single day the range of acoustic soundings is only 3 to 5 times greater than the range in rod measurements. However, between days the acoustic soundings give fluctuations which would result in large errors.

FIGURE 9 · COMPARISON OF ACOUSTIC SOUNDING AND REFERENCE RODS

at specific points in space and time. Then a choice of statistical procedures was made such that the information obtained could be utilized in the most efficient manner.

The natural limitations in this study, such as adverse weather conditions, did not permit the experimental design to be carried out to its best advantage. It is known, for example, that the data are somewhat biased because the measures could not be collected in a completely random manner. Nonetheless, although errors may result because the assumptions underlying the statistical theory have not been completely fulfilled, the statistical theory will not be rejected but will be applied with the condition that caution be used in its interpretation (Fisher, 1949, p. 16). If these limitations are kept in mind, the statistical applications are somewhat justified in that they may lead to the recognition of the causes of departures. As has been pointed out by Olson and Potter (1954), one must use judgment in passing over preliminary stages of analysis more rapidly than statistical rigor would permit in order to focus attention on later critical problems requiring statistical analysis.

With this point of view, several questions were asked in the study of sand level changes. These were: (1) what is the accuracy with which the sand level, at a given depth and at a given time, could be measured; (2) what are the monthly and seasonal variations of sand level; (3) is there a trend in the changes of sand level with time; and, finally; (4) is there a relation between the sand level changes at various depths? The choice of statistical tests used for obtaining estimated answers to these questions was the analysis of variance, the sign test of trend, and the estimate of correlation.

ANALYSIS OF VARIANCE

In a normally distributed population the form of the distribution is completely described by the mean μ and the variance σ^2. A random sample of n items from this population gives an estimate of these parameters in the form of the sample mean $\bar{x}$ and the sample variance s^2 (Dixon and Massey, 1951, pp. 32-35).

$$\bar{x} = \frac{\Sigma x}{n} \tag{1}$$

$$s^2 = \frac{\Sigma x^2 - (\Sigma x)^2/n}{n-1} \tag{2}$$

The fundamental principle upon which the analysis of variance is based is that the variance of the independent factors contributing to the population variance are additive. Therefore, the total population variance is the sum of the independent factors contributing to this variance. In the present study it was desired to separate the independent factors of sand level variability due to local irregularities

in sand level (plus measurement error) from those due to variations with time. By grouping the data in a hierarchical scheme, it was possible to separate and evaluate the magnitude of these variations.[1]

The assumptions underlying the analysis of variance are that the observations are from a normally distributed population and that the variance of each group is the same. Dixon and Massey (1951, p. 126) point out, however, that the results of the analysis are changed very little by moderate violations of the assumptions of normal distribution and equal variance.

Example of Analysis of Variance for the Station at 70-Foot Depth. For cases in which the random deviations are normally distributed, the computation of variance is conveniently made by utilizing the conventional form of the analysis of variance table (Eisenhart, et al., 1947, p. 311). The sources of variation are broken down into those (1) between k rods, (2) within j months, (3) between i months, and (4) between h seasons. Arranging the data so that these sources of variation could be separated results in the hierarchical order presented in table 1.

Computational procedure for the first step in the analysis of variance is indicated by the column headings and the summations following table I, part A. The summations are made with a calculating machine and entered directly in this part of the table. The sum totals of these columns are then entered in the analysis of variance table, part B, as lower level sum and higher level sum from the respective columns. The difference of these sums for each source is the sum of the squared deviations and corresponds to the numerator of equation (2). The sum of these squared deviations is then divided by the degrees of freedom to obtain the mean square or variance.

The number of degrees of freedom between rods is the number of rods per observation minus one (k-1), summed over all observations, and is equal to 199 for this station. Within months the number of degrees of freedom is the number of observations per month minus one (j-1), summed over all months and is equal to 11. This same procedure is followed for each successively higher level of sampling (Snedecor, 1946, p. 233).

The mean square between rods is assumed to be due to random sampling fluctuations around the true mean of the group with a true variance σ_k^2. The mean square within months, however, is dependent both upon the between rods variance, σ_k^2, affecting the individual measurements involved in the mean, and on the additional variance component, σ_j^2, due to the fluctuation of sample means around the group mean $\bar{X}_j$. The long-run expected value of the mean square within months is equal to

[1] The reader is referred to standard statistical texts such as Dixon and Massey (1951, Chaps. 8 and 10), Snedecor (1946, Chaps. 10 and 11) or the lucid digest of Olson and Potter (1954) for the complete details of this method of analysis.

Table 1. Analysis of variance for unequal samples of sand level change; station at 70 ft depth.

Part A. Data and computations.

Between Seasons (h)	Between Months (i)	Within Months (j)	Between Rods (k) 1	2	3	4	5	6	$\sum_k x$	$\sum_k x^2$	$\sum_j \left[\frac{(\sum_k x)^2}{N_k}\right]$	$\sum_i \left[\frac{(\sum_{jk} x)^2}{N_j}\right]$
1	1	4 May '53	+.01	-.02	+.01	+.04	+.01	+.04	+.09	.0039	.00270	.00432
		13 May '53	+.05	-.02	0	+.03	+.03	0	+.09	.0047		
	2	26 June '53	+.02	--	0	+.03	+.04	0	+.09	.0029	.00162	
2	3	22 July '53	+.03	+.01	0	-.04	+.03	-.01	+.02	.0036	.00006	.00008
	4	19 Aug. '53	+.04	0	+.03	-.06	0	0	+.01	.0061	.00001	
3	5	30 Oct. '53	-.02	-.03	+.03	-.01	+.02	-.04	-.09	.0043	.00135	.02575
	6	19 Nov. '53	-.02	-.06	+.04	-.01	0	-.02	-.07	.0061	.00081	
	7	28 Dec. '53	-.02	-.02	+.03	-.02	-.04	+.02	-.05	.0041	.00041	
	8	26 Jan. '54	+.02	-.03	+.04	+.01	-.02	-.01	+.01	.0035	.02175	
		27 Jan. '54	+.01	+.04	+.05	+.06	+.07	+.01	+.24	.0128		
		28 Jan. '54	+.04	+.02	+.06	+.06	+.05	+.04	+.27	.0133		
	9	29 Mar. '54	-.01	-.02	+.03	-.01	-.03	+.02	-.02	.0028	.00006	
	10	27 Apr. '54	-.03	-.04	+.03	+.02	-.05	+.01	-.06	.0064	.00060	
	11	14 May '54	+.04	+.03	+.02	+.07	+.02	+.02	+.20	.0086	.00666	
	12	18 June '54	0	-.03	+.03	+.03	+.02	+.02	+.07	.0035	.00081	
4	13	26 July '54	0	-.02	+.03	-.02	+.06	0	+.05	.0053	.00041	.00907
	14	16 Aug. '54	-.03	-.03	+.05	+.01	+.08	+.03	+.11	.0117	.00440	
		26 Aug. '54	+.03	-.04	+.01	0	+.07	+.05	+.12	.0100		
	15	1 Sep. '54	+.02	-.20	0	-.07	+.06	+.03	+.16	.0498	.00426	
5	16	4 Oct. '54	-.07	-.06	+.01	0	+.02	0	-.10	.0090	.00166	.09686
	17	15 Dec. '54	-.02	+.04	+.07	+.05	-.01	+.03	+.16	.0104	.01308	
		21 Dec. '54	+.03	+.05	+.06	+.02	+.05	+.02	+.23	.0103		
	18	25 Jan. '55	-.15	-.01	-.10	-.20	-.03	+.01	+.48	.0736	.03840	
	19	15 Feb. '55	+.06	0	+.07	-.20	+.02	+.07	+.02	.0538	.00006	
	20	10 Mar. '55	+.06	-.01	+.03	0	+.02	+.07	+.17	.0099	.00962	
		22 Mar. '55	+.03	0	+.06	-.01	+.02	+.07	+.17	.0099		
	21	6 Apr. '55	0	+.01	+.06	+.02	-.05	+.07	+.11	.0135	.02388	
		20 Apr. '55	+.04	+.02	+.07	+.08	-.05	+.01	+.17	.0159		
		29 Apr. '55	+.02	+.09	+.05	+.06	+.07	+.03	+.32	.0195		
	22	5 May '55	+.02	-.04	+.04	+.06	+.05	+.07	+.20	.0146	.00672	
		16 May '55	+.02	-.01	+.02	-.05	-.05	+.05	-.02	.0084		
	23	8 June '55	+.09	-.03	+.05	+.03	+.07	+.04	+.25	.0189	.01041	
6	24	6 July '55	+.14	+.01	+.03	+.10	+.08	+.07	+.43	.0419	.03081	.10087
	25	1 Aug. '55	+.11	+.01	+.04	+.05	+.06	+.12	+.39	.0343	.02535	
	26	10 Sep. '55	+.11	0	+.08	+.03	+.01	+.12	+.35	.0339	.02041	
	27	3 Oct. '55	+.06	0	+.09	+.02	+.05	+.03	+.25	.0155	.02442	
		12 Oct. '55	+.06	+.02	+.09	+.06	+.06	-.03	+.29	.0202		
	28	23 Nov. '55	+.05	-.02	+.06	+.07	+.05	+.05	+.26	.0164	.01126	
7	29	13 Jan. '56	+.06	+.01	+.03	+.04	+.06	+.09	+.29	.0179	.01508	.02266
		30 Jan. '56	+.03	-.01	+.02	+.04	+.01	-.01	+.08	.0032		
								Total	+4.43	.6144	.27707	.25961

$$\sum_h \left[\frac{(\sum_{ijk} x)^2}{N_i}\right] = \frac{(.27)^2}{17} + \frac{(.03)^2}{12} + \ldots \frac{(\sum_{ijk} x)^2}{N_i} = .14709$$

$$\frac{(\sum_{hijk} x)^2}{N_h} = \frac{(4.43)^2}{239} = .08211$$

Table 1. Analysis of variance for unequal samples of sand level change. Station at 70 ft depth.

Part B. Analysis of variance

	Between Rods	Within Months	Between Months	Between Seasons
Lower Level Sum	0.6144	0.27707	0.25961	0.14709
Higher Level Sum	0.27707	0.25961	0.14709	0.08211
Sum Squared Deviations	0.33733	0.01746	0.11252	0.06488
Degrees of Freedom	199	11	22	6
Mean Square	0.00169	0.00158	0.00511	0.01081
Lower Component		0.00169	0.00158	0.00511
Difference		-0.00011	0.00353	0.00570
Effective Subsample		5.97489	8.55393	31.52301
Component Estimate, s^2	0.00169	0	0.00041	0.00018
F		0.93491*	3.23417**	2.11545*

* Not significantly different from variations of shorter period.

** Significant at the 0.01 level; therefore highly significant.

($\sigma_k^2 + k\,\sigma_j^2$), where k is the known number of individuals which share in the particular sample value of this distribution (Olson and Potter, 1954, pp. 37-38).

The test of significance associated with the analysis of variance in this form is considered as the test of the hypothesis that the two mean squares σ_k^2 and $\sigma_k^2 + k\,\sigma_j^2$ are equal. This is equivalent to asking whether the contribution of σ_j^2 is equal to zero or whether there is a significant contribution to the mean square due to this factor σ_j^2. The F-test is used to test this hypothesis. The value of F is the ratio of the within months mean square to the between rods mean square; and, if this ratio exceeds the tabled values of F at a specified level of significance, then the hypothesis of equality of mean squares is rejected (Olson,and Potter, 1954, p. 38; Dixon and Massey, 1951, chap. 10; Snedecor, 1946, p. 231).

From the values in part B of table 1, the computed value of F is equal to 0.93491 and is less than 1. It is concluded therefore that the mean squares are approximately equal and that there is no significant contribution due to σ_j^2. If the F value exceeded the tabulated values (Dixon and Massey, 1951, pp. 310-313), the component estimate of σ_j^2

could have been computed from $\dfrac{(\sigma_k^2 + k\,\sigma_j^2) - \sigma_k^2}{k}$, as was the case for the other stations. In the present example, k is the effective subsample size. The reason for this is that the samples contributing to this mean square were not equal in size and k had to be computed as an effective subsample size. The effective subsample size is equal to 5.97489 and is computed from the following relation (Snedecor, 1946, pp. 234, 241):

$$k = \frac{\Sigma_j k - \dfrac{\Sigma_j (k)^2}{\Sigma_j k}}{(j-1)} \qquad (3)$$

This same procedure is followed for the next succeeding higher level of sampling to obtain the contribution in that group. In the present example, the difference happened to result in a minus value and it is concluded that the variance at this level, σ_j^2, is equal to zero. Since σ_j^2 should not be less than zero, this discrepancy may represent a sampling variation due to some unusually divergent measures in the rods (Olson and Potter, 1954, p. 40). The practical significance of this result is that the variability within months is negligible compared to the variability between rods.

Another way of estimating the variance among individual rod measures in an observation is to compute the pooled estimate or average variance.

The formula for this estimate is given by Dixon and Massey (1951, pp. 91-92) as

$$(4) \qquad s_p^{\,2} = \frac{(n_1-1)s_1^{\,2} + (n_2-1)s_2^{\,2} + \ldots + (n_k-1)s_k^{\,2}}{n_1 + n_2 + \ldots + n_k - k}$$

for k samples of variance. For the 70-foot station the pooled estimate of variance between rods is equal to 0.00169. The pooled estimate may be lower than that obtained in the analysis preceding because it is the average variance computed around each sample mean rather than around the mean of the group of all observations. It is felt that use of the pooled estimate is a better estimate for between rod variability. It is this value which has been used in obtaining the standard error discussed in the interpretation.

THE SIGN TEST FOR RUNS

The nonparametric sign test was used for determining the existence of a trend to the changes in sand level with time. This test does not involve the statistical parameters of the population distributions; it simply compares the distribution of values without specifying the form of the distributions (Dixon and Massey, 1951, p. 247). Furthermore, it is only necessary to assume that the observations represent a random sample.

For present purposes it was desired to know whether the changes in sand level with time had any cyclic trend or whether these changes were random. To test this, the procedure is to determine the median value of the ordered sequence of given mean values of sand level with time. All values above the median are then designated by a plus sign and all values below the median by a minus sign. The question is whether or not the positive values are distributed in a random manner among the negative values. The one or more adjacent similar signs are denoted as runs with the result that a run is started as soon as the sign changes (from plus to minus or from minus to plus). If the number of runs is very high, it would indicate a rapid periodic or cyclic trend. On the other hand, if the number of runs is very low, the sequence would be nearly monotonic (hence of slow period).

The test is based on the criterion for making one of three decisions for each group of observations (Hald, 1952, p. 749; Eisenhart, *et al.*, 1947, p. 419).

(1) Accept the hypothesis that there is a trend.
(2) Reject the hypothesis of trend.
(3) Continue sampling. There may be a trend but the observations are not sufficiently refined to bring this trend out in the analysis.

Example of Sign Test from Station at 70-Foot Depth. The median is determined by taking the mean values of the raw data from Appendix ID and arranging these in order of increasing values. This median value is +.02. Now taking the original serial sequence and designating the values above the median as plus and the values below as minus, the following is obtained:

Sign	Run	Sign	Run
-	1	+	6
-		+	
-		-	7
-		-	
-		+	8
-		+	
-		-	9
-		+	10
-		+	
+	2	+	
+		-	11
-	3	+	12
-		+	
+	4	+	
-	5	+	
-		+	
-		+	
md.		+	
-		+	
-		-	13
			13

Number of items = 40
Total number of runs = 13

Table 11 in Dixon and Massey (1951, p. 325) shows that the upper limit of runs to be expected from a pure random arrangement is 27, while the lower limit to be expected is 14. Values either above or below these tabled values would not be expected to occur more than 2.5 per cent of the time; i.e., there is a risk of being wrong in the decision 2.5 per cent of the time. Since the number of observed runs lies below the lower limit, it is concluded that there is a trend and the number of runs is abnormally smaller than would be expected from a simple random variation of data.

ESTIMATES OF CORRELATION

In order to determine the relationship between changes in sand level from one station to another, estimates of the correlation between stations were made. The test used is designed to determine whether or not the changes between stations are independent of each other. If these changes are not independent, it is desirable to know whether the correlation between them is positive or negative; i.e., either there is

simultaneous deposition of sand at the two stations in question, or there is deposition of sand at one station and erosion of sand at the other.

The relatively simple statistic used here to estimate the correlation is the tetrachoric r, described in Dixon and Massey (1951, pp. 233-235). The procedure is to make a graphic plot of one mean value of sand level against the other, for all mean values observed between the two stations on a simultaneous date. After all the values are plotted in this manner, the plotted points are divided equally by a horizontal line and a vertical line. If, then, the number of points in the upper right and lower left quadrants is greater than in the other two quadrants, the correlation is positive. If the opposite is true, the correlation is negative. This procedure is very useful in the exploratory stages of a study, such as the present one, to determine whether the factors are associated.

Example of Correlation between Stations at 70 and 30-Foot Depths. By taking the data from the table in Appendix IB and ID, and collecting all mean sand level changes observed on simultaneous dates for the 70 and 30-foot depth stations, the graphic plot of figure 10 is obtained. The total number of points in the plot is 26. Dividing these points by a horizontal and a vertical line so that an equal number of points lie above and below the horizontal line and an equal number of points lie on either side of the vertical line, the graphic plot is divided into four quadrants. The number of points in each of these quadrants is then designated by n_1, n_2, n_3, and n_4. The value of

$$\frac{n_1 + n_3}{n_1 + n_2 + n_3 + n_4} \quad \text{or} \quad \frac{n_2 + n_4}{n_1 + n_2 + n_3 + n_4}$$

is then computed, whichever is larger. In this case the second relation is the larger and results in a value of 0.769. This value is then entered in figure 2 of Dixon and Massey (1951, p. 234) to obtain the estimate of correlation. The estimate arrived at from this procedure is -0.75 for the correlation coefficient. It is concluded that these stations are not independent of each other. The correlation is negative; i.e., when the sand is built up at one station, there is a loss at the other.

INTERPRETATION

Of the several distinct factors involved in a study of this kind, it is difficult to isolate the local irregularities of the bottom from those changes which are significant. That is, the local irregularities of the bottom may be as large as the net changes in sand level one is trying to measure. Consequently such irregularities may completely mask the minor changes in sand level. The primary purpose of the present study was an attempt to evaluate the magnitude of the separate factors of variability involved in sand level changes.

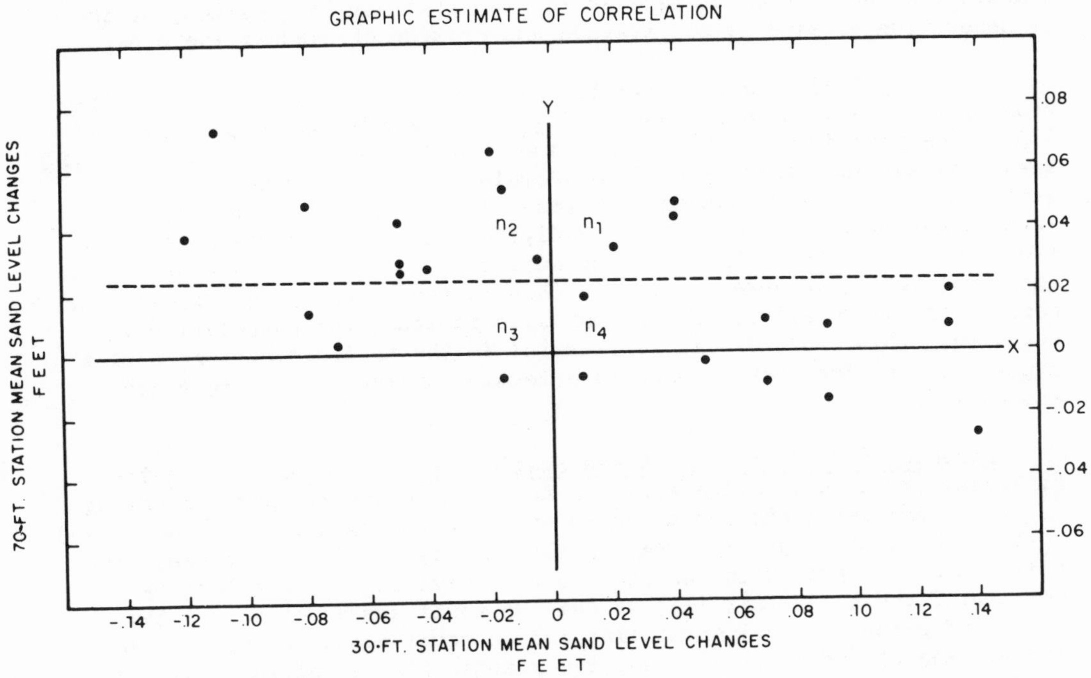

Graphic method of estimating the correlation between sand levels at the 30 and 70-foot depth stations on D range.

FIGURE 10 · CORRELATION BETWEEN SAND LEVEL CHANGES AT THE 30 AND 70-FOOT DEPTHS

Another serious consideration which could be made is the factor of costs. One might spend the entire time in a survey by making a large number of very precise measures from day to day. If, however, the changes of major importance are not day to day occurrences, but monthly or yearly occurrences, it would be less costly to take fewer day to day measures and still get the desired information. Before making such decisions, it is necessary that the sources of variation and their magnitudes be evaluated. In studies where this is done, it is possible to better allocate or budget sampling time so that in the future more information can be arrived at with a smaller cost in time and effort (Eisenhart, et al., 1947, p. 269). Suggestions are made for better budgeting of sampling in the discussion to follow and in the conclusions.

Earlier studies have indicated that the major changes in sand level on the beaches and in adjacent shallow waters tend to be systematic and can be related to the character of wave motion, tidal cycles, and to seasonal variations representing the combined effects of tides and waves. Initially it was intended that an attempt be made to correlate the sand level changes at the four stations on D range directly with local wave and tide characteristics. However, the changes were smaller than anticipated and it early became apparent that one must look to more recondite parameters than can be extrapolated from simple wave parameters. Therefore no attempt was made to correlate day to day wave action with sand level changes, but rather a gross seasonal distinction based on wave height was made. Summer was arbitrarily defined as those months or portions of months during which significant breaker height exceeded 5 feet less than one-third of the time, and winter was defined as the intervening period during which the breaker height exceeded 5 feet more than one-third of the time. For this purpose the breaker height at the wave convergence zone near D range was selected. The envelope of the significant breaker heights is plotted as a function of time in figure 7 and the summer seasons are indicated.

SAND LEVEL CHANGES ON D RANGE

The major part of this study was devoted to a study of sand level changes on D range. The large number of measurements and the control afforded by this area allowed for a relatively rigorous statistical evaluation of sand level changes. The statistical methods used have already been outlined in the previous section. The results of these analyses are presented in Appendix I and II and are summarized in table 2.

Reliability of Reference Rod Measurements. An estimate of sand level at a given depth is the mean value of a group of measures and this mean value is dependent upon the local irregularities at that station. The precision with which the mean value estimates the sand level is dependent upon the magnitude of these local irregularities. The standard error is a measure of this precision and can be computed for one observation. However, a better estimate is to compute an average standard

Table 2. Summary of statistical results, D range.

Station Depth (ft)	Total Number of Surveys	Duration of Survey in Months	Standard Error of Measurement $SE = \frac{S}{\sqrt{n}}$ (ft)	Range in Mean Sand Level (ft)	Standard Deviation of Sand Level Changes (ft)			Cyclic Trend
					Within Months	Between Months	Between Seasons	
18	11	4	.061	.62	0.055	0.040*	--	--
30	45	32	.067	.29	0*	0.048	0.042	Seasonal
52	49	35	.050	.16	0.015*	0.003*	0.033	Yes**
70	40	33	.033	.15	0*	0.020	0.013*	Yes**

* Not significantly different from variations of shorter period.
** Cycles more frequent than seasonal, but less frequent than monthly.

Correlation Between Stations		
Between Stations	Estimated Correlation Coefficient	Statistically Significant
30 52	+0.36	Weak
30 70	-0.75	Yes
52 70	+0.60	Yes

error pooled from all observations. The standard errors computed for each station in this manner, from 18 feet to 70 feet, decrease progressively away from shore (table 2). This indicates that either the local bottom irregularities or the error in measuring, or both, decrease away from shore. It is believed that while conditions nearer shore made measurements somewhat more difficult, these influences are minor compared to the actual bottom irregularities. The average standard error of observation for the four stations is approximately 0.05 foot and is a confidence limit within which the mean sand level falls. This means that if the changes in sand level which one is trying to measure are of this order of magnitude, they will be masked by this error.

Magnitude of Sand Level Changes. The magnitude of the significant changes in sand level decreases with increasing depth and is found to be very small in depths of 30 feet and greater. The changes in sand level range from 0.15 foot at depths of 70 feet to values in excess of 0.6 foot in water depths of 18 feet (table 2). The results of the analysis of variance for the station at the depth of 18 feet are not valid for statistical inference between seasons because of the limited data[2]. However, these results give an order of magnitude of the variability within and between months.

For the other three stations on D range, there is a general decrease in the magnitude of the between season variability from station to station seaward. The component estimate of the between months variability at the 30-foot depth shows that there is a fluctuation of approximately 0.05 foot around the average level of sand, and indicates a range of about 0.10 foot for the changes. This range is equivalent to twice the standard deviation (table 2), where the standard deviation is the square root of the variance component. The estimate of the between months variation at the 70-foot depth is about half that at the 30-foot depth and the variation at the 52-foot depth is negligible.

The practical significance of the zero within months variation at the 30 and 70-foot depths and the negligible variation at the 52-foot depth is that the within months variations here are very small relative to the between rods variation (table 2). Therefore these variations are masked by the variation which takes place between rods. Consequently, it is apparent that more frequent measures in time would have to be made before these within month effects could be isolated and evaluated.

The results in the analysis of trend indicate that the changes in sand level are not random but cyclic in nature (table 2). The test used,

[2]The reference rods at this station were lost following a series of high waves in November, 1953. Acoustic soundings at this station indicated an erosion of sand of 2 feet.

however, is by nature not efficient for indicating seasonal trends in the presence of pronounced variations of shorter period. Thus, for example, at the 30-foot station, there is a trend in sand level with season which could be observed on the graph (figure 1), and which is further indicated by the average sand level height for each season (table 3), but which did not show up in the trend test. For this station the sand level tends to be high (deposition) during the summer and low (erosion) during the winter. The deeper stations did not show measurable seasonal trends, of cyclic nature but they did show shorter period cycles.

Correlation of Sand Level Changes between Stations. It was desirable to know whether or not the changes in sand level between the various stations were related. The estimates of correlation previously described indicated (as in the example shown) that the correlation between the 70-foot station and the 30-foot station was inverse (or negative). That is, when sand was eroded at the 70-foot station, there was deposition at the 30-foot station; and, conversely, when there was deposition at the 70-foot station, there was erosion of sand at the 30-foot station. Between the 70-foot and the 52-foot depths there was a positive correlation; when sand was deposited at the 70-foot station, it was also built up at the 52-foot station. Between the 30-foot station and the 52-foot station, however, there was no strong correlation.

The fairly strong negative correlation between changes in sand level at the 30 and 70-foot depths, the somewhat weaker positive correlation between the 52 and 70-foot depths, and the very weak correlation between the 30 and 52-foot depths suggests that changes at the 52-foot station are transitional or borderline. That is to say, erosion at the 30-foot level is frequently accompanied by deposition at the 70-foot level and may or may not be accompanied by deposition at the 52-foot level.

No attempt was made to correlate the changes at the 18-foot depth with the other stations. The reason for this was that the number of observations from the 18-foot station was very limited in extent and did not allow for estimates of correlation.

Sand Size. A sample of the bottom sediment was obtained by the swimmers during each of the surveys on D range. The size distribution characteristics of the samples are listed in Appendix IIIA-B. The method of analysis and a detailed description of the character and composition of sediments obtained in this area during 1949 and 1950 are given in Inman, 1953.

The median diameter of the sediments averaged about 150 microns at the 18-foot depth, 120 microns at the 30-foot depth and 110 microns at the 52 and 70-foot depths. The size of the sand remained relatively constant at the three deeper stations during the first 18 months of the study, but fluctuated erratically during several periods beginning in

Table 3. Seasonal average of sand levels, D range.

Season	Station at 30-ft. depth		Station at 52-ft. depth		Station at 70-ft. depth	
	Number of Surveys	Average Sand Level (ft)	Number of Surveys	Average Sand Level (ft)	Number of Surveys	Average Sand Level (ft)
Winter 1952-1953	1	+.03	5	-.01	3	+.02
Summer 1953	9	+.02	5	-.01	2	0
Winter 1953-1954	12	0	14	+.04	10	+.01
Summer 1954	5	+.12	6	+.08	5	0
Winter 1954-1955	13	-.05	11	+.05	12	+.02
Summer 1955	2	+.01	5	+.07	5	+.06
Winter 1955-1956	3	-.04	3	+.08	3	+.03
Average Summer	16	+.05	16	+.05	12	+.02
Average Winter	29	-.02	33	+.04	28	+.02
Mean Sand Level	45	0	49	+.04	40	+.02

December, 1954. No simple correlation between sand size and sand level change was observed, although the large changes in sand size were commonly associated with greater than usual changes in sand level.

SAND LEVEL CHANGES ON B RANGE

The sand level changes associated with the tidal cycle were measured by placing a series of reference rods at intervals of 20 feet along a profile extending through the surf zone and onto the beach foreshore on B range. Three series of surveys, each consisting of a high-tide profile followed by a profile during the succeeding low tide, were made. The differences in sand level between tidal cycles within a day and the differences in level between days are shown in figure 8. The surveys were made during spring tides with ranges from high to low water level of about 6 feet. The significant breaker heights ranged from about 2 to 4 feet (Appendix IV).

Although these surveys represent a very small sample, there is some indication of systematic changes in profile related to the position of high and low tide breaker zones. All three survey series show deposition on the upper portions of the beach foreshore between the high and the following low tide, and the two October surveys show a tendency for a bar and trough to form in the zone of low tide breakers. It is interesting to note that for these surveys the magnitude of sand level change between high and low tide was one-half as great as the magnitude of the total change for the two-week period between the first and second surveys.

Several of the seaward reference rods have remained in position for over two years. They are completely covered with sand each winter and bared again for short periods during summer when the sand migrates landward and is deposited on the beach foreshore. The summer fluctuations in sand level at the deepest station on B range, approximately 6 feet below MLLW, are graphed together with the changes on D range in figure 7.

Experience indicated that accurate measurement of reference rods was limited to surf conditions where the breaking waves did not exceed 5 feet in height. For such conditions it is estimated that the accuracy of measurement is of the order of ±0.05 foot. This is the same order of magnitude as the standard error in level obtained for the deeper water stations on D range, which was based on the measurement of six rods. The similarity in the magnitude of error results from differences in the nature of the problem on the two ranges. The problem on D range was to determine the mean level for an area from six measurements of an irregular bottom, while on B range the interest was on the sand level at the position of a single reference rod.

ACOUSTIC SOUNDING

The evaluation of the accuracy of the acoustic method of hydrographic survey was one of the objectives of the investigation. It has

been demonstrated that depths obtained by echo sounding are reproducible when repeated soundings are made during the same day (Shepard and Inman, 1951; Saville and Caldwell, 1953). This finding resulted in optimism in the estimation of the reliability of hydrographic surveys. In the previous studies comparison of sounding data over periods longer than a day could not be used to evaluate the reliability of the method because of the uncertainty in the amount of true change in sand level.

The total or compounded errors involved in reducing an uncorrected sounding to some absolute datum can be roughly divided into errors incurred in (1) measuring the instantaneous depth of water, (2) positioning, and (3) resolving the corrected depth to datum. For convenience these errors are listed in outline form in table 4. Inspection of this table shows that many of the factors leading to error tend to remain constant during the course of a single survey; wave and tide conditions are similar, the same operator bias is in effect, etc. On the other hand, day to day observations may involve differences in personnel, state of the sea, and instrument characteristics. Near the surf zone, surf beat (Munk, 1949) may produce fluctuations of sea level, of the order of one-tenth of the wave height, which are not coherent over the area of the survey. Also, so subtle a factor as moisture of the fathogram paper can change the degree of halation and clarity of the trace and hence the accuracy of the fathogram.

A series of echo sounding surveys were conducted concurrent with the reference rod measurements on station D. In general, five echo sounding runs were made on each station and the data from these reduced in the manner described previously. The mean values of each day's observations from the station at the 30-foot depth are shown together with the results from the reference rod measurements in figure 9. For purposes of illustration, a value of the reference level was assumed such that the depths from the acoustic sounding and from the reference rod measurements coincided on the first day of the comparison, 23 July 1953.

It is apparent from figure 9 that the daily ranges in acoustic soundings and reference rods were relatively small, the scatter of acoustic sounding being in general three to four times greater than that of the reference rods. However, the day to day or the survey to survey variation in acoustic soundings is much greater than that for the reference rods. If it is assumed that the reference rods give a reliable measure of the real changes in sand level, then it can be concluded that the accuracy of the acoustic method of survey under these experimental conditions is of the order of $\pm \frac{1}{2}$ foot. Similar accuracy was obtained for the stations of 52 and 70-foot depths.

Table 4. Outline of factors leading to error in the reduction of acoustic sounding data to basic datum.

1. Errors in measuring depth.
    a. Calibration of instrument
    b. Stability of instrument
    c. Background noise: wave conditions, clarity of record
    d. Operator variables: control of gain, frequency and power of sounder, systematic errors in reading record.
2. Errors in positioning.
    a. Operator errors
    b. State of sea; pitch and roll of vessel
    c. Slope of bottom
3. Errors in correcting measured depth to absolute datum (from tide gage).
    a. Background noise of tide record: waves, clarity of record
    b. Anomalies in recorded level: water level change in tide well caused by water velocity against orifice, etc.
    c. Incoherent changes in sea level: surf beat, surges, etc.

SUMMARY AND CONCLUSIONS

1. A technique was developed for establishing a reference level on the bottom from which small net changes in sand level of the sand could be measured. The reference level was established by forcing six rods into the sandy bottom and then the changes in sand level were based upon the differences in lengths of rod exposed from survey to survey. Measurements, which were made by swimmers equipped with self-contained underwater breathing apparatus, resulted in a standard error of about ±0.05 foot per survey.

2. The total range in sand level probably exceeds 2 feet at the 18-foot deep station and changes in excess of 0.6 foot were measured before the reference rods were lost. The range decreases with increasing depth; ranges in net sand level of 0.29, 0.16, and 0.15 foot were measured at stations where the depth of water was 30, 52, and 70 feet respectively.

3. Statistical evaluation of sand level changes demonstrates that there is a general decrease in the magnitude of the variability in sand level from station to station seaward. This implies that the local bottom irregularities at each station (as measured by the variability between the six reference rods), as well as the frequency and magnitude of net change in sand level at the station, decrease with increasing depth.

4. A seasonal trend in sand level change was observed at the station at 30-foot depth where the sand level was high in the summer and low in the winter. At the deeper stations, 52 and 70 feet, seasonal trends were masked by fluctuations of shorter period.

5. There is a significant correlation between changes in sand level from stations at various depths. Changes at one depth are accompanied by changes at another; and, their relation may be direct or inverse, depending upon the stations involved.

6. A comparison of sand level changes between tidal cycles, obtained from reference rods placed along a traverse through the surf zone, showed systematic changes to occur in the beach profile near the position of high and low tide breaker zones. For the six surveys, the magnitude of sand level change between high and low tide was approximately one-half as great as the total change during the two week period.

7. Comparison of acoustic soundings with reference rod measurements indicated that the day to day accuracy of the acoustic sounding method was of the order of $\pm \frac{1}{2}$ foot, although soundings during one day of survey operation may be relatively precise in terms of reproducibility.

ACKNOWLEDGEMENTS

The field phase of the investigation was largely the work of Earl Murray, who was in charge of the boat and made all of the reference rod measurements. Assistance in the field work and analysis of data was given by Jean Short, and assistance with illustrations was given by Robert C. Winsett. Valuable suggestions and guidance during the course of the study were contributed by Robert S. Arthur, Jeffrey D. Frautschy, George F. McEwen, Robert L. Miller and J. S. Kahn.

LITERATURE CITED

1. Dixon, W. J. and F. J. Massey. 1951. Introduction to statistical analysis. McGraw-Hill, New York, 370 pp.

2. Eisenhart, C., M. W. Hastay, and W. A. Wallis. 1947. Techniques of statistical analysis. McGraw-Hill, New York, 473 pp.

3. Fisher, R. A. 1949. Statistical methods for research workers. Oliver and Boyd, London.

4. Hald, A. 1952. Statistical theory with engineering applications. John Wiley and Sons, Inc., New York, p. 749.

5. Inman, D. L. 1952. "Measures for describing the size distribution of sediments". Jour. Sed. Petrol., Vol. 22, pp. 125-145.

6. Inman, D. L. 1953. "Areal and seasonal variations in beach and nearshore sediments at La Jolla, California." Beach Erosion Board, Corps of Engineers, Tech. Memo. No. 39, 134 pp.

7. Munk, W. H. 1949. "Surf beats." Trans. Amer. Geoph. Union, Vol. 30, pp. 849-854.

8. Olson, J. S. and P. E. Potter. 1954. "Variance components of cross-bedding in some basal Pennsylvanian sandstone of the eastern interiör basin; statistical applications." Jour. Geology, Vol. 62, pp. 26-48.

9. Saville, Thorndike, Jr. and J. M. Caldwell. 1953. "Accuracy of hydrographic surveying in and near the surf zone." Beach Erosion Board, Corps of Engineers, Tech. Memo. No. 32, 28 pp.

10. Shepard, F. P. and D. L. Inman. 1951. "Sand movement on the shallow inter-canyon shelf at La Jolla, California." Beach Erosion Board, Corps of Engineers, Tech. Memo. No. 26, 29 pp.

11. Snedecor, G. W. 1946. Statistical methods. The Iowa State College Press, Ames, Iowa, 485 pp.

18

Reprinted from *Mar. Geol.* **13**:31–55 (1972)

FIELD OBSERVATIONS OF SAND TRANSPORT BY SHOALING WAVES

DAVID O. COOK* and DONN S. GORSLINE

Department of Geological Sciences, University of Southern California, Los Angeles, Calif. (U.S.A.)

ABSTRACT

Cook, D. O. and Gorsline, D. S., 1972. Field observations of sand transport by shoaling waves. *Mar. Geol.*, 13: 31–55.

A study of wave-generated currents and associated sand transport in the offshore zone of southern California has modified several previously-recognized theories concerning sedimentary processes in this region.

A set of passing waves generates oscillating currents at the sea floor which have a spectrum of velocities whose average strengths over broad shelves are predicted accurately by classical wave theory. Long period swells and offshore breezes cause a net transport of bottom water towards the beach, while short period waves and onshore winds are associated with neutral or seaward flow.

Ripple marks on the shallow sea floor, whose dimensions are a function of sediment size and current velocity, allow all available sand sizes to be moved with equal ease. A ripple remains in the same position even though sediment and water may be differentially transported.

Most bottom sand transport occurs in the lowest few centimeters of the water column at a rate which varies with the square of current strength. Fine sand-grain size populations migrate in the direction of the greatest current velocities as a unit without being sorted.

During winter in southern California, rip currents transport beach sand to the offshore where it is confined by predominant seaward oscillations caused by steep waves and strong winds. Rips are not active in calm summer months, and long period swells produce shoreward surge asymmetry which replenishes the beach with sediment.

INTRODUCTION

General statement

Literature concerning the transportation of sand by shoaling waves is replete with exhaustive laboratory investigations, but complementary field observation has only begun in recent years. For this reason, it was proposed to conduct a comprehensive in situ study of the hydrodynamics of the nearshore environment and attendant sediment motion in

*Present address: Raytheon Company, Environmental Research Laboratory, P.O. Box 360, Portsmouth, Rhode Island 02871.

the area off southern California. SCUBA methods were employed to obtain direct observations and to position test equipment.

Most sediment from the continent must pass through the zone of wave action before reaching depositional sites in the marine environment. The sorting action of the waves separates sand from the silt and clay and moves the sand alongshore in the inner shelf to depths dependent on the characteristics of the surface waves impinging on the coast. Thus the sand is not restricted to the surf zone alone, but moves in a complex path passing outside the breaker line in rip currents and general diffusion, and then moving inshore again as a result of bottom drift.

The sand is concentrated at convergences in the drift system and, depending primarily on shelf width, either forms seaward-extending shoals, which eventually reach the shelf break, or moves down submarine canyons to deeper water.

Silt and clay are separated in the sorting process and move out of the coastal drift system in suspension.

Previous work

The conventional model for water circulation in the littoral zone was established by Shepard and Inman (1950). The principal currents are: (1) coastal currents, which flow parallel to shore outside the breakers at low velocities; (2) wave drift, the mass transport of water towards shore associated with the oscillatory motion caused by passing waves; (3) longshore currents, which move parallel to the coast between the breakers and shore; and (4) rip currents by which the water transported towards the beach by wave drift is returned to the offshore.

According to theory, energy is continually lost from a passing wave because of bottom friction, internal turbulence, and permeability, causing onshore velocities to predominate over the offshore component. A net onshore flow of water, wave drift, occurs near the sea floor and increases exponentially towards shore (see Bagnold, 1947; Longuet-Higgins, 1953; and Russell and Osario, 1955). Longinov (in Zenkovich, 1967) found that for short-period waves in the Black Sea, shoreward mass transport of water occurred at depth and increased to a maximum just beyond the breakers. Inman and Nasu (1956), working with moderate-to long-period waves in southern California, reported that drift usually was towards shore, but occasionally flowed seaward.

Attenuation of wave-induced motion with depth may be predicted from classical wave theory. Attempts to relate theory to observed currents (Inman and Nasu, 1956) and pressures (Draper, 1957; Longinov, in Zenkovich, 1967) at the ocean bottom, however, have produced uncertain and often conflicting results.

Most effort in studies of nearshore sediment motion has been devoted to the net migration of sand normal to the coast. Three models for net sand movement have been proposed:

(1) Grant (1943) assumed that because of the theoretical inequality of onshore and offshore surge velocities, all bottom sand should be transported toward the beach. A

balance is maintained by seaward-flowing rip currents laden with sediment. Sato et al. (1963) lent support to this model by observing that near the coast of Japan large amounts of irradiated sand were moved in the direction of wave advance.

(2) The null point theory presumes that both onshore wave drift and an offshore gravitational component produced by bottom slope affect sediment grains on the shallow ocean floor. For each grain of a certain size, shape, and density there exists a position of equilibrium between seaward and shoreward motion. Using a wave tank with a plane bottom, Ippen and Eagleson (1955) confirmed the existence of null points for individual grains. Miller and Zeigler (1958) incorporated the null point concept in a model which they proposed for sediment transport in the shallow neritic environment. Vernon (1966) made field measurements on the transportation of dyed sand on the sea floor off southern California which supported this theory.

(3) Several recent investigators concluded that most transportation takes place by diffusion of suspended sediment. Nagata (1961) performed an experiment on the coast of Japan and observed that suspended sand seemed to be moved in a direction opposite to that of wave drift. Murray (1967) decided that the motion of fine dyed sand off a low energy beach conformed to the predictions of diffusion theory. Passega et al. (1967) inferred from subtleties in the texture of sediment off the Adriatic coast that it had been deposited from a suspension.

METHODS

Individual observations were accomplished over periods of one-half to six hours at six diverse bottom sites (Fig.1) using SCUBA equipment. The study entailed over 80 days in the field and 350 dives.

Limitations

The coast of southern California may be classified as one of moderate energy level. Swells causing the water and sediment motion under consideration had periods of 6–12 sec and heights between 0.2 and 1.5 m. Observations were performed at depths of 3–30 m, or in the range of *relative wave height* (wave height/water depth) between 0.05 and 0.08. Average maximum current velocities ranged from 5 to 35 cm/sec. Bottom sediment at the test sites consisted entirely of sands with mean diameters of 0.10–1.00 mm.

Wave measurement

The first event during a day of underwater observation was an evaluation of the average height and period of waves reaching the coast. This was accomplished visually with the use of a wave staff. Measurements were performed in water depths exceeding 25 m so that swells could be described by their deep water characteristics.

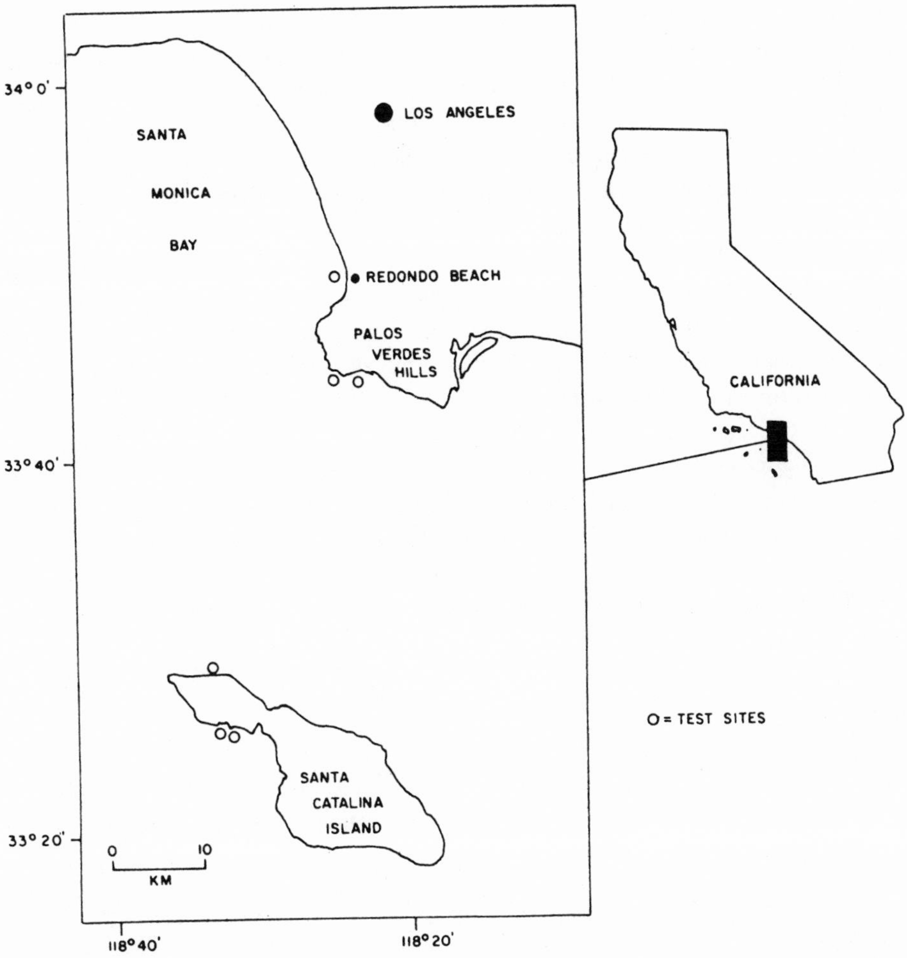

Fig.1. Location map.

Current determination

Oscillating currents at the sea floor were recorded with a mechanical current meter (Summers et al., 1971). Two paddles, which move in response to currents 15 cm above the bottom, scribe on a drum which completes one rotation in 14 min. A stationary pendulum on the opposite side of the drum defines the null point. The meter is oriented normal to the oscillations, and trials were made in locations where tidal currents are minimal. The velocity range is from 0 to 60 cm/sec. Calibration was made by field tests and comparison with a strain gauge current meter of the U.S. Navy Underwater Warfare Center in San Diego, California.

Records from the instrument permit both an evaluation of average current strength

and a comparison of the relative magnitude and duration of onshore and offshore pulses for each trial. The degree of surge asymmetry is indicated by onshore–offshore ratios of velocity and duration of flow. In computing these ratios the larger of the two quantities is placed in the numerator so that the value is always greater than 1.0. A plus sign denotes asymmetry in the shoreward direction, and a negative sign a predominance of seaward pulses. By combining ratios of current strength and duration, the direction of wave drift may be appraised.

Sediment traps

Three varieties of traps (Fig.2) to intercept moving sediment were used in the study. They were designed to sample the total load, the distribution of suspended sediment near

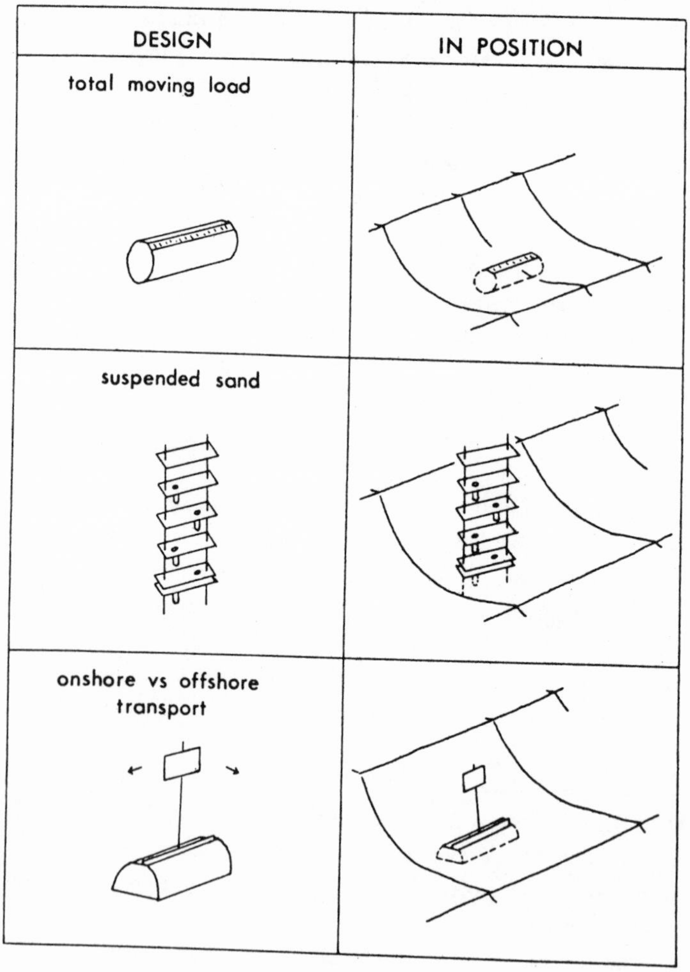

Fig.2. The sediment traps used in the present study.

the bottom, and the volume of sand moving onshore compared to that moving offshore. Thus, the competency and capacity of the oscillating currents could be determined.

Total load

The device for sampling all material being transported is extremely simple in design, consisting of a number 107A tin can with a slit 1.5 cm wide. This slit size is consistent with the dimensions of bottom ripple marks at the study sites. In practice, the slit is covered with masking tape, and the can is carefully buried in a ripple trough so that only the tape is exposed. After an interval during which equilibrium is restored, the tape is removed and moving grains settle into the can. Sand is allowed to accumulate for 1–4 h before the trap is recovered.

The advantage of this design is that nothing protrudes above the sea floor to obstruct current flow. Several funnel-like traps were tested, but these caused scour and anomalous sand motion. However, a problem inherent in the sub-surface sampler is that rolling particles may be intercepted while suspended grains by-pass the slit. This difficulty is minimized by the manner in which sand is moved. Low-velocity oscillations cause transportation by traction and most sand reaching the trap is caught. Stronger currents sweep sediment past the slit until the wave passes, and then all particles settle into the can.

In the laboratory the contents of the can were dried and weighed. Dividing this weight by the time of exposure yields a sediment transport rate which is expressed in terms of grams per hour. The samples were then split and analyzed for hydraulic equivalent grain size distribution using an automatically-recording settling tube (Cook, 1969; Felix, 1969).

Suspended sediment

A trap consisting of vials arranged in tiers was designed to sample sand suspended near the bottom. The device is buried in a ripple trough so that the lip of the lowest vial is even with the sediment–water interface. After burial, the containers are uncapped and material settles into them as the surges causing suspension wane. Suspended sand is sampled at the following intervals: 0–1.2, 1.3–7.6, 7.7–15.2, 15.3–22.8, and 22.9–30.5 cm above the sea floor. These intervals were selected on the basis of preliminary experiments which provided estimates of suspended sand distribution. One to four hours are allowed to elapse before the trap is removed.

A funnel-type of sampler was used by Nagata (1961) in his work on the suspension of sediment by waves. When a similar model was tested in this study, it was observed that the return surge washed out some sand previously carried into the funnel. Although the vial design does not sample all sediment passing the trap, it is designed to collect representative aliquots over a continuous range.

Sand contained in the vials was weighed and analyzed for grain size distribution. The amount of sediment suspended in a height interval has been expressed in terms of its rate of accumulation.

Onshore and offshore migration

The buried can principle was employed again in the design of a trap to intercept sand undergoing shoreward and seaward transport. In this case, however, a longitudinal baffle separates the container into two halves. Hinged to the baffle is a gate which is moved back and forth by the force of the surge on a sail located several centimeters above the bottom. The sampler is buried in the sea floor with its axis normal to oscillatory surges. The gate keeps accumulating sand segregated in "onshore" and "offshore" compartments. In an effort to minimize the possibility of error in correlations of sand migration with water motion, sediment was collected only while the current meter was recording. This design is subject to the same advantages and limitations as the sampler for the total moving load and was selected only after extensive experimentation.

The sediment was later weighed and sized. Preferential movement is expressed in terms of onshore–offshore transport ratios.

Ripple mark sampling

Oscillation ripple marks cover the shallow sea floor and have been examined in the field (e.g., Inman, 1957; Risk, 1965) and laboratory (e.g., Bagnold, 1946; Scott, 1954; Manohar, 1955). In order to ascertain their bearing on sediment transport, these features were scrutinized during each day of observation. A ruler was used to determine the wave length. Previous research indicated that ripple marks segregate sediment by size with extremes located at the crest and trough. Therefore, channel samples were taken from several crests and troughs with small vials and later subjected to size analysis.

Ripple marks were also photographed using still and motion picture cameras. Some time lapse studies were also made.

TEST SITES

Three locations on the southern California coastline (Fig.1) served as sites for underwater research. Wave-generated currents and sediment motion were studied off Palos Verdes and Catalina Island, regions which are characterized by steep narrow shelves. Redondo Beach, which has a low gradient and smooth topography, was selected for additional current measurements.

RESULTS OF HYDRODYNAMIC STUDIES

Distribution of current velocities

Although wave motion is conventionally envisioned as a regular phenomenon, waves approaching a beach vary in height and possess a range of periods. Water motion at depth is irregular, and currents have a spectrum of velocities.

Typical distributions of surge velocity are presented on Fig.3. The graphs show the percent of onshore and offshore currents in increments of 10 cm/sec. The currents

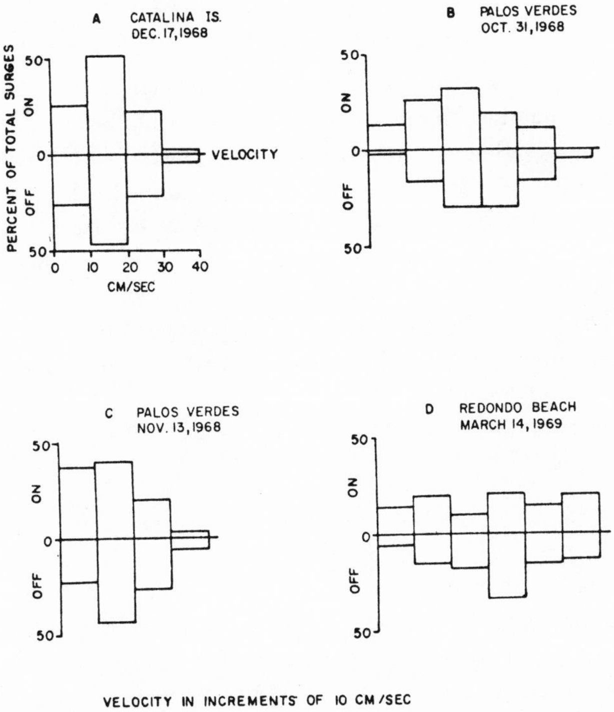

Fig.3. Typical distributions of current velocity.

measured in the first example are weak and demonstrate no directional asymmetry (Fig.3A). The quasi-normal distribution of surge strength, which is marked here, is a feature of most recordings. The second and third examples reveal a disparity between onshore and offshore flow (Fig.3B, C). The directional inequality of surges will be discussed at length. It is evident that with stronger currents the range of velocities is broader. The complex pattern displayed by the last example (Fig.3D) was caused by different waves arriving from southern California's two generating areas. Usually one set of swells is dominant, but occasionally such interaction occurs in exposed locations.

Comparison of observed and theoretical velocities

Equations for predicting the velocity of oscillating currents at the ocean floor have been derived from classical wave theory. Deep water swells may be approximated by a trochoidal profile and the following expression applies (Inman, in Shepard, 1963):

$$u_m = \frac{\pi H}{T \sinh (2\pi h/L)} \qquad (1)$$

where u_m = maximum horizontal velocity; H = wave height; T = wave period; h = water depth; L = wave length.

As waves shoal, several factors cause their properties to change. Simply stated, the period remains constant, height increases while wavelength and velocity diminish. When the *relative wave height* exceeds 0.04 to 0.06, swells are no longer trochoidal but consist of isolated crests separated by flat troughs. Water motion may be better described by solitary wave theory, and a new equation obtains for bottom velocities (Inman, in Shepard, 1963):

$$u_m = \frac{H\sqrt{gh}}{2h} \qquad (2)$$

where terms are as in eq.1, and g = acceleration of gravity.

A comparison of velocities calculated from wave staff measurements using both trochoidal and solitary wave theory was made with average current strengths recorded on the current meter. Data for Catalina and Palos Verdes are characterized by a high degree of scatter and are not presented. The scatter is probably a result of wave refraction over the irregular bathymetry off Palos Verdes and Catalina. Wave staff readings were performed offshore from underwater test sites, and swell dimensions at the adjacent areas can differ appreciably.

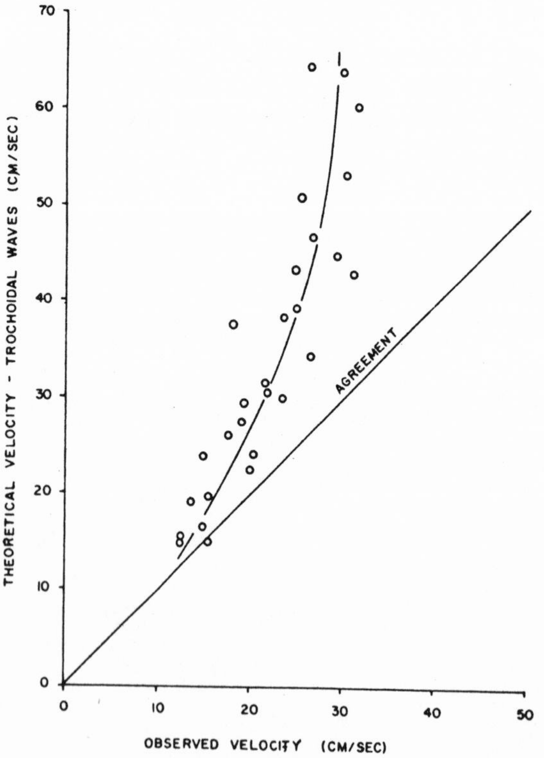

Fig.4. Comparison of surge velocities calculated using trochoidal wave theory with values recorded over a gentle continental shelf (Redondo Beach).

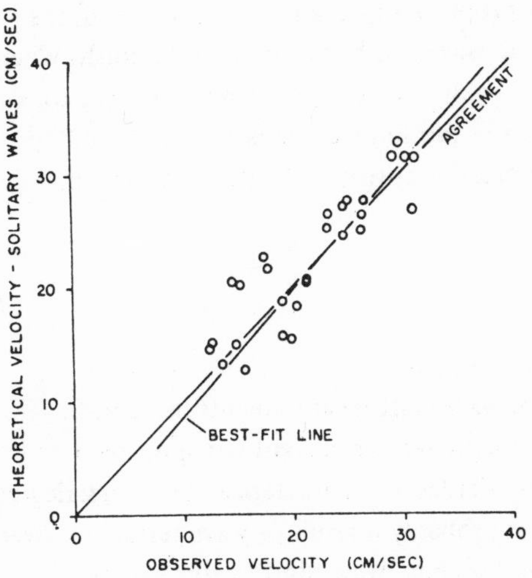

Fig.5. Comparison of surge velocities calculated using solitary wave theory with values recorded over a gentle continental shelf (Redondo Beach).

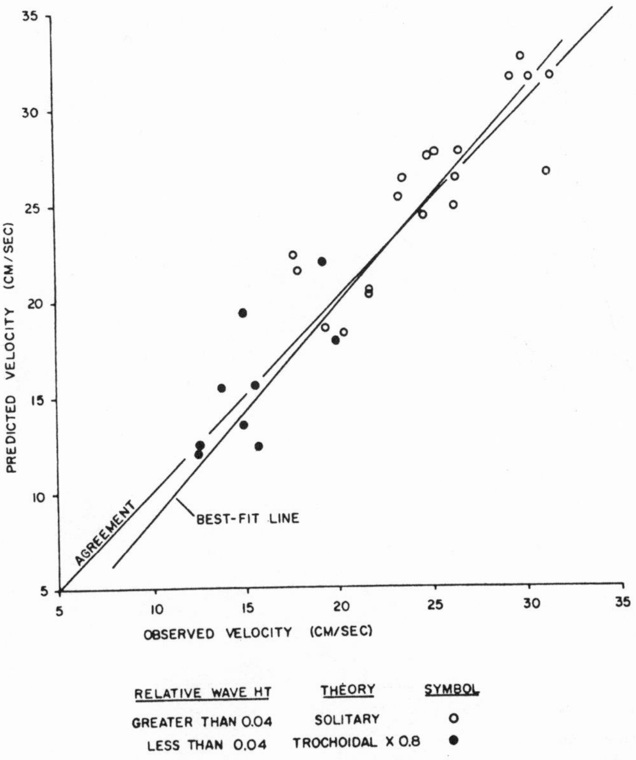

Fig.6. Use of two theories to predict surge velocities over gentle continental shelves. Solitary theory is used with relative wave heights greater than 0.04 and trochoidal theory in deeper water.

Results from Redondo Beach are considerably more revealing, and these are demonstrated on Fig.4 and 5. Theoretical velocities for trochoidal waves are consistently a few cm/sec larger than recorded currents in the 10 to 25 cm/sec range. This association is not true at higher surge strengths, probably because swells are transformed to the solitary form in shallow water.

Predictions of solitary wave theory agree well with observations of strong currents. They show less correlation with lower readings taken in deeper water where waves are still trochoidal.

These results indicate that the change of swell form takes place at a relative wave height of 0.04, and this value is used to integrate the two theories (Fig.6). Observations taken where relative wave heights were lower than 0.04 are plotted against trochoidal theory, and these predicted currents have been reduced by an empirically-derived factor of 0.80. Solitary theory is compared with velocities recorded in shallower water. In this form the accuracy of calculated velocities is significant in comparison to the studies of Inman and Nasu (1956), Draper (1957) and Longinov (in Zenkovich, 1967).

Fig.6 lends confidence to predictions of bottom currents generated by waves with moderate heights and periods. Computed velocities are within ± 5 cm/sec of the correct value. However, caution must be exercised when applying the equations in areas of irregular bathymetry where bedrock crops out on the sea floor.

Surge asymmetry

Net water transport superimposed on wave-induced oscillations is expressed on a current meter record as asymmetry of onshore and offshore pulses. Either an inequality of average velocity or duration of flow may cause drift along the sea floor. As the bottom shoals, theoretical considerations suggest that onshore surges should become stronger, while offshore flow occupies a longer time interval. Results from the study did not completely support this concept but did reveal some interesting relationships.

The distribution of onshore–offshore ratios for average current velocity and period of flow show both shoreward and seaward asymmetry. They are characterized by appreciable variability, and the scatter of values overshadows any correlation between directional surge disparities and depth. When onshore–offshore ratios are combined to create a factor representing net water drift, this parameter is also variable both in direction and magnitude at all depths.

The surge asymmetries observed in this investigation are thus by no means a simple function of wave shoaling. The only other extensive field examination of oscillating currents, performed by Longinov, yielded results which coincided with theoretical expectations. However, Longinov's work was undertaken in an area of low wave energy, which was not subjected to the rigors of an exposed coast.

Non-parametric statistics were used to inspect the data, which consisted of 45 individual observations, for possible relationships between current asymmetry and wave period, the wind, and tide. A Kruskal-Wallis one-way test for variance by ranks (Siegel,

TABLE I

Results of Kruskal-Wallis tests concerning surge asymmetry at Redondo Beach, California

Surge asymmetry factor	Confidence Levels of Correlations	
	wind direction (%)*[1]	wave period (%)*[2]
Average onshore and offshore velocities	82.0	88.0
Average onshore and offshore duration of flow	99.5	95.0
Bottom drift	99.5	97.5

*[1] Wind direction relating inversely to surge inequalities.
*[2] Long period swells (10–11 sec) causing more shoreward asymmetry than short period waves (7–8 sec).
Correlations of tidal flux with surge inequalities were lower than the 80% confidence level.

1956) was employed for the purpose. The correlations having a confidence level of 80% or more are listed on Table I. The absence of relationships in data from Catalina Island and Palos Verdes may be explained by the fact that anomalous currents are often encountered near irregular shorelines. The following inferences are drawn from current measurements at Redondo Beach:

(1) Tidal flux does not have a significant bearing on surge asymmetry. The range between high and low water near Los Angeles is 2 m, and different results may occur in regions with greater tidal flux.

(2) The wind affects ratios for duration of current flow and net bottom drift. An onshore breeze causes shoreward transport of surface water and compensatory offshore flow at depth. As a result, seaward surges occupy the largest amount of time. An offshore wind, on the other hand, generates sub-surface drift towards the beach.

(3) Swell characteristics also play a role in determining water drift. With long period waves, shoreward pulses prevail temporarily, causing net motion in the same direction. This conclusion has interesting implications concerning the annual cycle of erosion and deposition on southern California beaches and will be discussed later.

The asymmetry of surge velocities showed similar possible correlations with wind direction and wave period but at a lower level of confidence.

The term "wave drift" implies a genetic relationship between net transport superimposed on oscillating surges and shoaling swells. Because other factors influence these currents, the name "bottom drift" may be more appropriate.

RESULTS OF SEDIMENT TRANSPORT STUDIES

Sand motion and ripple marks

Dimensions of ripple marks

Investigators of oscillation ripple marks in the laboratory and field have generally found that their dimensions depend primarily on sediment size and current velocity. This study substantiates these conclusions. A direct relationship between ripple wavelength and grain diameter is demonstrated on Fig.7. The largest forms, megaripples, occur in coarse sand and have wavelengths of a meter or more. In fine sand, ripple crests are spaced several centimeters apart. The subordinate effects of current velocity are illustrated by data for a limited grain size plotted on Fig.8. Ripple dimensions and current intensity are shown to be related inversely, and this causes an increase in sea floor microtopography in the offshore direction.

Circulation and grain paths over a rippled bed

Close-up cinematography was used to examine water circulation over ripple marks and the resulting motion of sand grains. Distinct sediment transport patterns were developed for conditions of weak and strong surge, and these obtain with ripples of all sizes. Typical grain paths observed under low-velocity currents are illustrated on Fig.9. As the oncoming surge flows over a ripple crest, particles on the stoss side are swept over the top and cascade down the lee face. A large eddy is established in ripple troughs. Where the eddy impinges against the sediment, surface grains are carried up the ripple lee in a direction opposite to general current flow. The subsequent departure of the eddy is marked by a slight inflection of the ripple surface below the crest. Occasionally, eddies collide with the bottom and cause an apparent explosion of trough grains in all directions. Small horizontal eddies resembling desert dust devils suspend wisps of sediment near the ripple crest.

Observed transport patterns caused by high-velocity surges are shown on Fig.10. In this case grains moving over the crest are launched into suspension. The vertical eddies are well developed and dominate circulation in ripple troughs. Suspended particles both from crests and troughs are whirled about and eventually projected high enough to be carried over the next crest. As the surge wanes, eddies increase in diameter and decay, releasing a cloud of grains which are carried over the sea floor. These are deposited during succeeding pulses.

Allen (1968) analyzed circulation and sediment transport for ripples in a flume. The patterns he observed with unidirectional flow are similar to those recorded on the sea floor where strong oscillating currents were present. Because wave-generated currents change direction every few seconds, insufficient time is available for the development of turbulence with low velocities.

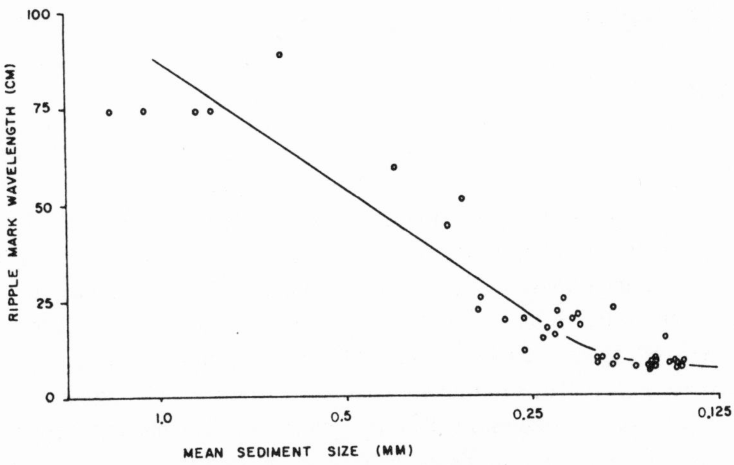

Fig. 7. Ripple wavelength plotted against mean diameter of bottom sediment.

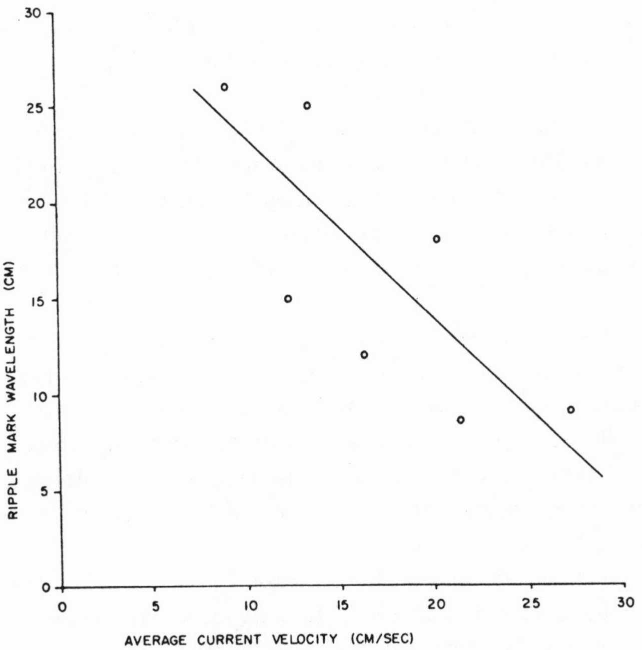

Fig. 8. Effects of surge velocity on ripple wavelength. The size of sand is approximately constant with mean diameters lying between 0.177 and 0.250 mm.

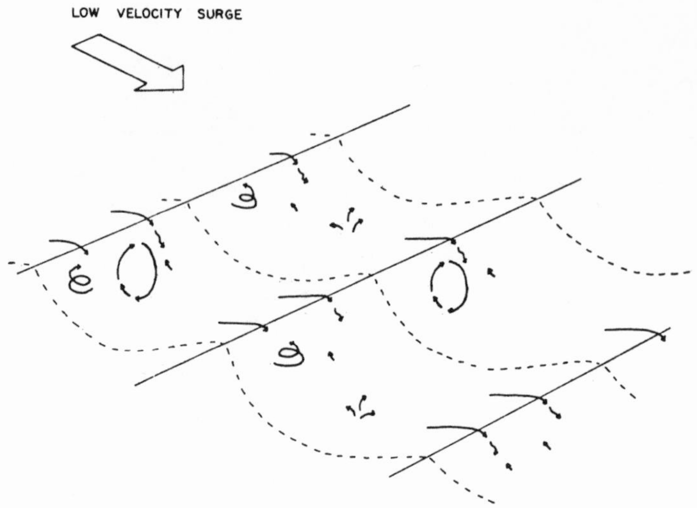

Fig.9. Grain paths over ripple marks during a low-velocity surge.

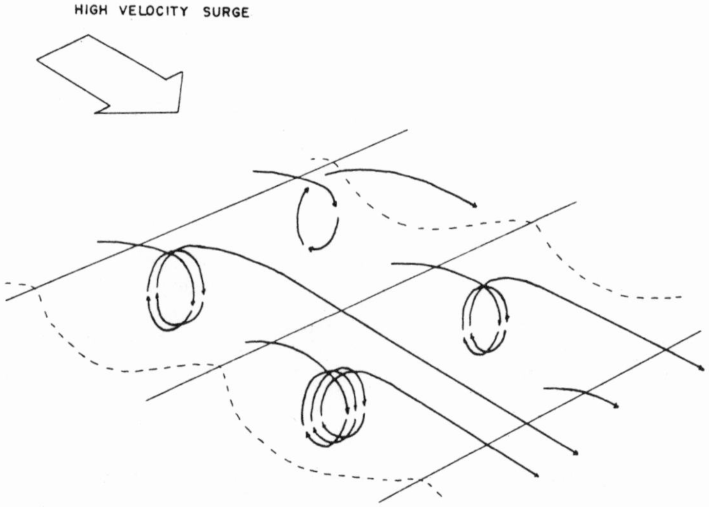

Fig.10. Grain paths over ripple marks during a high-velocity surge.

Sediment distribution over ripple marks

Differences in the intensity of current flow over the ripple form cause a segregation of grains by hydraulic size. Extremes of mean diameter exist at the crest and trough. This range in grain size is illustrated on Fig.11 where it is graphed against average diameter. With fine sand in which small ripples develop, crest sediment is slightly coarser than that in the trough. This may be explained by winnowing of small particles from the exposed ripple peak. On megaripples consisting of coarse sand, however, the trough grains are the

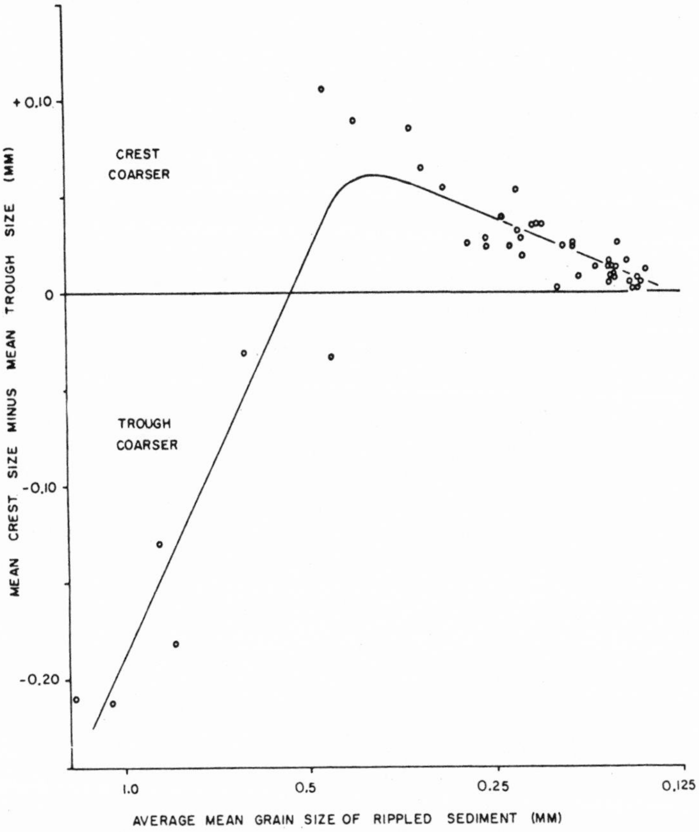

Fig.11. Contrast in the mean size of grains concentrated at ripple crests and troughs.

largest. Inman (1957) also observed this tendency and attributed it to insufficiency of available medium-size sand to completely cover pre-existing gravel pavements.

Ripple mark migration

Changes in the position of oscillation ripple marks may be considered from two temporal standpoints. During a period of seconds, a crest can migrate as much as 2 cm in the direction of surge flow. This is a result of the aforementioned transfer of grains from the stoss to the lee and imparts a degree of asymmetry to the ripple form. When the current changes direction, the asymmetry is reversed as the crest returns to its former position.

One might assume that a directional inequality of surges would cause gradual and permanent migration of ripple marks in the direction of net water motion. Onshore movement of ripples has been observed in a wave tank by Bagnold (1947), and Scott (1954), and in Lake Huron by Risk (1965). Inman and Bagnold (1963) proposed a formula for sub-littoral sediment transport based on ripples shifting under the influence of supposedly shoreward wave drift.

The migration of ripple marks on the shallow sea floor was examined both by Vernon (1966) and in the present study using time-lapse photography. The position of ripple crests was monitored over periods ranging from 1 to 48 hours. In all cases it was found that aside from short term "chattering" of crests, no significant movement of the ripples occurred. This stability was maintained in spite of surge inequalities and net transportation of bottom sediment. Thus, it appears that the concept of shoreward-marching ripple marks is not necessarily valid in the shallow marine environment.

The migratory nature of ripples observed in the laboratory and the lacustrine environment may be explained by noting that shoreward surge velocities generated by small waves exceed the offshore component by a factor of at least 2 (Scott, 1954). With oceanic swells the ratio of strengths is usually less than 1:1.3. Apparently a larger velocity differential is required to cause a shift in position of the entire wedge of sand composing a ripple crest.

Total moving load

The slitted can trap is designed to intercept the total load moving at all levels and in any direction. The grain size and quantity of sand recovered from this trap forms the basis for the following discussion.

It was previously noted that oscillation ripple marks segregate sediment by size with extremes located at the crests and troughs. Thus, if grain size distributions from these portions of ripples are plotted as cumulative curves, an envelope is created which defines sizes available for movement. A cumulative curve representing the size distribution of the total load moved by waves characteristically lies within this envelope. This relationship holds regardless of average current velocity and extends to the tails of the curve. Fig.12 shows typical examples. One might expect that finer sizes would be moved with greater facility than coarse material. That all available sediment is transported with equal ease may be explained as follows. Nearshore sands are well-sorted, and variations in threshold velocity for a given grain population are small. Also, ripple marks create a spectrum of energy niches so that particles of different hydraulic size may be at equilibrium at the same bottom location. It would seem that ripple dimensions adjust to surge velocity so that the entire range of available sand may be placed in motion.

Information on the relationship between transported load and sediment available for motion is lacking in studies of deposition in other natural environments (Bagnold, 1968). In the laboratory Bagnold (1954) and Eagleson et al. (1961) respectively allowed wind and small waves to affect a mixture of grain sizes. The transported load was better sorted than the parent material, as might be expected. The situation in the offshore zone is not analogous, however, because sand reaching this area by means of rip currents already is confined to a narrow size range (Cook, 1970).

The rate of sand transport may be expressed by the weight of particles intercepted by a sediment trap per unit of time. Values of this rate are plotted against average current velocity on Fig.13. Three observations warrant special consideration. In the first place,

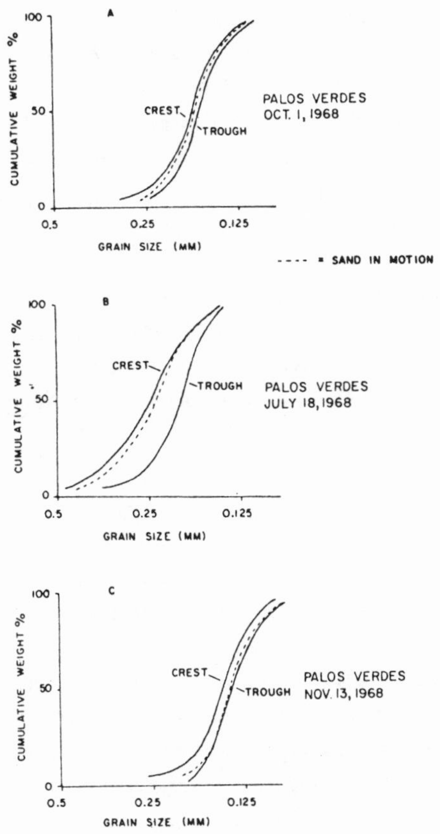

Fig.12. Examples of the relationship between the grain size of sand transported and sediment available for motion. The data represent 3 individual observations.

transportation appears to occur with currents much lower than the experimental threshold velocities. This apparent anomaly may be resolved by noting that velocities on the graph represent averages in a spectrum. Secondly, considerable variation in the rate of sand transport with a given current is evident, reflecting several undefined factors. These include surge acceleration, inequalities of ripple size, "cementation" by organic matter, and sediment porosity. Thirdly, an empirical curve indicates that the rate of sediment movement varies as approximately the square of excess current velocity above the average surge strength at which sand begins to move. However, theoretical considerations, experiments in flumes (Gilbert, 1914), and information for aeolian transport (Bagnold, 1954) indicate that capacity increases with the cube of velocity of the transporting agent. This difference may result from incomplete development of turbulence with oscillatory flow and the inverse relationship between current strength and boundary roughness in the form of ripple marks.

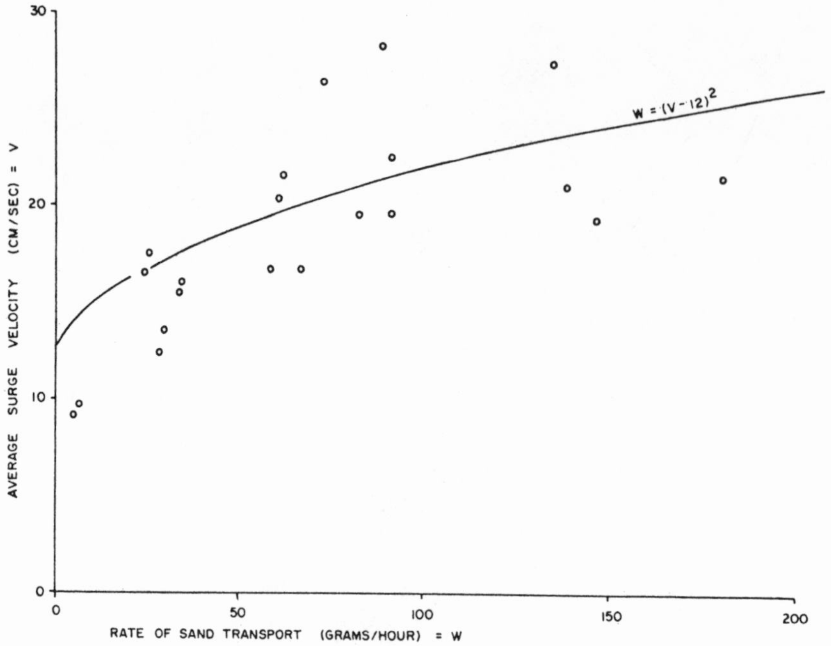

Fig.13. Average surge velocity versus rate of sand transport. Data is for sands with mean diameters ranging between 0.125 and 0.250 mm.

Suspended sediment

It was observed that surges with velocities exceeding 30 cm/sec cause temporary suspension of sand grains at the sea floor. Whenever the sediment transport rate is moderate to high, a significant quantity of material is moved in this manner. The sediment trap, consisting of vials arranged in tiers, intercepts particles carried at heights up to 30 cm above the bottom.

Typical examples of the vertical distribution of suspended sediment are illustrated on Fig.14. Plotted in this manner, it is obvious that the concentration of particles diminishes exponentially with increasing height. At least 95% of the transport occurred within 15 cm of the bottom, and the percent transported at levels greater than 30 cm is negligible. Hom-ma and Horikawa (1963) sampled suspended sediment at intervals of 60 cm in high energy conditions off the coast of Japan and reported a similar rate of decrease in concentration upward from the bottom.

The grain size of sediment carried by surges is finer with distance above the ocean floor. Typical cumulative curves representing bed load and different levels of suspension are demonstrated on Fig.15. This relationship obtains regardless of current strength.

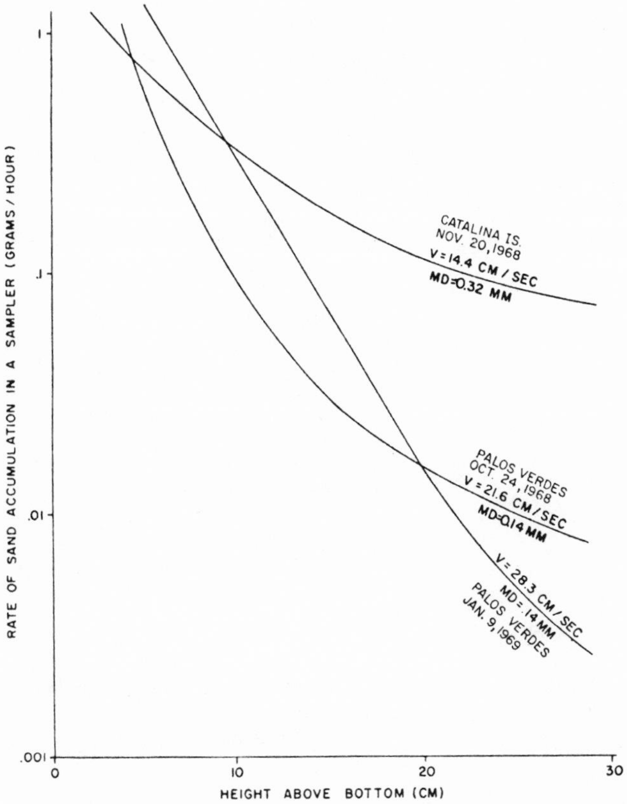

Fig.14. Examples of the vertical distribution of suspended sand. The curves represent 3 individual observations.

Sand migration

The net transportation of sea floor sediment by waves was investigated with directional traps. Using current meter records, onshore–offshore ratios for sediment transport may be compared with inequalities in surge velocity, duration of flow, and bottom drift. The slope of the ocean floor varied only between 3 and 5° and can be considered uniform. Experiments were performed exclusively in fine sand.

Net transport of the sediment does not correlate closely with any of the aforementioned hydrodynamic parameters, but agrees best with the ratio of onshore and offshore current strengths. In an effort to improve agreement, onshore–offshore velocity ratios were calculated using only the pulses sufficiently strong to move sand. This parameter (Fig.16) shows a convincing relationship with the direction of sediment transport. A downslope gravity force appears to have little influence on small particles carried over a rippled surface.

The grain size distributions of fine sand preferentially transported onshore and offshore

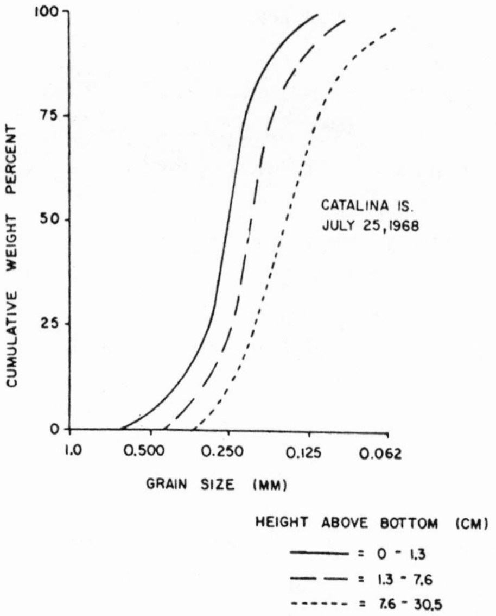

Fig.15. Typical grain size distribution of sand suspended to different heights above the sea floor.

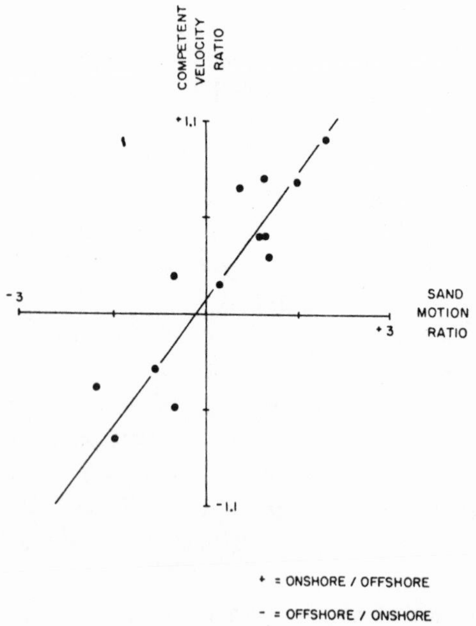

Fig.16. Relationship between medium to fine sand migration and asymmetry of those velocities capable of transporting grains. The threshold velocity for particle motion is 20 cm/sec.

were identical in all cases. The absence of sorting normal to the beach complements the following observations:

(1) Surges of varying strengths are competent to move all available sand-size particles with equal facility on a rippled bottom.

(2) Directional inequalities in current strength are small in relation to the total spectrum of velocities.

(3) The effect of the offshore component of gravity is reduced by the presence of ripple marks.

(4) Sand reaching the offshore is confined to a narrow size range.

(5) The hydraulic behaviour of particles in the fine to medium sand size category is relatively uniform (Inman, 1949).

Models for nearshore sand transport

The results of this investigation permit an evaluation of previous theories proposed to explain the exchange of sand between the beach and offshore. It appears that all three of the models are founded upon questionable assumptions.

Grant (1943) presumed that all nearshore bottom sediment is moved towards the beach under the influence of shoreward wave drift. However, current meter measurements have revealed that bottom drift is variable in direction.

The null point concept is also predicated on a net onshore transport of water over the sea floor. If a heterogenous sediment load were dumped on a steep shelf at a time when bottom drift happened to be shorewards, a null point type of sorting might occur. But such an event is unlikely because sediment reaching the offshore zone is well sorted and usually lies in the fine sand range, apparently reducing its susceptibility to the offshore gravitational component.

The confinement of most moving bottom sand to the lowest few centimeters of the water column precludes extensive diffusion of these particles. However, rip currents which introduce sediment to the offshore are dense suspensions in which concentration gradients cause grains to disperse. Thus diffusion influences the original distribution of nearshore sands, but is probably not a part of the reworking process.

The results of this study and related investigations of rip currents (Cook, 1970) lead to the construction of a relatively simple model for the exchange of sand between the beach and nearshore. During the winter season, local storms cause high, short period waves and strong onshore winds off the southern California coast. Large rip currents carry sand from the beach to the inner shelf (Cook, 1970). After the grains are deposited on the bottom, they either tend to remain in place or migrate offshore under the influence of seaward surge asymmetry. The result is an accumulation of sediment outside the breakers at the expense of the beach.

In summer, the rigorous conditions ameliorate, winds blow with less force, and long period swells generated at great distances reach the coast. Bottom sand is moved towards land by dominant onshore surges and re-enters the littoral zone. Rip currents are weak,

and the beach is replenished while the offshore reservoir of sand is depleted.

The shoreline of southern California undergoes an annual cycle of winter erosion and summer deposition which is explained by this model. Shepard and LaFond (1940) monitored sediment level on the beach and inner shelf at La Jolla, California. They concluded that the distribution of material between these adjacent environments was in dynamic equilibrium with wave characteristics.

CONCLUSIONS

(1) Oscillating currents at the sea floor generated by a set of passing waves are not uniform but have a spectrum of velocities. The average wave-induced current strength over gently-sloping continental shelves is predicted accurately by trochoidal theory where the relative wave height is less than 0.04, while solitary theory is more accurate in shallow water.

(2) The conventional concept of wave drift is inadequate since a net transport of water frequently takes place towards the offshore. Seaward bottom drift is associated with onshore winds and short period waves, while long period swells cause shoreward pulses to predominate.

(3) The dimensions of oscillation ripple marks in bottom sediment are directly related to sand size and vary inversely with current strength. The sediment is segregated according to hydraulic size with extremes concentrated at ripple crests and troughs. Despite the occurrence of both net water and sand transport, oscillation ripples do not migrate along the sea floor. Ripple dimensions seem to approach an equilibrium with bottom sediment and current velocity such that all available sand-size grains are transported with equal facility.

(4) The rate at which nearshore sand is moved increases with the square of current velocity, and practically all transport of sea floor sand by oscillating currents occurs within less than 1 m from the bottom. Wave-generated currents do not sort fine sand but cause the entire grain population to migrate as a unit. Fine sand is preferentially transported in response to inequalities in the velocity of onshore and offshore surges.

(5) During winter storms, rip currents carry sand from the beach to the offshore zone in southern California. Seaward surge asymmetry caused by short period waves and onshore wind temporarily prevents this sediment from returning to shore. Long period summer waves eventually replenish the beach with sand from the offshore.

ACKNOWLEDGEMENTS

The authors are indebted to Dr Harold D. Palmer who performed an associated project on scour in the offshore zone using time-lapse photography and collaborated on much of the field work and laboratory analysis. Captain Herbert J. Summers constructed the current meter used extensively in the study and provided another underwater movie camera. Many students assisted in diving operations which were conducted from the University of

Southern California research vessel "Ahoyoha III". The project was generously supported by the Office of Naval Research under contract number N00014-67-A-0269-0002. Stipends were provided from Office of Naval Research contracts number 228(17) and number N00014-67-A-0269-0009, and from the National Aeronautics and Space Administration.

REFERENCES

Allen, J. R. L., 1968. *Current Ripples – Their Relation to Patterns of Water and Sediment Motion.* North-Holland, Amsterdam, 433 pp.

Bagnold, R. A., 1946. Motion of waves in shallow water. Interaction between waves and sand bottoms. *Proc. R. Soc. Lond.*, A187: 1–18.

Bagnold, R. A., 1947. Sand movement by waves: some small-scale experiments with sand of very low density. *J. Inst. Civ. Eng., Pap.*, 5554: 447–469.

Bagnold, R. A., 1954. *Physics of Blown Sand and Desert Dunes.* Methuen, London, 265 pp.

Bagnold, R. A., 1968. Deposition in the process of hydraulic transport. *Sedimentology*, 10: 45–56.

Colby, B. R., 1964. Discharge of sands and mean velocity relationships in sand-bed streams. *U.S. Geol. Surv., Prof. Pap.*, 462A: 1–47.

Cook, D. O., 1969. Calibration of the University of Southern California automatically-recording settling tube. *J. Sediment. Petrol.*, 39: 781–786.

Cook, D. O., 1970. Occurrence and geologic work of rip currents in Southern California. *Mar. Geol.*, 9: 173–186.

Draper, L., 1957. Attenuation of sea waves with depth. *La Houille Blanche*, 12: 926–931.

Eagleson, P., Glenne, B. and Dracup, J., 1961. Equilibrium characteristics of sand beaches in the offshore zone. *U.S. Army Corps Eng., Beach Erosion Board, Tech. Memo.*, 126: 1–66.

Evans, O. F., 1942. The relation between the size of wave-formed ripple marks, depth of water, and the size of generating waves. *J. Sediment. Petrol.*, 12: 31–35.

Felix, D. W., 1969. Design of an automatically recording settling tube for analysis of sands. *J. Sediment. Petrol.*, 39: 777–780.

Gilbert, G. K., 1914. The transportation of debris by running water. *U.S. Geol. Surv., Prof. Pap.*, 86: 1–263.

Grant, U. S., 1943. Waves as a sand transporting agent. *Am. J. Sci.*, 241: 117–123.

Hom-Ma, M. and Horikawa, K., 1963. Suspended sediment due to wave action. *Proc. Conf. Coastal Eng., Counc. Wave Res., 8th, Univ. Calif., Berkeley, Calif., 1963:* 168–193.

Inman, D. L., 1949. Sorting of sediments in the light of fluid mechanics. *J. Sediment. Petrol.*, 19: 51–70.

Inman, D. L., 1952. Measures for describing the size distribution of sediments. *J. Sediment. Petrol.*, 22: 125–145.

Inman, D. L., 1957. Wave-generated ripples in nearshore sands. *U.S. Army Corps Eng., Beach Erosion Board, Tech. Memo.*, 100: 1–42.

Inman, D. L. and Bagnold. R. A., 1963. *Littoral Processes. The Sea.* Interscience, New York, N.Y., 3: 529–553.

Inman, D. L. and Nasu, N., 1956. Orbital velocity associated with wave action near the breaker zone. *U.S. Army Corps Eng., Beach Erosion Board, Tech. Memo.*, 79: 1–43.

Ippen, I. and Eagleson, P., 1955. A study of sediment sorting by waves shoaling on a plane beach. *U.S. Army Corps Eng., Beach Erosion Board, Tech. Memo.*, 63: 1–83.

Longuet-Higgins, M. S., 1953. Mass transport in water waves. *Philos. Trans. R. Soc., Lond.*, 245: 535–581.

Manohar, M., 1955. Mechanics of bottom sediment movement due to wave action. *U.S. Army Corps Eng., Beach Erosion Board, Tech. Memo.*, 75: 1–121.

Miller, R. L. and Zeigler, J. M., 1958. A model relating dynamics and sediment pattern in equilibrium in the region of shoaling waves, breaker zone, and foreshore. *J. Geol.*, 66: 417–441.

Murray, S. P., 1967. Control of grain dispersion by particle size and wave state. *J. Geol.*, 75: 612–634.
Nagata, Y., 1961. Balance of transport of sediment due to wave action in shoaling water, surf zone, and foreshore. *Rec. Oceanogr. Works Japan*, 6: 53–62.
Passega, R., Rizzini, A. and Borghetti, G., 1967. Transport of sediment by waves, Adriatic coastal shelf, Italy. *Bull. Am. Assoc. Pet. Geologists*, 51: 1304–1319.
Risk, M. J., 1965. *Shallow Water Ripple Marks in Lake Huron.* Thesis, University of Western Ontario, London, Ont., 120 pp.
Russell, R. C. and Osario, J. D., 1955. An experimental investigation of drift profile in a closed channel. *Proc. Conf. Coastal Eng., Counc. Wave Res., 6th, Univ. Calif., Berkeley, Calif; 1953*: 171–193.
Sato, S., Ijima, I. and Tanaka, N., 1963. A study of critical depth and mode of sand movement using radioactive glass sand. *Proc. Conf. Coastal Eng., Counc. Wave Res., 8th, Univ. Calif., Berkeley, Calif.*: 304–323.
Scott, T., 1954. Sand movement by waves. *U.S. Army Corps Eng., Beach Erosion Board, Tech. Memo.*, 48: 1–37.
Shepard, F. P., 1963. *Submarine Geology.* Harper and Row, New York, N.Y., 557 pp.
Shepard, F. P. and Inman, D. L., 1950. Nearshore circulation. *Proc. Conf. Coastal Eng., Counc. Wave Res., 1st, Univ. Calif., Berkeley, Calif.*: 50–59.
Shepard, F. P., and LaFond, E. C., 1940. Sand movement along Scripps Institution Pier. *Am. J. Sci.*, 238: 272–285.
Siegel, S., 1956. *Non-parametric Statistics for the Behavioral Sciences.* McGraw-Hill, New York, N.Y.. 312 pp.
Summers, H. J., Palmer, H. D. and Cook, D. O., 1971. Some simple devices for the study of wave-induced surges. *J. Sediment. Petrol.*, 41: 861–866.
Vernon, J. W., 1966. *The Shelf Sediment Transport System.* Thesis, Univ. Southern California, Los Angeles, Calif., 135 pp.
Zenkovich, V. P., 1967. *Processes of Coastal Development.* Interscience, New York, N.Y., 738 pp.

19

Reprinted from pp. 365–380 of *Estuarine Research: Vol. II. Geology and Engineering,* L. E. Cronin, ed., Academic Press, 1975, 587 pp.

TIDAL CURRENTS, SEDIMENT TRANSPORT, AND SAND BANKS IN CHESAPEAKE BAY ENTRANCE, VIRGINIA

by

John C. Ludwick[1]

ABSTRACT

Taking the mean over all stations in the entrance area, and with reference only to a level 100 cm above the bed, tidal currents at ebb and flood strength are nearly equal in speed and average 42 cm/sec. Taking individual stations, ebb maximum is 79 cm/sec; flood maximum is 66 cm/sec; however, at more than half the observation stations, maximum flood speed and duration exceed maximum ebb speed and duration.

If transport rate of bed sediment is proportional to stream power, $\tau_0 u_{100}$, where τ_0 is shear stress at the bed and u_{100} is current speed 100 cm above the bed, then at 19 of 24 stations, ebb-directed sediment transport exceeds flood-directed sediment transport. Much of the longer flood is incompetent and the greater ebb shear stress more than offsets larger flood speeds. Net sediment transport is directed landwards only in parts of channels and in dead-end sinuses re-entrant into shoals and open to flood currents.

In Chesapeake Bay entrance, a wide entrance of moderate tidal range, net sediment transport at the headlands is directed seaward. Individual shoals within the entrance are bounded on one side by a net sediment transport to seaward and on the other side by a net sediment transport to landward. Ebb deltas and flood deltas occur in alternate succession across the entrance with four to five of the former and three to four of the latter. There is also an interior flood delta found where ebb-directed sediment is interdicted and swept landwards in a flood-dominated channel.

1. Institute of Oceanography, Old Dominion University, Norfolk, Virginia 23508.

INTRODUCTION

In this paper a summary is given of sediment transport phenomena relating to shoals and sand banks in the tidal entrance to Chesapeake Bay, Virginia (Fig. 1). The entrance to this body of water is wide, the distance between the two entrance capes amounting to 17 km. Mean water depth is 11.3 m in the entrance section. Tidal range is 1 m, wave action is not intense; hence, the estuary is properly classified as one of moderate energy.

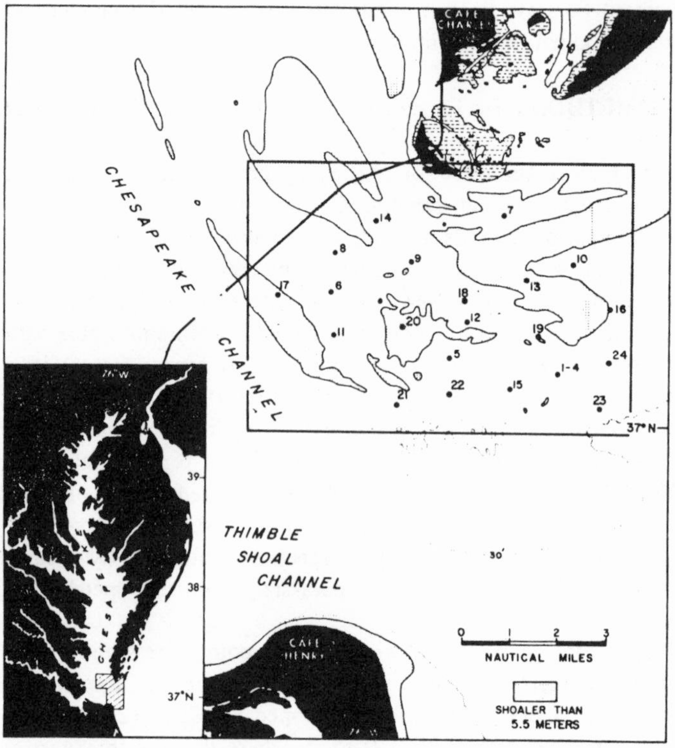

Figure 1. Index chart to the location of tidal current stations.

Sand banks, shoals, and tidal channels are prominent features of the entrance area, particularly in the northern part where sand-sized sediment is debouched into the estuary mouth by the littoral drift system along the ocean side of the Delmarva Peninsula. In contrast with smaller inlets, Cheseapeake Bay entrance lacks a typically developed single ebb-tidal delta and single flood-tidal delta. Instead there is a complex arrangement of sand banks separated by interdigitating tidal channels. It is this pattern of banks and the associated tidal

currents that is the subject of this investigation.

Previous work on estuarine shoals in other areas is extensive and has been summarized elsewhere (6, 12, 13). Study of the entrance area to Chesapeake Bay has been conducted by the present author for several years and the results obtained to date have also been reported elsewhere (7, 8, 9). In the sand bank area, water depth at bank top is approximately 2.5 m; water depth in tidal channels is approximately 10 m. Tidal currents at the surface at strength reach speeds of 132 cm/sec on ebb and 105 cm/sec on flood; at the bottom, speeds reach 79 cm/sec on ebb and 66 cm/sec on flood. At all locations peak speeds are competent to move bottom sands which range from fine-grained (MD = 0.16 mm) to coarse-grained (Md = 0.66 mm).

FIELD METHODS, DATA REDUCTION, AND TRANSPORT CALCULATION

Current speed and direction in the entrance were measured at 24 stations with a Kelvin Hughes Direct Reading Current Meter suspended from an anchored ship. Observation periods at each station averaged 27 hours and thus included two ebb cycles and two flood cycles. The meter was calibrated in a flume and found to be accurate within 3 cm/sec. At each station the meter was raised successively through 11 different levels with a 4-minute observation period at each. Individual readings at each level were taken every 15 seconds.

In the office, averages of current speed for each 4-minute period for each depth were obtained. These speeds were then plotted versus the logarithm of the corresponding distance above the bed. Values for 11 standard depths were obtained by interpolation. These depth-corrected values of speed were then plotted for each standard level versus time. At standard times corresponding to every hour on the hour, readings of speed were obtained by linear interpolation. By this means a pseudo-synoptic data set for current speed was produced for each station.

The 11 time- and depth-corrected data points for speed versus depth were fitted by the method of least squares with a parabolic function. This function is an empirical velocity-defect law developed by Hama (2) and pertains to outer flow at distances greater than 0.15 d, where d is the thickness of a turbulent boundary layer, or water depth in the case of fully-developed flow in a uniform channel. The equation is

$$\frac{U - u}{u_*} = 9.6\,(1 - \frac{y}{d})^2 \qquad (1)$$

whre U is the surface speed, u is speed at a distance y above the bed, and u_* is the friction velocity.

From the least squares curve fitting to the 11 data points, an estimate is obtained of τ_0, the boundary shear stress, according to the definition, $u_* =$

$(\tau_0/\rho)^{1/2}$, where ρ is fluid density.

An algorithm proportional to bed sediment transport rate is taken from the work of Bagnold (1). His analysis of the energetics of sediment transport leads in the present study to a computation of flow power per unit area, $\tau_0 u_{100}$, where τ_0 is derived from u_* as explained above and u_{100} is the observed current speed at a point 100 cm above the bed. The units of this quantity are dyne-centimeters per second per square centimeter. Another publication (10) shows in detail how this quantity is reduced to mean tidal range, further corrected by subtracting 150 dyne-cm/sec/cm^2, a threshold value for initiation of sediment motion, and finally integrated separately over each ebb half-cycle and each flood half-cycle. The results are then averaged for ebb and flood. After integration and averaging, the units of the measure are dyne-centimeters per square centimeter per average ebb (or flood) half-cycle. Table 1 gives the magnitude and direction of the sediment-transport index, $\tau_0 u_{100}$, for 24 stations in the entrance area of Chesapeake Bay.

THE CHARTS OF BED-SEDIMENT MOTION

From the sediment-transport index described above at each of the 24 stations, it is possible to develop charts showing streamlines of near-bottom sediment transport for ebb and for flood half-cycles, and for the vector sum of ebb and flood sediment transport. This development is best viewed as a problem in vector field interpolation. A simple chart plot of the vectors for $\tau_0 u_{100}$ at the 24 stations for, say, ebb shows the need to obtain interpolated values between stations, values consistent in both vector direction and magnitude.

The first step in this procedure is to prepare a chart plot of the north-south component of $\tau_0 u_{100}$ for, say, ebb, and a separate chart plot of the east-west component of $\tau_0 u_{100}$. Isopleths are drawn on each chart. The two charts are then superimposed and at each intersection of isopleths the resultant direction is obtained and plotted. At contour intersections the resultant direction is obtained by trigonometry; the resultant magnitude is obtained by the pythagorean relationship. Any desired number or density of resultant directional arrows can be obtained by decreasing the isopleth interval of the component charts. Each directional resultant is plotted as a vector of unit length; only the directional information is used initially.

The unit vectors of the final dense field agree exactly with observed data at the 24 stations and are interpolated values elsewhere. Finally a streamline chart is prepared by drawing lines that are everywhere tangent to the unit vectors. Streamline charts for ebb, flood, and the vector sum of ebb and flood are shown in Figures 2-A, 3-A, and 4-A.

It is evident that lines of divergence and convergence occur in the sediment-transport field, particularly during flood flow. In the case of planar

TABLE 1.

Integrated Effective Flow Power Adjusted To Mean Tidal Range

Station Number	Water Depth m	Ebb $\tau_0 u_{100}$[1]	Ebb Dir °T	Flood $\tau_0 u_{100}$[1]	Flood Dir °T	Vector Sum $\tau_0 u_{100}$[2]	Vector Sum Dir °T
1-4	7.7	560	083	365	272	207	079
5	7.8	558	113	462	292	97	118
6	8.5	543	108	15	318	530	107
7	10.4	1003	069	660	271	462	037
8	8.3	582	095	294	273	288	097
9	10.1	1658	109	650	310	1077	097
10	7.2	92	089	1360	266	1268	266
11	6.1	1879	112	1038	318	1050	086
12	8.2	6352	097	1296	274	5058	098
13	3.3	1655	090	446	290	1245	083
14	10.7	751	134	1208	315	457	317
15	7.3	4004	078	562	281	3494	074
16	6.7	1846	079	828	283	1140	062
17	6.4	2538	129	28	327	2511	129
18	13.7	1707	125	2246	308	549	317
19	6.8	1514	101	76	285	1438	101
20	5.6	3253	119	262	322	3014	117
21	5.7	700	134	893	312	195	305
22	6.4	1243	118	380	299	864	118
23	10.0	2094	063	93	274	2015	062
24	8.4	145	081	538	266	394	269

1 Units are dyne-cm/cm^2/half cycle x 10^{-4}

2 Units are dyne-cm/cm^2/tidal cycle x 10^{-4}

potential flow of an imcompressible fluid, it is not possible for streamlines to join. However, in the case of three-dimensional flow, merging or joining of streamlines indicates ascending fluid motion; divergence of streamlines indicates descending fluid motion. A pattern of alternating zones of convergence and divergence is characteristic of spiral flow in a set of cells with horizontal parallel axes.

Sediment, unlike liquid, can be lost or gained during transport; hence sediment input in a stream tube does not necessarily equal sediment output. Sediment can be lost from transport by deposition or gained by erosion. Spacing

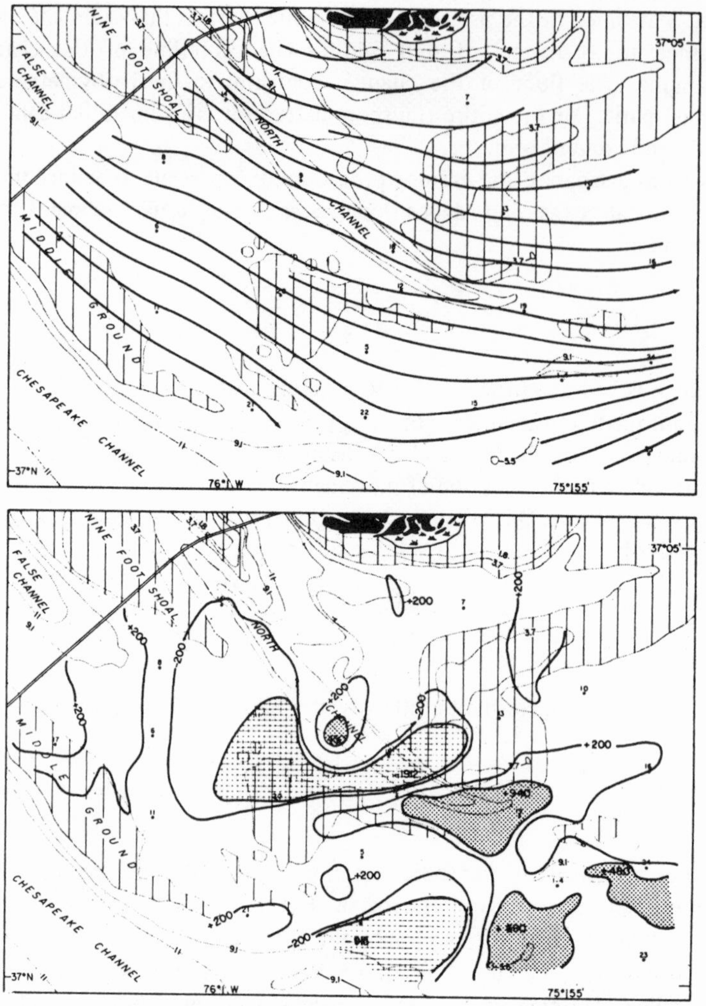

Figure 2. Ebb-directed sediment transport at the bed. (A). Streamlines of the sediment-transport vector $\tau_0 u_{100}$; depths are in meters; vertically ruled areas are shoaler than 5.5 m. (B). Erosion-deposition chart on which erosion is positive (+) and deposition is negative (-); units are dyne-cm per sq. cm per ebb half tidal cycle per 463 m of transport x 10^{-4}; cross-hatched areas indicate erosion intensity greater than -400 units; stippled areas indicate deposition intensity greater than +400 units.

between sediment-transport streamlines is not a measure of sediment-transport flux. There is no underlying stream function in the development presented here. What is assumed in the use of the streamline charts is that sediment transport is confined to a channel of unit width. The centerline of this channel is a streamline. The floor of the channel conforms to the bathymetry along the channel path. For the entire transport pattern, it is assumed that conditions are steady and non-uniform.

Now as a separate and ensuing procedure, it is possible to deduce the location of erosional areas and depositional areas. The usual sediment continuity equation is

$$\frac{\partial \eta}{\partial t} = -\varepsilon \frac{\partial Q}{\partial x}, \tag{2}$$

where η is bed elevation relative to a datum plane, t is time, ε is a dimensional constant related to sediment porosity, Q is weight rate of bed-sediment transport per unit width of channel, and x is distance along a sediment-transport pathway or streamline in the direction of movement.

In what follows, Q is taken as proportional to $\tau_0 u_{100}$, the right hand partial derivative is approximated by a finite difference,

$$\frac{\partial Q}{\partial x} \approx \frac{\Delta Q}{\Delta x} = k \frac{(\tau_0 u_{100})_2 - (\tau_0 u_{100})_1}{x_2 - x_1}, \tag{3}$$

and $x_2 - x_1$ is held constant arbitrarily at a value of 463 m, whence

$$\frac{\partial \eta}{\partial t} \propto -\Delta\ \tau_0 u_{100}. \tag{4}$$

Thus a decrease in transport rate along a transport path requires deposition; whereas, an increase requires erosion. Returning to the component charts and the streamline chart for, say, ebb, the three are finally superimposed and the magnitude of $\tau_0 u_{100}$ is determined at equispaced points along each streamline. Differences between adjacent pairs of values are taken and the distribution of erosion and deposition intensity is then mapped over the area of study. Positive values correspond to inferred deposition; negative values correspond to inferred erosion. The size of the positive or negative difference is proportional to the rate of deposition or erosion. Charts of this type have been prepared for ebb (Fig. 2-B), flood (Fig. 3-B), and the the vector sum of ebb and flood (Fig. 4-B).

RESULTS

With reference to the ebb streamline chart for bottom-sediment transport (Fig. 2-A), it is seen that the motion is not markedly deflected by local shoals probably because of low declivity of bottom slopes. Transport is, however, directed distinctly across North Channel. Transport is also up and across the end of Nine Foot Shoal, an ebb spit. In the southeast corner of the chart, the pronounced change in transport direction is a manifestation of the James River jet, the edge of which has spread more than half-way across the entrance width and into the study area. It is noteworthy that many of the elongated features of the area are more or less parallel to the ebb transport streamlines.

With reference to the erosion-deposition chart for ebb (Fig. 2-B), it is seen that there are two erosion areas characterized by intensity indices greater than -400. Maximum erosion on ebb is inferred to occur in a part of North Channel. The large erosion area may be an extension of False Channel. The other erosion area is positioned so as to suggest an escaping or easement flow from Chesapeake Channel, perhaps as a result of the blocking action of a large shoal at the mouth of that channel. Continued action in this pattern should cause an ebb bifurcation channel to form. The principal depositional areas on ebb are situated so as to suggest that eroded sediment is soon deposited to seaward in a patch. The area of maximum deposition at the seaward end of north Channel corresponds to an area where seaward-facing sand waves up to 4 m in height are prominently developed. In this area these waves migrate seaward at rates of 35 to 150 m/year (9).

With reference to the flood streamline chart for bottom-sediment transport (Fig. 3-A), it is seen that there is much less transport parallelism than for the ebb case. There are features of the pattern suggestive of lines of transport convergence and lines of transport divergence. Moreover, there is some indication that lines of convergence and divergence occur alternately as one moves across the chart from northeast to southwest. Flood transport in North Channel is parallel to the channel axis and into a prominent sinus in a flood delta. Ebb and flood transports are not opposed exactly; most of the intersection angles open to the north.

With reference to the erosion-deposition chart for flood (Fig. 3-B), it is seen that the intensity of both these processes is lower than for ebb conditions. Maximum erosion is -483 units; maximum deposition is +348 units. Some deposition occurs on Middle Ground and in North Channel. Erosion of note also occurs in the same channel.

With reference to the streamline chart for the vector sum of ebb and flood transport (Fig. 4-A), it is seen that most of the area is characterized by net ebb-directed transport. This is because much of the longer flood flow is

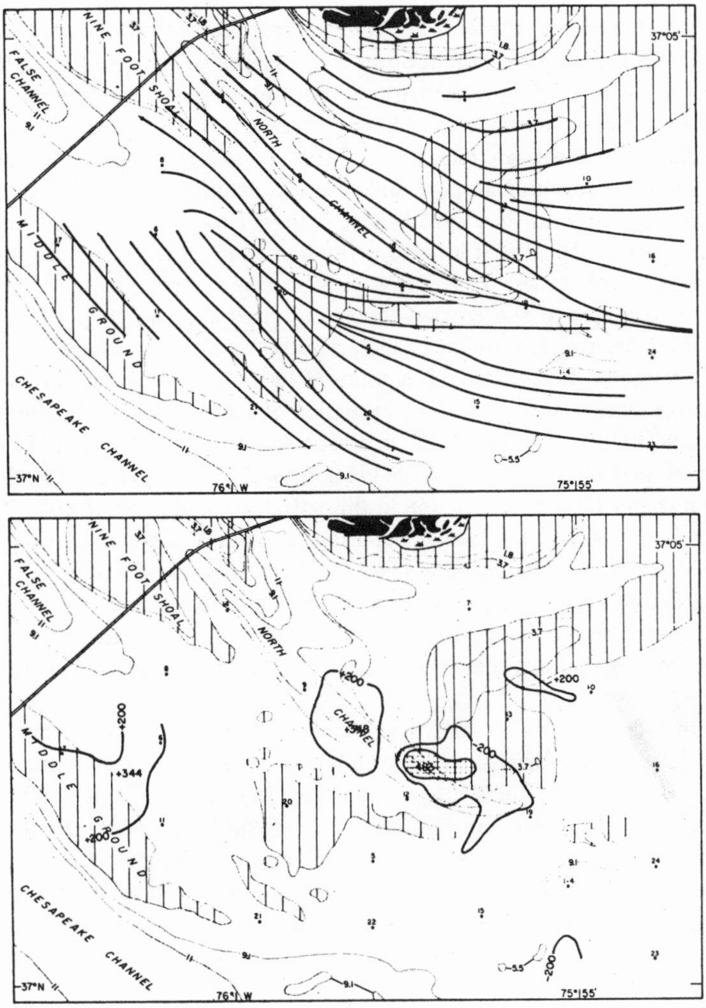

Figure 3. Flood-directed sediment transport at the bed. (A). Streamlines of the sediment-transport vector $\tau_0 u_{100}$; depths are in meters; vertically ruled areas are shoaler than 5.5 m. (B). Erosion-deposition chart on which erosion is positive (+) and deposition is negative (-); units are dyne-cm per sq. cm per flood half tidal cycle per 463 m of transport x 10^{-4}; cross hatched areas indicate erosion intensity greater than -400 units; stippled areas indicate deposition intensity greater than +400 units.

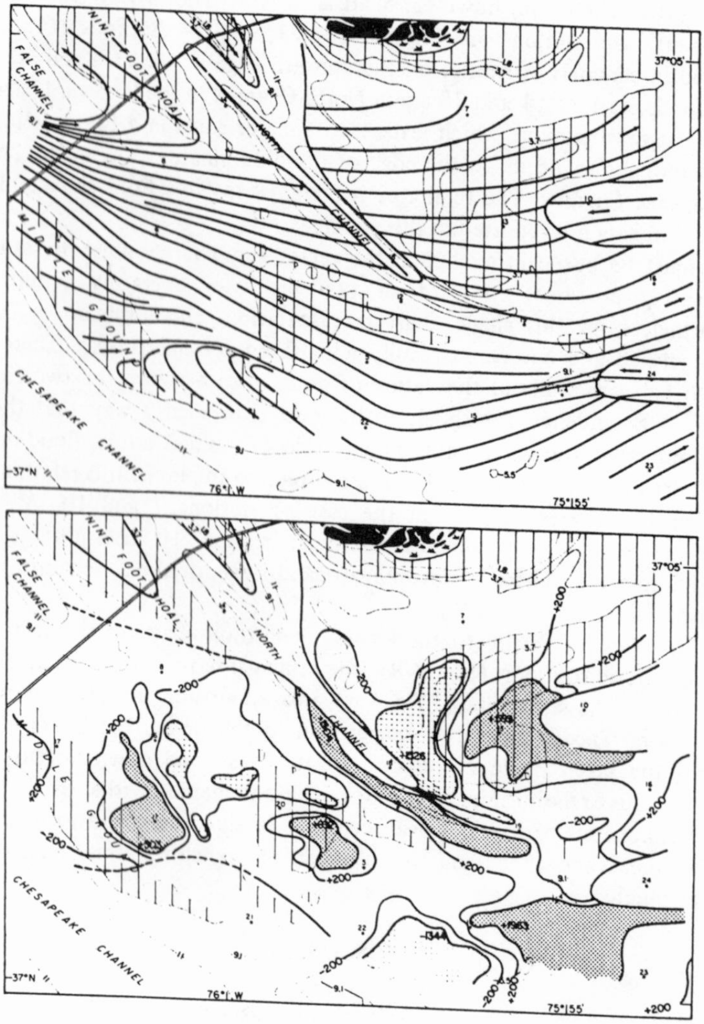

Figure 4. Vector sum of ebb and flood sediment transport at the bed. (A). Streamlines of the resultant sediment transport vector $\tau_0 \; u_{100}$; depths are in meters; vertically ruled areas are shoaler than 5.5 m. (B). Erosion-deposition chart on which erosion is positive (+) and deposition is negative (-); units are dyne-cm per sq. cm per tidal cycle per 463 m of resultant transport x 10^{-4}; cross hatched areas indicate erosion intensity greater than -400 units; stippled areas indicate deposition intensity greater than +400 units.

incompetent and because ebb shear stress is usually greater than flood shear stress (10). There are, however, 5 areas in which flood-sediment transport is dominant. The stations are 10, 14, 18, 21, and 24. Stations 10 and 24 are located in channels or sinuses on the seaward edge of the area and are open to the flood. Stations 14 and 18 are in North Channel in the landward part and are on the same streamline which leads into a major sinus in a flood delta. Station 21 is thought to occur on the edge of flood-dominated flow in the bottom of Chesapeake Channel. The fifth area is marginal to Nine Foot shoal and is based on station data not reported in this paper.

It must be borne in mind that the vector sum chart represents resultants of alternating processes and not of two processes operating simultaneously. Nevertheless, in this single chart representation, ebb-dominant areas can be distinguished and compared readily with flood-dominant areas. There is much yet to be learned about the nature of boundaries between the two flow areas. There is a tendency for shoals to be located in such a way that there is an ebb-dominated transport path on one side of a shoal and a flood-dominated transport path on the other. There are several examples of this relationship, but perhaps the clearest is seen in the case of stations 7 and 10. At station 7 transport is strongly ebb-dominated; at station 10 transport is strongly flood-dominated. There is a prominent sand ridge, or shoal, separating the two stations.

With reference finally to the erosion-deposition chart for the vector sum of ebb and flood transport (Fig. 4-B), one principal motif is evident. Moving in the direction of the streamlines, (Fig. 4-A), areas of erosion are followed by areas of deposition. The former often correspond to the up-current backs of existing shoals; the latter often correspond to the crests and down-current slopes of shoals. This relationship holds for ebb-dominated paths as well as for flood-dominated paths. It is also seen that erosion areas tend to be in existing deeper areas and to extend, in the form of branching arms, around existing higher shoal areas.

DISCUSSION

An essential, but not necessarily obvious, aspect of tidal flow in estuaries is that, at a fixed observation point, ebb and flood flows are not usually equal and not always oppositely directed. The concept of flow inequality has become embodied in terms such as net flow, time-velocity asymmetry, flow dominance, residual flow, and non-tidal flow. Various measures are in use to describe the inequality; direction and magnitude of the difference between peak ebb and peak flood velocity; difference in duration of flow; and difference in the areas under the ebb and flood curves on a time-velocity plot of tidal currents. This last measure has units of length and has been referred to as the difference in tidal

excursion.

The causes of inequality in tidal flow are probably many, but at least three seem to be fundamental: (a) flow inequality caused in essence by mixing that occurs between river water and sea water, i.e., vertical entrainment due to breaking of internal waves at density interfaces, and vertical mixing due to turbulent exchange; (b) preferred flow location associated with coriolis deflection; and (c) flow inequality associated with the local mutual evasion of ebb and flood tidal current pathways.

The first-listed cause produces a characteristic gross circulation pattern in partially-mixed estuaries of moderate salinity in temperate climates. In this pattern net flow at depth is directed into the estuary; net flow in near-surface layers is directed seawards.

The second-listed cause is manifested in the northern hemisphere as flood-dominant flow on the left-hand bank and ebb-dominant flow on the right-hand bank looking down an estuary toward the sea.

The third-listed cause may show itself as ebb-dominated flow in a given channel and flood-dominanted flow in an adjacent channel or area. Several common channel patterns of mutually evasive tidal flows are known: the forked pattern; the flanking pattern; and the parallel offset pattern (8).

The three above-listed causes of flow inequality can co-exist; however, coriolis effects are usually overshadowed by other processes. The flow inequalities associated with gross estuarine circulation tend to be depth-related, at least in partially-mixed estuaries. In deep channels, headward net flow is expected; in shallow areas net flow to seaward is expected. In mutual evasion, it is the depositional history and present configuration of shoals and channels that are important factors along with tidal range and relation between tidal elevation and current speed, i.e., the presence of a progressive tidal wave in some estuaries and a standing tidal wave in others.

When bed-sediment transport dominance, as distinct from fluid-flow dominance, is considered, conclusions reached often depend on the assumptions made about threshold velocity for sediment transport and transport rate. Net transport of bed sediment is not simply correlated with the net transport of fluid near the bed. In the case of Chesapeake Bay entrance, although the net fluid flow at depth is directed headwards, taking the entrance section as a whole, the net sediment transport in some local areas is headward, and in other areas, at the same depth, is seaward. The pattern of bed-sediment-transport dominance near entrance shoals appears to be related more closely to local mutual evasion of tidal flows than to gross estuarine circulation or coriolis deflection.

Models of shoal configuration and flow dominance have been developed by others for inlets (3, 4). There are similarities and contrasts when these models are compared with shoaling patterns in a wide entrance. Near the seaward end of North Channel there is an arrangement of banks and channels similar in some

respects to the ebb-tidal model of Hayes (3, 4). There are two prominent flood-dominated areas corresponding to channels heading landward from the sea. In between is a small shoal corresponding perhaps to the ebb bar; however, this present feature is bounded on one side by ebb-dominated flow and on the other side by flood-dominated flow, and not by flood-dominated flow on both sides, as in the Hayes model.

North Channel corresponds to the major channel in a tidal inlet and like such a channel is ebb-dominated in seaward parts and flood-dominated in landward parts. Nine Foot shoal and the immediately adjacent sand banks exhibit many if not all of the features of Hayes' flood-tidal delta including the flood ramp, flood channels, high areas, and ebb spits.

In the present instance, however, the sediment forming this major flood delta apparently first experiences a net motion across North Channel which interdicts some of the transported material and serves as a chute for sediment movement landward. The flood delta is formed where spreading of the current occurs and deposition takes place. The final completed flood delta is probably a quasi-stable form in balance amongst mutually evasive ebb-dominated and flood-dominated flows.

The finding of flood-dominated bed transport in North Channel and ebb-directed transport in the shallower marginal areas is similar to results reported by Hubbell, Glenn, and Stevens (5) for the Columbia River estuary, Washington, and by Wright, Sonu, and Kielhorn (14) for East Pass near Destin, Florida. Middle Ground of the latter report is a flood delta situated between two branches of the main channel. Bed transport is flood-dominated in both branches. Their flood delta lacks a central flood-dominated sinus when compared to Nine Foot shoal of the present study.

Among the charts in the present paper, including bathymetric, streamline, and erosion-deposition charts for ebb and flood, certain geomorphic forms, sediment-transport directions, and erosion-deposition patterns appear to be interrelated. A barchan-like, or parabolic-shaped, shoal is a recurring form usually seen compounded with other similar, but opposite-facing, parabolic shoals. The prominent flood delta of the area is an obvious example. In addition, the line of somwhat discontinuous shoals that begins at the north headland and extends southward three-fourths of the distance across the entrance is an example of linked alternate-facing parabolas. Flood-dominant transport occurs in those sinuses open to the sea; ebb-dominated transport occurs in the seaward end of channels leading out from the estuary. Characteristically, these latter channels leading out from the estuary. Characteristically, these latter channels display the ramp-to-the-sea shape described by Oertel (11) from the Georgia coast.

If a comparison is made of the principal features of a major interior flood delta, Nine Foot shoal, and features of the smaller flood deltas in the exterior

line of shoals, it is evident that there are only minor differences, except for scale. Thus the exterior line of zig-zag, or alternate-facing, parabolic shoals is regarded as a single feature comprised of alternating ebb deltas and flood deltas linked together. Across the entire entrance there are four or five ebb-dominated and three or four flood-dominated pathways of bed-sediment transport. At the headlands, net flow is directed seaward.

It was seen that bed-transport streamlines are not deflected by shoals, and that the up-current backs of shoals are often the site of erosion. A likely conclusion is that it is the decrease in water depth with distance up the back of a shoal that causes a velocity and shear stress increase, an increase in flow power, a corresponding increase in sediment transport rate, and hence erosion. These relations are similar to those occurring on the backs of sand waves. Present findings thus indicate that the sand banks and shoals of the area may be formed by processes quite similar to those that form sand waves and cause them to migrate. It is noteworthy that the down-current sides and the crests of the entrance shoals are often sites of deposition. It is thus also inferred that the sand banks are migrating in the direction of bed-sediment-transport dominance.

With reference to the development of shoals, perhaps the essential difference between wide estuaries and inlets is the stability of flow through the opening. When the width of an opening exceeds some unknown limit, it may well be that the flow alters from frontal to fingered and a linked series of ebb and flood deltas are formed as a consequence, rather than single features as in an inlet. The development of spiral flow in long cells with parallel horizontal axes may be favored in flood flow because of weaker density stratification. The relationship of the proposed fingered flow and spiral flow is presently unknown.

SUMMARY AND CONCLUSIONS

The measurement of tidal currents, their velocity profiles, inferred shear stress at the bed, and flow direction, provide a basis for a deduction of relative bed-sediment-transport rate over an area of shoals in the wide entrance to Chesapeake Bay. Streamlines of ebb- and flood-directed bed-sediment-transport are little deflected by shoals. During flood there is some evidence of convergent and divergent transport suggestive of spiral flow. Ebb transport is across North Channel; flood transport is along the axis of the channel.

When the vector sum of ebb and flood transport is plotted, five areas of flood dominance appear on a chart that shows largely ebb dominance of bed-sediment transport. The five areas are mostly in or related to channels.

A decrease in transport rate with distance requires deposition; an increase requires erosion. Using this continuity concept, areas of deposition and erosion can also be deduced from data on tidal currents. The intensity of erosion and

deposition is greater on ebb than on flood. It is seen that erosion on the upcurrent side of a shoal is soon followed, down the streamline, by deposition on the shoal crest or down-current side. The process is strikingly similar to that by which sand waves migrate. The parabolic, or barchan-shaped, shoal with a pronounced sinus between the arms appears to be a common quasi-stable form under reversing tidal flow. Best agreement with present shoal and channel configuration, and bedform-facing direction, is given by the streamline and erosion-deposition charts for the vector sum of ebb and flood transport.

Across the entrance to Chesapeake Bay there is an outer line of shoals and an inner major flood delta. The outer line is comprised of alternate-facing, ebb parabolas, or deltas, and flood parabolas, or deltas, linked together into a crude zig-zag pattern. The sinuses are alternately ebb-dominated and flood-dominated. The major inner flood delta is a simple barchan form.

ACKNOWLEDGMENT

The author appreciates the assistance of those who have helped with this work, particularly graduate students John T. Wells and Mitchell A. Granat. The effort was supported by the Office of Naval Research, Geography Programs Branch, under Contract Number N00014-70-C-0083.

REFERENCES

1. Bagnold, R. A.
1965 Beach and nearshore processes. Part I, Mechanics of marine sedimentation. In **The Sea,** Vol. 3, p. 507-528.

2. Hama, F. R.
1953 Discussion: The frictional resistance of flat plates in zero pressure gradient (by L. Landweber). **Trans. Soc. Naval Arch. and Marine Engineers,** 61: 27-28.

3. Hayes, M. O., Goldsmith, V., and Hobbs, C. H.
1970 Offset coastal inlets. **Proceedings of 12th Coastal Eng. Conf.,** p. 1187-1200.

4. Hayes, M. O., Owens, E. H., Hubbard, D. K., and Abele, R. W.
1973 The investigation of form and processes in the coastal zone. In **Coastal Geomorphology,** p. 11-41. (ed. Coates, D. R.) State Univ. New York, Binghamton, New York.

5. Hubbell, D. W., Glenn, J. L., and Stevens, H. H.
1971 Studies of sediment transport in the Columbia River estuary. In **Proceedings of 1971 Tech. Conf. on Estuaries of the Pacific Northwest,** Circular 42: p. 190-226. Eng. Expt. Station, Oregon State Univ., Corvallis, Washington.

6. Lauff, G. H.
1967 **Estuaries.** Am. Assoc. Adv. Sci., Publ. 83, Washington, D. C.

7. Ludwick, J. C.
1970a Sand waves in the tidal entrance to Chesapeake Bay: preliminary

observations **Chesapeake Ści.,** 11: 98-110.

8. Ludwick, J. C.
1970b Sand waves and tidal channels in the entrance to Chesapeake Bay. **Virginia Jour. Sci.,** 21: 178-184.

9. Ludwick, J. C.
1972 Migration of tidal sand waves in Chesapeake Bay entrance. In **Shelf Sediment Transport,** p. 377-410. (eds. Swift, D. J. P., Duane, D. B., and Pilkey, O. H.) Dowden, Hutchinson, and Ross, Stroudsburg, Pennsylvania.

10. Ludwick, J. C.
1973 Tidal currents and zig-zag sand shoals in a wide estuary entrance. **Old Dominion Univ., Institute of Oceanography,** Tech. Rept. No. 7: 1-23.

11. Oertel, G. F.
1972 Sediment transport of estuary entrance shoals and the formation of swash platforms. **Jour. Sed. Pet.,** 42: 857-863.

12. Robinson, A. H. W.
1956 The submarine morphology of certain port approach channel systems. **Jour. Inst. Nav.,** 9: 20-40.

13. Shepard, F. P. and Wanless, H. R.
1971 **Our Changing Coastlines.** McGraw-Hill, New York.

14. Wright, L. D., Sonu, C. J., and Kielhorn, W. V.
1972 Water-mass stratification and bed form characteristics in East Pass, Destin, Florida. **Mar. Geol.,** 12: 43:58.

20

Reprinted from *Geophys. Res. Lett.* **3**(2):97–100 (1976)

PRELIMINARY RESULTS OF COINCIDENT CURRENT METER AND SEDIMENT TRANSPORT OBSERVATIONS FOR WINTERTIME CONDITIONS ON THE LONG ISLAND INNER SHELF

J.W. Lavelle, P.E. Gadd, G.C. Han, D.A. Mayer, W.L. Stubblefield, and D.J.P. Swift

NOAA/Atlantic Oceanographic and Meteorological Laboratories,
15 Rickenbacker Causeway, Miami, Florida 33149

R.L. Charnell

NOAA/Pacific Marine Environmental Laboratory, 3711 15th Avenue N.E.,
Seattle, Washington 98105

H.R. Brashear, F.N. Case, K.W. Haff, and C.W. Kunselman

Oak Ridge National Laboratory, P.O. Box X, Oak Ridge, Tennessee 37830

Abstract. We have observed late fall and winter bedload sediment transport and the overlying current field in ridge and swale topography on the inner continental shelf south of Long Island, and can report movement of bed material at a water depth of 20 m to a distance of approximately 1500 m after several storm events. Movement over an 11-week observation period was longshore and oblique to the ridge crest at the experimental site. Currents were also predominately longshore, but long term averages demonstrate that a vertical shear existed in the fluid motion. Although the number of sediment transport "events" suggested by the current meter data is nearly balanced in eastward and westward directions, both estimates of transport from current speeds and sand tracer dispersion patterns show that several westward flowing events dominated the transport during a two and one-half month period. A quantitative upper bound of 31 cm/sec on the threshold velocity for sediment movement in this size range is also set by the data.

Introduction

Increasingly widespread interest in the characterization and quantification of shelf sediment transport stems from the requirements of the growing number of shelf and nearshore users to understand the dynamics of an area on which they may have potential impact. The uses and interests are myriad, but more common expressions of concern are phrased in terms of recovery rates of contaminated sediments by replacement, the stability of the substrate for offshore structures, the influence of offshore work on beach and nearshore features, and the temporal variability of sediment transport as an influence on faunal habitats. While considerable efforts have been made in observing and describing water-sediment coupling in the laboratory, under riverine flow, and in the nearshore area, few direct measurements of offshore sediment movement and the associated near-bottom water velocity field have been made. For these reasons, we are reporting preliminary results of an experiment recently completed in the New York Bight to directly measure offshore cohesionless sediment movement and its immediate forcing mechanism, the overlying water velocity field.

The Long Island Near-Shore (LINS) Study (Figure 1) was centered at 40°33'N and 73°25'W, halfway between Jones and Fire Island Inlets, Long Island, New York, some 9 km offshore. The study area, an 8 x 10 km rectangle, was located in an area of undulating morphological features described in Duane et al. (1972) as ridge and swale topography. Bedforms at the study site have wavelengths of approximately 1 km with wave heights of 4-7 m, intersect the shoreline obliquely, and are composed of relatively clean, medium to fine sands; the ridges are asymmetrical with steeper southwest facing flanks. The experimental design was twofold: to gather sediment dispersion and current meter data which could be used to aid in quantifying sediment transport; and to gather qualitative data on the construction and/or the maintenance mechanism of ridge and swale features which are widespread on the Atlantic continental

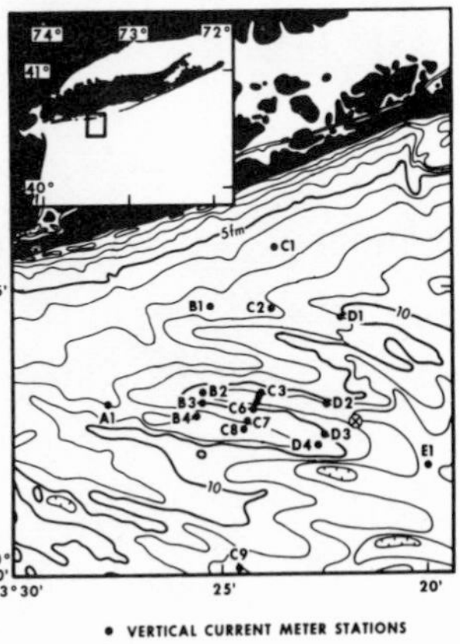

Fig. 1. Bathymetry and current meter station locations for the Long Island Nearshore Study (LINS). Stations C1, C6, C7, and C8 returned no usable data.

shelf (Swift et al., 1973). Field work was divided into two concurrent operations: a sediment tracer experiment and a current meter array of high spatial resolution. We present here a qualitative, preliminary view of the data collected in those efforts.

Current Meter Observations

During the first six weeks of the current meter operation (October 16 to December 4, 1974), nineteen stations (Figure 1) were occupied. A single current meter string was retained in the area during the remainder of the experiment. Aandaraa RCM-4 Savonius rotor current meters which record instantaneous direction and integrated average speed at 10-minute intervals were used throughout. Measurement emphasis was placed on a well-defined ridge and trough; meters were located on a crest, flank, and trough on each of three transects (B, C, and D of Figure 1) as well as the adjacent flank of transect C. Additional meters were set outside the central study area to measure far-field velocities.

Flow during the observation period trended both east and west, parallel to the coast. Figure 2 is a vector time series of velocities at station 2C (1.5 m above the bottom) and is representative of near-bottom water movement during one of the most active periods of flow. The data presented here have been subjected to a 40-hr low pass filter and then resampled at hourly intervals. Although east is the dominant flow direction during this sampling interval, the most intense flow was westward during a three day period near the end of this period.

Predominance of eastward flow is consistent with

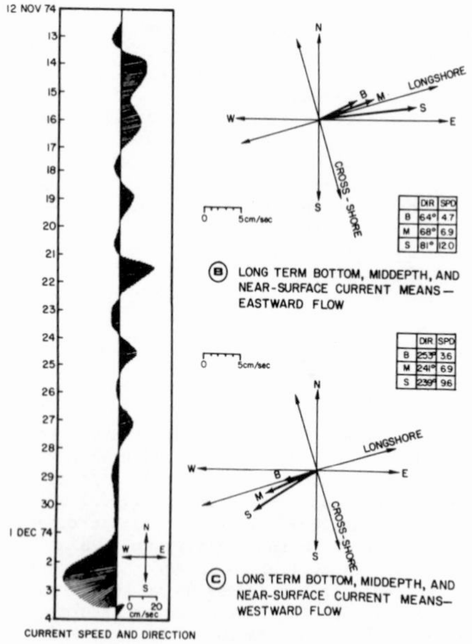

Fig. 2. Near bottom current vector time series and velocity averages. Shoreline direction represents the trend of the 5 fathom isobath between Jones and Fire Island Inlets.

the observation of Charnell and Mayer (1975) who reported the existence, in the statistical sense, of a clockwise gyre in the long term mean flow within the New York Bight apex during the fall and winter of 1973. The strong westward flow (Figure 2A) occurred during the storm of December 1 - December 4, 1974, an event which was reported to have been the most damaging northeaster since the Ash Wednesday storm of 1962 (C. Galvin, Coastal Engineering Research Center, personal communication). Winds from the east-northeast up to 16 m/s were recorded at John F. Kennedy International Airport during the initial 36 hrs of this period; winds from the northwest at an average speed of 10 m/s followed on December 3 and 4. The second most important sustained flow during the observation period, that which began during December 16, also followed east winds. Periods of high speed winds from the west and northwest cause less intense near-bottom water movement. The asymmetry of the fluid response to easterly and westerly winds in this area has been noted by Beardsley and Butman (1974).

Vertical shear in current velocities was unmasked in the data (Figures 2B and 2C) when long term velocity averages were made on data from meters grouped by position in the water column. Flow recorded by meters 1.5 to 4 m from the bottom (B), 5 and 6 m from the bottom (M), and 6 to 11 m from the surface (S), were averaged separately in time over periods when flow had eastward and westward components. Water depths at the stations varied from 15 to 22 m. These data show that near surface water flow had an offshore component for both eastward and westward flow, while bottom flow tended to be more inshore, parallel to the isobaths during westward flows and more strongly inshore during eastward flows. Speeds decreased in a relatively uniform fashion from the upper to the bottom meters. Wind records document a northerly wind component throughout much of the observation period; the observed shore-normal components may be an indication of upwelling contributions to fluid motion.

Sand Tracer Measurements

In order to directly assess the flow response of the sediment to the observed water movement, we employed the Radioisotope Sand Tracer (RIST) system developed at Oak Ridge National Laboratory (Duane, 1970; Case et al., 1971). Indigenous sand was sorted to produce a fraction whose size distribution was roughly Gaussian, with a mean diameter of .15 mm (fine to very fine sand), a standard deviation of .03 mm, and no material larger than .25 mm or smaller than .06 mm. Approximately 500 cm^3 of this material was surface coated with 10 Curies of the isotope ^{103}Ru ($T_{½}$ = 39.6 days). On November 12, equal portions of the tagged sand in water soluble bags were released at three points at the east end of the main trough (Figures 1 and 3). The injection points formed an equilateral triangle with sides roughly 100 m in length. The ensuing dispersal pattern of labeled sand was surveyed at intervals by scintillation detectors mounted in a cylindrical vehicle which was towed across the bottom. Raydist precision navigation with 10 m resolution was used. Four post-injection surveys were made during the 11-week tracer experiment.

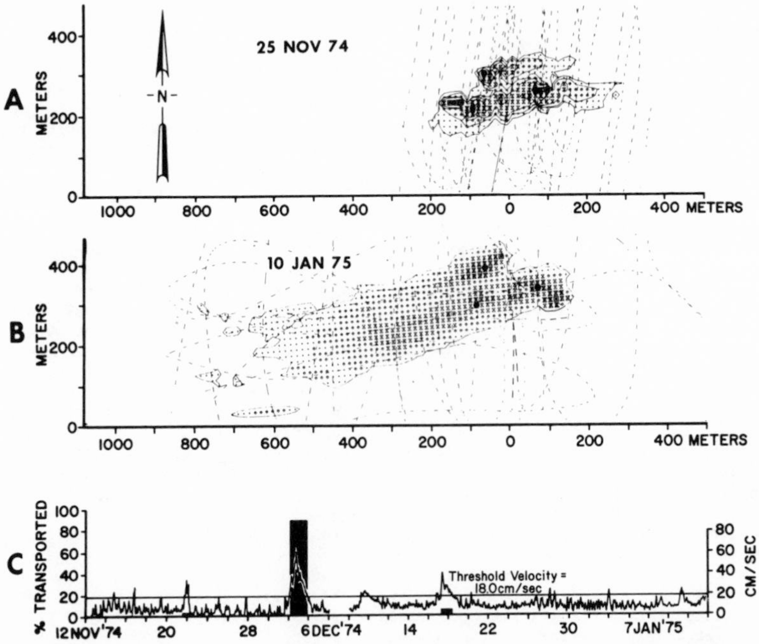

Fig. 3. a and b) Dispersion patterns measured 13 and 59 days after injection of tagged sand. Point sources are represented by dots. Broken line is the survey trackline; stippling represents radiation intensity. c) Near bottom current speed record over the length of the experiment, and calculated sediment transport information (see text).

Dispersion patterns mapped two and eight weeks after injection are shown in Figure 3; each has been corrected for background radiation and decay. After two weeks (November 25) roughly ellipsoidal smears trended east from each of the three injection points (Figure 3a). Each smear could be traced for about 200 m before the signal was lost in the background radiation. After eight weeks (January 10), the three eastward smears had been replaced by a single, more extensive pattern extending 700 m to the west (Figure 3b). The reversal of the patterns from east to west was more markedly demonstrated by preliminary data from a survey made during mid-December (December 17-19). Although those data have not yet satisfactorily been processed, the data at the time of that survey were sufficient to indicate that the reversal of the dispersion pattern of Figure 3a had already occurred, and in fact extended approximately 1,500 m to the west. We should point out that the patterns of Figure 3 must be regarded as minimum transport patterns, in the sense that tagged material which has been buried or has diffused downward into the bed is attenuated in signal strength by the overburden. For this reason, the signal measured is an underestimate of the true signal and the observations must be regarded as a lower bound to the true transport.

The temporal pattern of sediment transport over a 60-day period may be inferred from Figure 3c. The basic record is current speed, measured 1.5 m from the bed, versus time. The horizontal line at 18 cm/sec is an estimated threshold, based on the work of Shields and subsequent workers for shear velocity (Graf, 1971) and a choice of 3.0×10^{-3} for the drag coefficient (Sternberg, 1972). We believe that this choice of threshold velocity is in part verified by empirical evidence obtained during the course of the experiment (see below). We have made estimates of the relative role each transport event played in the overall transport record, based on the concept of the proportionality of frictional energy expenditure to the transport volume (Bagnold, 1963). For each event where velocities exceeding threshold were recorded, we have calculated a transport volume:

$$Q_i = \frac{\alpha}{T_i} \int_i (|V| - |V_{TH}|)^3 \, dt \qquad (1)$$

where $|V|$ is measured current speed, $|V_{TH}|$ is threshold speed, α is a constant of proportionality, and T_i is the duration of the transport event. Expression of sediment transport as a power of the difference of measured and threshold velocity is supported by analysis of stream transport data (Kennedy, 1969). Without assigning a value to α, one may calculate relative rates of transport, one event to the next, or one event to the total transport evidenced by the current meter record. We have taken the second of these options, and have represented relative sand transport by solid bars superimposed on the current record (Figure 3c). Despite the exceedence of the sediment transport threshold at many points in the record, only the solid bars centered on December 2 and December 16-

17 are visible in the figure, bearing witness to the dominance of the calculated transport by these two events. Furthermore, the figure also shows that most of the calculated transport occurred during the early December storm. While this calculated transport index may be biased by the choice of threshold speed as well as the functional dependence on velocity, we believe any other reasonable parameterization is likely to lead to the same general conclusion: the storm event of December 1 - December 4 moved more sand at 20 m water depth than the combination of all other transport events.

The reversing nature of sediment flow during the observation period provides a constraint on the entrainment velocity. A threshold speed greater than approximately 31 cm/sec at 150 cm off the bottom would eliminate transport during the first 14 days of the record, in contradiction to the observation of eastward transport (Figure 3a). Setting the threshold much below 12 cm/sec would result in more eastward transport during the entire tracer experiment than was the case. Based on the relative extent of the dispersion patterns in Figures 3a and 3b, we believe that the calculated threshold velocity of 18 cm/sec is realistic.

Summary

Water movement on the Long Island Inner Shelf at depths of 10 to 20 m and at frequencies below 1/40 hr^{-1} was predominately alongshore with a net flow over the observation period to the east. The non-tidal flow reversals at these depths suggest domination by winds associated with frontal passages; the net eastward flow likely reflects the average winds from the north and west through the fall and winter months. Vertical shear of the flow is observable in long term averages of the current records; small offshore mid-depth flows and some onshore bottom flow may reflect as an upwelling circulation the net offshore component of the wind. The most intense water movements recorded during the experimental period followed high northeasterly and easterly winds.

Sediment is transported both eastward and westward parallel to the shoreline, and oblique to the ridge and trough system. Current speeds recorded 150 cm from the bed show that the sediment entrainment threshold is exceeded only intermittently; sand transport occurs only during storm events, separated by periods of quiescence. Mean water movement was to the east over the observation period in sharp contrast to the observed mean westward sediment transport. Some eastward sediment transport was observed, but the most intense water movement and resultant sand movement were associated with several "northeaster" storm events. Asymmetry of the ridges (steeper southwest facing flanks) suggests that westward flows associated with such storms constitute the primary sediment flow mechanism in this ridge and swale topography.

Acknowledgements.

Support for this work has come from NOAA's New York Bight Marine Ecosystems Analysis (MESA) Project, NOAA's Environmental Research Laboratories, and ERDA's Division of Biomedical and Environmental Research. Oak Ridge National Laboratory is operated by Union Carbide Corporation for the U.S. Energy Research and Development Administration.

References

Bagnold, R.A., Beach and near-shore processes, part I, mechanics of marine sedimentation, In: The Sea, vol. 3, pp. 507-528, Interscience Pub., New York, 1963.

Beardsley, R., and B. Butman, Conditions on the New England continental shelf: response to strong winter storms, Geophys. Res. Letters, 1, 181-184, 1974.

Case, F.N., E.H. Acree, and H.R. Brashear, Detection system for tracing radionuclide-labeled sediment in the marine environment, Isotopes and Radiation Technology, 8, 412-414, 1971.

Charnell, R.L., and D.A. Mayer, Water movement within the apex of the New York Bight during summer and fall of 1973, Tech. Memo., National Oceanic and Atmospheric Administration, Boulder, Co. (in press).

Duane, D.B., Tracing sand movement in the littoral zone: progress in the Radio Isotopic Sand Tracers (RIST) study, July 1968-February 1969, Coastal Eng. Res. Center Misc. Paper, Washington, D.C., 1970.

Duane, D.B., M.E. Field, E.P. Meisburger, D.J.P. Swift, and S.J. Williams, Linear shoal on the Atlantic inner continental shelf, Florida to Long Island, In: Shelf Sediment Transport: Process and Pattern, pp. 447-498, Dowden, Hutchinson and Ross, Stroudsburg, Pa., 1972.

Graf, W.H., Hydraulics of Sediment Transport, p. 96, McGraw Hill, New York, 1971.

Kennedy, J.F., The formation of sediment ripples, dunes, and antidunes, In: Annual Review of Fluid Mechanics, vol. 1, pp. 147-168, Annual Reviews, Inc., Palo Alto, Calif., 1969.

Sternberg, R.W., Predicting initial motion and bedload transport of sediment particles in the shallow marine environment, In: Shelf Sediment Transport: Process and Pattern, pp. 61-82, Dowden, Hutchinson and Ross, Stroudsburg, Pa., 1972.

Swift, D.J.P., D.B. Duane, and T.F. McKinney, Ridge and swale topography of the Middle Atlantic Bight, North America: secular response to the Holocene hydraulic regime, Mar. Geol., 15, 227-247, 1973.

AUTHOR CITATION INDEX

Abbott, R. T., 35
Abele, R. W., 324
Acree, E. H., 329
Akers, W. H., 123
Allen, J. R. L., 168, 308
Amein, M., 192
Angas, W. M., 50
Avignone, N., 193

Bagnold, R. A., 79, 308, 324, 329
Baines, W. D., 212, 216
Barkov, L. K., 13
Barrell, J., 5, 71
Bass, N. W., 35
Batchelor, G., 216
Beardsley, R., 329
Bernard, H. A., 168
Blanton, J. O., 249
Borghetti, G., 309
Bradley, W. H., 35, 71, 79
Brand, D. D., 168
Brashear, H. R., 329
Brehmer, M. L., 192
Bretschneider, C. L., 89
Bruun, P., 43, 47, 191
Bumpus, D. F., 219
Butman, B., 329
Byrne, J. V., 123

Caldwell, J. M., 284
Carson, R. M., 232
Case, F. N., 329
Caston, V. N. D., 219, 232
Chadwick, G. H., 123
Charnell, R. L., 329
Clark, T., 71
Clifton, H. E., 95
Coch, N. K., 192
Colby, B. R., 308
Coleman, J. M., 89
Colquhoun, D. J., 191, 192
Cook, D. O., 191, 308, 309
Cooper, G. A., 123
Craft, B. B., 124, 168
Cram, I. H., Jr., 71
Csanady, G. T., 249
Curray, J. R., 71, 79, 85, 123, 125, 168, 191, 193, 232

Daly, R., 71
Davies, J. L., 89
De Beaumont, E., 168
De Cserna, Z., 168
Dietz, R. S., 71, 79, 85, 129
Dill, C. E., Jr., 193
Dillard, W. R., 35
Dillon, W. P., 12, 191
Dixon, W. J., 284
Dolan, R., 5, 12, 191
Dorf, E., 47
Dowling, G. B., 72
Dracup, J., 308
Drake, C. L., 85, 123
Draper, L., 232, 308
Duane, D. B., 207, 219, 232, 329
Dunbar, C. O., 71, 123

Eagleson, P. S., 12, 89, 192, 308
Eisenhart, C., 284
Ekman, V. W., 216
El-Ashry, M. T., 191
Elias, M. K., 35
Emery, K. O., 12, 71, 79, 85, 86, 96, 123, 191
Evans, O. F., 308
Ewing, J., 85
Ewing, M., 71, 85, 123

Fairbridge, R. W., 71, 79, 85, 123
Felder, W., 12
Felix, D. W., 308
Fenneman, N. M., 71
Field, M. E., 207, 219, 232, 329
Fischer, A. G., 191
Fisher, J., 5, 12

Fisher, J. J., 191, 192
Fisher, R. A., 284
Fisk, H. N., 35, 123
Flint, R. F., 72, 85, 168
Frankenberg, D., 192
Frerichs, W. E., 168
Frey, D. G., 168
Friedrich, B., 168
Frye, J. C., 36, 124
Furst, P. T., 168

Galvin, C. J., 216
Garrels, R., 71
Garrett, J. R., 249
Gerritsen, F., 47
Gienapp, H., 232
Gierloff-Emden, H. G., 168
Gilbert, G. K., 5, 71, 79, 308
Gilluly, J., 71
Glenn, J. L., 324
Glenne, B., 308
Goldsmith, V., 324
Gorsline, D. S., 129
Gould, H. R., 71, 124, 168
Grabau, A. W., 124
Graf, W. H., 329
Grant, U. S., 308
Greenman, N. N., 124
Guilcher, A., 41
Gulliver, F., 71
Gunnerson, G. G., 71

Hadley, L. M., 232
Hald, A., 284
Hama, F. R., 324
Harbaugh, J. W., 35
Hardin, G. C., 85
Harris, R. L., 192
Harrison, W., 192, 216
Hastay, M. W., 284
Hayden, B., 12
Hayes, M. O., 324
Hedgpeth, J. W., 36
Heezen, B. C., 71, 85
Heinter, R. E., 95
Hersey, J. B., 79
Hey, R. W., 168
Hill, H. W., 232
Hirshman, J., 85
Hjulström, F., 79
Hobbs, C. H., 324
Hodges, G. F., 232
Hodgson, W. D., 124
Holck, A. J. J., 123
Holliday, B. W., 193
Hom-Ma, M., 308
Hopkins, T. S., 232
Horikawa, K., 308
Houbolt, J. J. H. C., 232
Howard, J. P., 96
Howe, H. V., 124, 168
Howe, W. B., 35
Howell, J. V., 71
Hoyt, J. H., 168, 192
Hubbard, D. K., 324

Ijima, I., 309
Inman, D. L., 42, 71, 79, 192, 284, 308, 309
International Hydrographic Bureau, 232
Ippen, A. T., 192, 308

Jeffries, H., 212
Jelgersma, S., 124
Jennings, J. N., 89
Jewett, J. M., 36
Johnson, D. W., 12, 36, 71, 85, 168
Johnson, J. W., 12, 89
Johnson, M. A., 232
Judson, S., 72

Kay, M., 85, 124
Kennedy, J. F., 329
Kenyon, K. E., 216
Keulegan, G. H., 192, 216
Kielhorn, W. V., 325
Kinahan, H. C., 24
Kinsman, B., 192
Klovan, J. E., 192
Knapp, D. J., 212, 216
Knopf, A. K., 72
Koczan, J., Jr., 96
Kofoed, J. W., 208
Komar, P. D., 13
Kraft, J. C., 208
Kuenen, P. H., 13, 71
Kuhn, T. S., 96
Kumar, N., 96

Ladd, H. S., 36
LaFond, E. C., 309
Lahee, F. H., 124
Langfelder, J., 192
Lankford, R. R., 125
Larras, J., 89
Lauff, G. H., 324
Lawford, A. L., 232
Lawson, A. C., 71
Leatherock, C., 35
Le Blanc, R. J., 124
Lee, A. J., 232

Lee, W., 36
Leet, L. D., 72
Le Pichon, X., 85
LeRoy, D. O., 123
Lewis, W. V., 89, 168
Logvinenko, N. V., 13
Longuet-Higgins, M. S., 308
Longwell, C. R., 72, 85
Lovely, H. R., 124
Lucke, J. B., 36
Ludwick, J. C., 324, 325
Luskin, A., 85
Lynde, R. E., 219

McArthur, D. S., 216
McCave, I. N., 232
McClennen, C. E., 219
McCloy, J. M., 216
McFarlan, E., Jr., 35, 71, 124, 168
McGuirt, J. H., 124, 168
McKee, E. D., 124
McKinney, T. F., 329
McManus, D. A., 232
McMaster, R. L., 192
Major, C. F., Jr., 168
Manohar, M., 308
Marmar, H. A., 40
Massey, F. J., 284
Mayer, D. A., 329
Meisburger, E. P., 207, 219, 232, 329
Melton, F. A., 124
Menard, H., 71, 79
Miller, D. J., 72
Miller, H., 6
Miller, J. C., 208
Miller, R. L., 13, 308
Mitchell, H., 16
Moody, D. W., 13, 192, 208
Moore, D. G., 72, 79, 85, 124, 125, 168
Moore, R. C., 36
Morgan, W. H., 47
Mortimer, C. H., 249
Munk, W. H., 284
Munsart, C. A., 192
Murray, G., 85
Murray, H. H., 36
Murray, S. P., 192, 216, 309
Murty, T. S., 249

Nagata, Y., 309
Nasu, N., 192, 308
Neumann, G., 216
Newman, W. S., 192
Noda, E. K., 13
Norcross, J. J., 219
Northrup, J., 85

Oaks, R. Q., Jr., 192
O'Conner, H., 36
Oertel, G. F., 325
Olson, J. S., 284
Oostdam, B. L., 208
Osario, J. D., 309
Owens, E. H., 324

Palmer, H. D., 309
Pannekoek, A. J., 124
Parker, R. H., 124
Parrot, B. S., 168
Passega, R., 309
Payne, L. H., 192
Payne, R. H., 72
Pettijohn, F., 72
Philips, R. L., 95
Pierce, J. W., 192
Pierson, W., Jr., 216
Pilkey, O. H., 129, 192, 232
Pitt, E. G., 232
Post, R., 36
Postma, H., 125
Potter, P. E., 284
Powers, M. C., 192
Price, W. A., 13

Ramster, J. W., 232
Rao, D. B., 249
Reid, R. O., 89, 216
Reineck, H. E., 96
Resio, D., 12
Rich, J. L., 36, 72, 86
Richards, H., 36
Riley, C. M., 123
Risk, M. J., 309
Rizzini, A., 309
Roberts, A., 85
Robinson, A. H. W., 325
Rodgers, J., 71, 123
Ronai, P. H., 36
Russell, R. C., 309
Russell, R. J., 124, 168

Salsman, G., 72
Sanders, J. E., 96, 192
Sanford, R. B., 192
Sato, S., 309
Saulsbury, F. P., 208
Saunders, P. M., 249
Saville, T., Jr., 284
Schlee, J., 192
Schureman, P., 232
Schwartz, M. L.., 6, 13, 96, 192
Scott, J. T., 249
Scott, T., 193, 309

Scruton, P. C., 124, 168
Sears, P., 208
Shaw, D. M., 219
Shepard, F. P., 13, 36, 72, 79, 86, 123, 124, 168, 284, 309, 325
Shideler, G. L., 193
Shumway, G., 72
Siegel, S., 309
Singh, I. B., 96
Smith, J. D., 232
Snedecor, G. W., 284
Sonu, C. J., 216, 325
Spieker, E. M., 125
Stafford, D., 192
Stahl, L., 96
Stanley, D. J., 3, 6, 193, 232
Stanley, E. M., 219
Stearn, C., 71
Stephenson, M. B., 124, 168
Sternberg, R. W., 232, 329
Stetson, H. C., 72, 86, 125, 129
Stevens, H. H., 324
Stommel, H., 40
Stone, R. B., 192
Stride, A. H., 219, 232
Summers, H. J., 309
Sundborg, Ä., 79
Sutton, G. H., 85, 123
Swain, F., 125
Swift, D. J. P., 3, 6, 13, 96, 192, 193, 207, 208, 219, 232, 329

Tanaka, N., 309
Tanner, W. F., 193
Tharp, M., 71, 85
Thomson, M. R., 124
Trask, P., 42
Tucker, M. J., 232
Twenhofel, W., 72

Uchupi, E., 86, 193, 219
U. S. Department of Commerce, ESSA, Weather Bureau, 216

Van Andel, T. H., 79, 125
Van Dorn, W. G., 216
Van Reenan, E. B., 208
Van Straaten, L. M. J. U., 36, 96, 125
Veley, V. F. C., 232
Vernon, J. W., 309
Vincent, L., 12
Visher, G. S., 193
Von Engeln, O., 72

Wagner, K. A., 192
Wallis, W. A., 284
Wanless, H. R., 36, 191, 325
Waters, A., 71
Watts, G., 46
Weller, J. M., 36
Wells, A. J., 36
Wells, D. R., 13
Wexler, H., 48
Wheeler, H. E., 36
Wiegel, R. L., 89, 216
Williams, S. J., 207, 219, 232, 329
Willman, H. B., 124
Woodford, A., 71
Wrath, W., 72
Wright, H. E., Jr., 168
Wright, L. D., 89, 325

Young, R. E., 36

Zeigler, J. M., 13, 308
Zenkovich, V. P., 89, 309
Zenkovitch, V. P., 13, 168, 193

SUBJECT INDEX

Age dating, 173, 204, 206, 207
Alluvium, 144
  pre-transgressive, 144, 150
API project, 51, 92
Ash Wednesday storm, 10
Assateague Island, Maryland, 217
Attenuation ratio, 88

Barrier Island, 170, 173, 174, 178, 188–191, 199, 207
Bar, longshore, 138
Barnegat Bay, New Jersey, 27
Barriers, 20, 22, 28, 106
  formation, 94, 166
  over topped, 121
Basal transgressive sand, 112
Base level, 14, 55
Beach, 173, 178, 182, 200, 201, 203, 205, 255, 256, 280, 283, 306, 307
  ridges, 94, 113, 135, 138, 146, 156, 173
  rock, 156
Beds
  bottom-set, 60, 75
  foreset, 52, 60, 67, 75
  forms, ridges, 179, 186, 187, 190, 200, 201, 204, 205, 207, 217–221, 226, 227, 231, 317–324, 326–329
  forms, ripples, 181, 186, 290, 291, 297–299, 300–302, 304, 307
  forms, sand banks, 311, 323
  forms, sand waves, 201, 203, 231
  prograded paralic, 70
  topset, 60, 75
Bethany Beach, Delaware, 199, 201, 205–207
Bioturbation, 185
Black Sea, 286
Bottom friction, 236, 246, 286, 312–320
Boundary layer, 245, 247, 312–317
Bruun's hypothesis, 10, 189

California, Southern, 286–288, 291, 292, 307
Cape Hatteras, North Carolina, 170, 179, 182, 184
Cape Henlopen, Delaware, 199, 200, 201, 206
Cape Henry, Virginia, 170, 182, 186
Central shelf
  Delaware, 199–207
  European, 221
  Gulf of Mexico, 213–217
  Maryland, 217–221
  North Carolina-Virginia, 169–198
  North Sea, 222–231
  Southern California, 285–307
Chenier plain, 113, 166
  Southwestern Louisiana, 107, 120
Chesapeake Bay, 186, 190, 310–312, 315, 317, 318–324
Chicago, 242
Coast
  deltics, 165
  Florida, 44
  Georgia, 126
  Gulf, 80
  Mexico, Northwest, 92
  Mexico, Pacific, 131
  Nayarit, Mexico, 76, 112
  New Jersey, 95
  North America, Pacific, 108
Coastal
  boundary layer, 243
  deposits, 92
  erosion, 70
Continental
  embankment, 67, 69, 78
  shelf, origin of, 42
  shelves. *See* Shelf
  slopes, 238, 239
  slopes, Atlantic, 67
Coriolis effect. *See* Coriolis force
Coriolis force, 222, 238, 241, 242, 244, 246, 321
Currents
  alongshore, 21
  bottom, 213–216, 217–221, 222–231, 241, 285–307, 326–329

coastal jets, 240–243
frictional stress, 2, 214
geostrophic. *See* Geostrophic flow
measurement, 213, 217–221, 224–226, 249, 288, 289, 312, 327
rip, 180, 286, 287, 301, 306, 307
storm, 213–216, 229, 327–329
thermohaline, 233, 234
tidal, 186, 200, 203, 207, 214, 222, 223, 225, 226, 228, 229
wave-induced, 223, 292–307
wind-driven, 217–231, 233–237, 243–246
zones, 3, 4
Currituck Spit, North Carolina, 173, 174, 182

Delaware Bay, 200, 201, 203, 207
Delta, 76
building, destructive phase of, 105
Catskill, 120
Danube, 88
development, destrucitve phase of, 122
Ebro, 88
Mississippi, 78, 88, 105, 107, 120, 121
Mississippi, destructive phase of, 119
Niger, 88
Nile, 88
Rio Grande de Santiago, Mexico, 112
river, morphology, 87
San Francisco, 88
Senegal, 88
Deposition, rate of. *See* Sedimentation
Destin, Florida, 322
Diamond Shoals, North Carolina, 176, 183, 184, 186, 190
Discharge effectiveness ratio, 88

Ekman flow, 246–248
Eperic seas, sedimentation in, 118
Epicontinental seas, environmental conditions, 118
Equation of motion, 2, 4
Erosional retreat, 9, 10
Estuary, 310–324
Eustatic oscillations, 39
Eustatic sea level change, 32

Faunas
False Cape, Virginia, 173, 176, 187
Fenwick Island, Maryland, 207
fetch, wave, 118
flow. *See* Currents
fluid dynamics, 5
selective preservation, 30
Fronts, fluid, 243–246, 249

Geomorphology, 173, 207
inductive, 11
Victorian, 9
Georgia, 322
Geostrophic flow, 4, 238, 240–242
Great Lakes, 234, 243–245
Gulf of Maine, 238
Gulf of Mexico, 116
Northwestern, 101, 108, 116
Gulfport, Mississippi, 213

Hatteras Island, North Carolina, 173
Hen and Chicken shoal, 201, 203, 307
Hurricanes, 213–216
Hydraulic processes, 179, 180, 181, 185–200, 207, 213–231, 297, 306, 311–324. *See also* Currents

Intertidal zone, 256
Isostacy, 69

Japan, 287, 303

La Jolla, California, 255, 256, 307
Lagoons, 28, 144
Lake Bonneville, 2, 59
Lake Huron, 300
Lake Meade, 60
Lake Michigan, 22, 24, 242
Lake Ontario, 237–239, 242, 243, 246
Littoral drift, 41–43, 179, 180(fig.)
Long Island, 234, 235, 239, 247, 326, 329
Longshore
currents, 179, 214, 236–239, 241, 242, 244–246, 286
gradients, 87

Middens, Indian, 144, 159
Mississippi River, mud load, 116
Mixing, 30
Moray Firth, 1
Muds shelf, 146

Nags Head, North Carolina, 182
Natural cyclic period, 74, 83
Nearshore sand level, 58
New Jersey, 9, 26, 243
North Carolina, 172, 173
North Sea, 222–231
Null line, 181, 287, 306
model, 4, 12

Ocean City, Maryland, 206
Offlap, 98
Old Red Sandstone, 1
Onlap, 98, 120
Oregon Inlet, North Carolina, 173, 174, 179, 182

Overlap
regressive, 98, 100
transgressive, 98
Overstep, 122

Palos Verdes, California, 202, 288, 291, 293
Peat, 30, 204, 207
Peneplanation, 53
Platform
abrasive, 67, 70
of marine abrasion, 9, 73
Processes, equilibrium, 8
Profile
equilibrium, 5, 8, 9, 11, 12, 14, 15, 43, 52, 57, 58, 67, 70, 73, 75, 92, 93
subaqueous, 2
Progradation, 113, 167
Pycnocline, 223, 240

Quartz, iron-stained, 127
Quaternary
history, 172–174
sediments, 173

Redondo Beach, California, 288, 291, 293–296
Regression, 97, 98, 113. *See also* Shoreline, retreat
depositional, 104, 105, 107, 114
discontinuous depositional, 120
erosional, 119
Holocene, Nayarit Coast, 157
Late Wisconsin regression, chronology, 101
mixed erosive and regressional, 120
Rehoboth Bay, Delaware, 199, 200, 201, 203, 204
Relict
sediments, 74
topography, 75
topography, channels, 201–207
Ripple marks, 42
River
Albermarle, North Carolina, 182, 186
Altamaha, Georgia, 128
Brazos, Texas, 105
Colorado, Texas, 105
Columbia, 322
Indian, Delaware, 206
Mississippi, 80, 116
Rio Grande de Santiago, 131, 156, 157, 161, 162
St. Johns, 128

Salinity, 246–248
Sandbridge formation, 172, 173, 178, 179
Sands
basal, transgressive, 103, 119, 150
littoral, 119, 120, 144
Recent, 127
regression latteral, 150
relict, 127
ridge topography, 11
sheet, 118, 119
sheet regression, 167
silty, 146
Santa Catalina Island, California, 288, 291, 293
Scientific revolution, 93
Scotland, 1
SCUBA, 186, 256, 262, 286, 287
Sea level, 42
changes, 38, 156
changes, eustatic, 99, 102
changes, relative, 99
erosion by rise of, 43
Holocene transgression, 189, 201, 203, 204, 206, 207, 217
Pleistocene, 170, 172, 201, 204, 205
rate of change, 204
relative, 114
rise, Holocene, 57
rise and warming trend, 47
Sediment
analyses, 171, 172, 184–188,194–198 (table), 278, 290, 301
analyses, factor, 176–178, 183, 184, 186
bedload transport, 177, 178, 180, 181, 223, 231, 263, 285, 286, 290, 291, 297, 298, 299, 300–305, 307, 311–321, 326–329
Continental Shelf, 115
defined, 4, 5
Emery classification, 74, 84, 93
grain size, 175–177, 182, 184, 185, 188, 190, 286, 296, 297–307
grain size, clay, 178–179, 205, 207, 286
grain size, gravel, 182, 190, 199, 201, 202, 204, 205, 207
grain size, mud, 187, 200, 201, 203, 204, 207, 217
grain size, sand, 170, 171, 176, 180, 182, 190, 201, 203, 204, 207, 217, 222, 277, 278, 280, 286, 287, 290–307
grain size, silts, 201, 203, 286
grain size, trends, 182, 183, 184
peat, 30, 204, 207
Pleistocene, 178, 182
porosity, 302
problem, 115
rate of supply, 105
recent, 127
relict, 74, 84, 93, 115, 126, 127, 170, 190–191, 201, 202, 204

shelf, 116
shell, 178, 179
slope, 116
sorting, 180, 181, 185, 186, 188, 189, 190, 286, 297, 299, 300, 301, 306
suspended transport, 180, 181, 287, 289, 297, 298, 303, 305
tracers, 287, 376–379
traps, 289, 290, 301, 303
winnowing. *See* Sorting

Sedimentation
Pleistocene, 116
Recent, 126
seasonal rates (erosion and deposition), 99, 273–283
tidal flux, 313–324
wave-transported, 290, 304

Seismic profiling, 199, 201–206

Shelf
Atlantic coast, 122
break, 55, 59, 234, 239, 243, 245, 246, 249
Florida, 66
Georgia, 126
needs, 103, 114
Northwestern Gulf, Mexico, 92, 116
Palos Verdes, California, 110, 121
relict, 11
Southern California, 62, 108, 116
Texas, 102, 105, 120, 121

Shoreface, 10, 12, 174, 176, 178, 179, 181, 182, 184, 185, 189, 201, 207, 217–221, 256, 275–280
erosion, 95
retreat, 94

Shoreline
inspection of, 97, 119
movements, 41
retreat, 173, 182, 188–201, 207

Slope
minimum, 43
Rhode Island, 83
stability, 78

Statistical analyses, 266, 267–277, 281, 282, 295, 296

Strand plain, 113, 135, 138
Nayarit, 94
width, 158

Surf
base, 84
beat, 281
zone, 179, 181, 256, 280, 286

Temperature, 234

Terrace
Continental, 52, 76, 81, 118
Delta, 60, 67, 80
low-tide, 138
surf-cut, 58
wave-built, 52, 55, 57, 59, 67, 70, 73, 75, 80
wave-cut, 23, 59, 60, 70

Thermocline, 240–242, 244–246

Thermohaline circulation, 233–237, 244, 245, 247

Threshold velocity, 302, 313–316, 328, 329

Tidal currents, 186, 200, 203, 207, 214, 222, 311–324. *See also* Currents

Tidal faults, 30

Tidal marshes, 30

Tide, 226–228, 230, 233, 295, 296. *See also* Tidal currents

Transgression, 31, 32, 97, 98, 131, 191
depositional, 120
discontinuous depositional, 121
erosional, 105, 122
rapid erosional, 122
sequence, 31, 33

Transport, longshore, 157

Turbulence, 240, 246, 286, 297, 302

Uncomformities
Holocene, 201, 203
Holocene-Pleistocene, 201, 204

Undertow, 10

Uniformitarianism, 117

Upwelling, 240, 242
downwelling, 241, 242

Virginia, 172, 173

Virginia Beach, 170

Wave
base, 1, 23, 24, 70, 73, 74, 83, 84
base-defined, 52, 53
breaking, 213, 214, 216, 286
climate, 12, 87, 88
climate, North Carolina, Virginia, 185, 186
deep-water, 292
drift, 286, 289, 295, 306, 307
fairweather, 180, 218
height, 275, 287, 291, 293
internal, 56
Kelvin, 242, 243
length, 293
measurement, 287–289, 293
orbital velocity, 74, 181
period, 287, 291, 293, 295, 307
Poincare, 243
power, 88
refraction, 179
relative height, 287, 295
shoaling, 293
significant, 275, 280

slope, 180
storm, 181, 213, 306, 307
surge, 2, 180, 291, 292, 295–298, 300–303, 306, 307
swell, 180, 287, 295, 296, 306
theory, solitary, 93–95, 307
theory, trochoident, 292–295, 307
topographic, 238–240, 242, 243
Wimble Shoals, North Carolina, 187
Wind, 213, 216, 218, 222–225, 227–230, 295, 296, 306, 327
mixing, 245
stress, 233, 234, 236, 240, 246, 247
Wisconsin Glacial, 38

About the Editors

DONALD J. P. SWIFT is currently Research Oceanographer at the Atlantic Oceanographic and Meteorological Administration-National Oceanographic and Atmospheric Association, in Florida. Dr. Swift, who has held professional positions at several universities in both the United States and abroad, received his Ph. D. in Sedimentology from the University of North Carolina in 1964. He completed his M.S. degree at the John Hopkins University and his A.B. at Dartmouth College.

HAROLD D. PALMER presently serves as Technical Coordinator for Marine Services with Dames & Moore in Washington, D. C. Dr. Palmer received his B. S. in Geology from Oregon State University and his M. S. and Ph. D. from the University of Southern California. He is a Fellow of the Geological Society of America and has served as both Chairman and Secretary of the American Society of Civil Engineers, Committee on Hydrographic and Oceanographic Surveying and Charting. Currently, he serves on the evening faculty at the Johns Hopkins University.